Fundamentals
of
Dual-Phase Steels

Fundamentals
of
Dual-Phase Steels

Proceedings of a symposium sponsored by the Heat Treatment Committee of The Metallurgical Society of AIME, and the ASM/MSD Structures Activity Committee at the 110th AIME Annual Meeting, Chicago, Illinois, February 23-24, 1981.

Edited by
R.A. KOT
Republic Steel Research Center
Cleveland, Ohio

and

B.L. BRAMFITT
Bethlehem Steel Research Laboratories
Bethlehem, Pennsylvania

A Publication of The Metallurgical Society of AIME

A Publication of The Metallurgical Society of AIME
P.O. Box 430
420 Commonwealth Drive
Warrendale, Pa. 15086
(412) 776-9000

Printed in the United States of America.
Library of Congress Card Catalogue Number 81-83951
ISBN Number 0-89520-383-9

Foreword

The 25 papers in this volume were given at the TMS-AIME symposium "Fundamentals of Dual-Phase Steels" held at the AIME Annual Meeting in Chicago, Illinois, February 23-24, 1981. The symposium was cosponsored by the Heat Treatment Committee of the Metallurgical Society of AIME and the ASM-MSD Structures Activity Committee. It was the third major international conference dealing with dual-phase steels, the first held in Chicago in October 1977 and the second in New Orleans in February 1979. The proceedings were published as Formable HSLA and Dual-Phase Steels, edited by A. T. Davenport, and Structure and Properties of Dual-Phase Steels, edited by R. A. Kot and J. M. Morris. However, as is apparent after a review of these previous proceedings, significant gaps remained in our understanding of the physical and mechanical metallurgy of dual-phase steels. To this end, the current symposium deals specifically with the fundamental aspects of the metallurgy of dual-phase steels and is thus aimed at closing some of the existing gaps in our knowledge. The symposium was divided into four sessions with the following session chairmen:

> Session I: J.M. Rigsbee, University of Illinois-Urbana
> Session II: R.A. Kot, Republic Steel Corporation
> Session III: B.L. Bramfitt, Bethlehem Steel Corporation
> Session IV: R. Stevenson, General Motors Corporation

The physical and mechanical metallurgy of dual-phase (ferrite-martensite) steels is quite complex. This volume is subdivided into these two topics, with keynote papers providing state-of-the-art reviews in each area. However, examination of the various papers demonstrates that it is often difficult to clearly separate the two disciplines. Although the overall structure-property relationships in dual-phase steels have been well characterized, a number of specific questions remain unanswered. For example, the effects of retained austenite and epitaxial ferrite on the continuous yielding behavior and the improved ductility of dual-phase steels are still unresolved. However, a clearer understanding of other phenomena is fairly well developed. For example, the various models describing continuous yielding in dual-phase steels presented in this volume represent a significant advance in closing one of the important gaps remaining in our understanding of dual-phase steels.

Thanks are due to the members of the TMS-AIME Heat Treatment Committee and the ASM-MSD Structures Activity Committee for their support and to the numerous people who assisted in reviewing the manuscripts. In particular, we acknowledge the efforts of Dr. Robin Stevenson of General Motors and Professor Mike Rigsbee of the University of Illinois-Urbana.

R. A. Kot
Republic Steel Research Center
Cleveland, Ohio

B. L. Bramfitt
Bethlehem Steel Research Laboratories
Bethlehem, Pennsylvania

August, 1981

Table of Contents

Section I-Physical Metallurgy of Dual-Phase Steels

PHYSICAL METALLURGY OF DUAL-PHASE STEELS

G. R. Speich
United States Steel Corporation
Research Laboratory
Monroeville, Pa 15146

Dual-phase steels are a new class of high-strength low-alloy sheet steels characterized by a microstructure consisting of a dispersion of hard martensite particles in a soft, ductile ferrite matrix. These steels have a combination of strength, ductility, and formability that makes them attractive for weight-saving applications in automobiles. As a result, large-scale research programs on these steels have been conducted recently in many industrial laboratories and universities, leading to a rapid increase in our understanding of the production, metallurgy, and properties of these steels. The physical metallurgy of these steels is reviewed in the present article, with emphasis on (1) intercritical heat treatment, (2) continuous-cooling transformations, (3) structure-property relationships, and (4) tempering and strain-aging reactions.

Introduction

Dual-phase steels are a new class of high-strength low-alloy (HSLA) steels characterized by a microstructure consisting of a dispersion of about 20 percent of hard martensite particles in a soft, ductile ferrite matrix. The term "dual-phase" refers to the presence of essentially two phases, ferrite and martensite, in the microstructure, although small amounts of bainite, pearlite, and retained austenite may also be present. These steels have a number of unique properties (sometimes called dual-phase properties), which include (1) continuous yielding behavior (no yield point), (2) a low 0.2 percent offset yield strength (~340 MPa, 50 ksi), (3) a high tensile strength (~690 MPa, 100 ksi), (4) a high work-hardening rate, and (5) an unusually high uniform and total elongation. The high work-hardening rate results in a yield strength of 550 MPa (80 ksi) after only 3 to 5 percent deformation. As a result, in formed parts dual-phase steels have a yield strength comparable to that of other 80 ksi HSLA steels. More importantly, the high work-hardening rate, combined with the high uniform elongation of these steels, gives them a formability equivalent to that of much lower strength sheet steels. As a result these steels are an attractive material for weight-saving applications in automobiles.

Although research on dual-phase steels began more than a decade ago (1-4), the intense interest in these steels is of much more recent origin, beginning with the work of Hayami and Furakawa (5), in 1975 and of Rashid (6,7) in 1976. Hayami and Furakawa showed that continuous annealing of cold-rolled Si-Mn sheet steels in the intercritical temperature range produced steels with a ferrite-martensite microstructure similar to that shown in Figure 1, which had ductility superior to that of normal precipitation-hardened or solid-solution-hardened HSLA sheet steels. Some of their results are shown in Figure 2. Rashid subsequently showed that intercritical heat-treatment of V-N HSLA steel to produce ferrite-martensite microstructures also resulted in a large increase in ductility with no change in tensile strength, as shown in Figure 3 [the ductility of the dual-phase steel is actually comparable with that of a much lower strength 345 MPa (50 ksi) steel]. Rashid also demonstrated that this ductility increase was accompanied by an increased formability so that automotive parts could easily be formed from dual-phase steels, as shown in Figure 4. Prior to this time, HSLA steels were considered to have poor formability.

It was at this point that large-scale research efforts were begun in many steel research laboratories to develop new steel compositions and alternative processing paths for producing the large tonnages of dual-phase sheet steels predicted for use in automobiles in the 1980's. Much of this research effort has already been reported in three major international conferences (8-10) and in several recent review papers (11-13). The more recent advances in our understanding of the physical metallurgy of dual-phase steels is discussed in the present paper, with emphasis on (1) intercritical heat treatment, (2) continuous-cooling transformations, (3) structure-property relationships, and (4) tempering and strain-aging reactions.

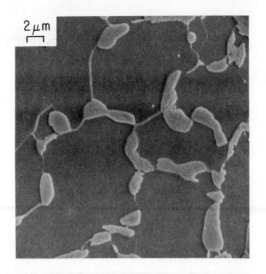

Fig. 1 - Ferrite-martensite microstructure of dual-phase steel (0.06C, 1.5Mn; water-quenched from 760°C).

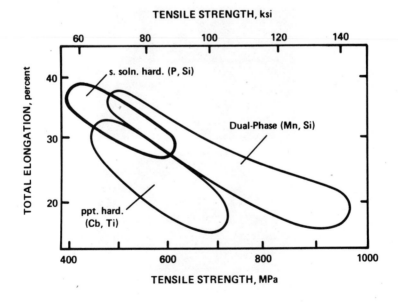

Fig. 2 - Relation between tensile strength and total elongation for various HSLA sheet steels (5).

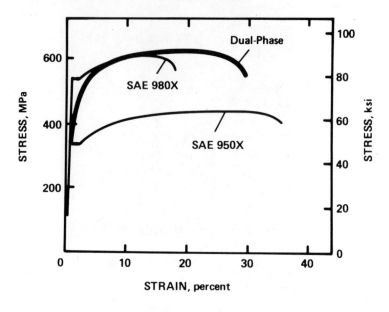

Fig. 3 - Stress-strain curves for HSLA and dual-phase steels (6).

Fig. 4 - Typical automotive parts produced from dual-phase steel.

Intercritical Heat-Treatment

Annealing Techniques and Steel Compositions

Dual-phase sheet steels can be produced by intercritical heat treatment with either continuous-annealing (14-19) or box-annealing techniques (19,20). Most present production has concentrated on using
continuous-annealing techniques because of higher production rates, better
uniformity of properties, and the possibility to use lower alloy steels.
However, box annealing has also been considered where continuous-annealing
facilities are not available.

In the continuous-annealing technique, the steel sheet is heated for
a short time (~2 min.) into the intercritical temperature range to form
ferrite-austenite mixtures, followed by accelerated cooling (~10°C/s) to
transform the austenite phase into martensite. The actual cooling rate is
dependent on sheet thickness and the quenching conditions on a given
annealing line. As a result, steel compositions must be adjusted to
obtain the hardenability needed for the cooling rate (sheet thickness).
For hot-rolled grades (thickness greater than 0.070 in.), 1.5Mn-Si-V
(14-17) or 1.2Mn-Si-Mo-Cr (18) steels have been proposed. For thinner,
cold-rolled grades, simpler 1.2Mn steels have been proposed. The carbon
content of these steels is generally below 0.15 percent to be sure that
the sheet steels are weldable. Rare-earth additions may also be made for
sulfide-shape control to improve ductility. Some typical chemical compositions are given in Table I (21).

In the box-annealing technique, a similar heat treatment is performed, but the annealing times are much longer (~3 h) and the cooling
rates are much slower (~20°C/hr). Because of the slow cooling rate, much
higher alloy steels are required to achieve the desired hardenability.
For this annealing technique, 2.5Mn steels (19,20), sometimes containing
appreciable amounts of Si and Cr, have been proposed (Table I) (21).

From a fundamental viewpoint, the formation of dual-phase microstructures in these different steels by intercritical annealing techniques can
be understood if the formation of austenite during intercritical
annealing, the subsequent transformation of this austenite, and changes in
the ferrite after intercritical annealing can be explained. These three
areas are discussed in the following three sections for low-carbon, 1.5Mn
steels, which are the base steels for most dual-phase-steel development.

Formation of Austenite During Intercritical Annealing

The formation of austenite during intercritical annealing has been
studied recently in low-carbon 1.5Mn steels by a number of investigators (22-26). Speich et al. (22) have shown that the formation of
austenite during intercritical annealing can be separated into several
steps: (1) almost instantaneous nucleation of austenite at pearlite or
grain-boundary cementite particles, followed by very rapid growth of
austenite until the carbide phase is dissolved; (2) slower growth of
austenite into ferrite at a rate that is controlled by carbon diffusion in
austenite at high temperatures (~850°C, 1562°F) and by manganese diffusion
in ferrite (or along grain boundaries) at low temperatures (~750°C,
1382°F); and (3) very slow final equilibration of ferrite and austenite at
a rate that is controlled by manganese diffusion in austenite.

An example of austenite formation in an 0.06C-1.5Mn steel at 740°C
(1364°F) is shown in Figure 5. The kinetics of austenite formation at

Table I. Typical Dual-Phase Steel Compositions, weight percent (21)

Production Technique	C	Mn	Si	Cr	Mo	V	Al	N	S	P	Other
Continuous Annealing, Hot-Rolled gages	0.12	1.55	0.61	-	-	0.06	0.05	0.007	0.006	0.015	Rare Earth
	0.11	1.43	0.58	0.12	0.08	-	0.04	0.007	0.012	0.015	-
Continuous Annealing, Cold-Rolled gages	0.11	1.20	0.40	-	-	-	0.04	-	0.005	0.015	-
Box Annealing	<0.13	<2.20	<1.50	<1.00	-	-	<0.08	-	<0.020	<0.02	Rare Earth
As-Rolled	0.06	0.90	1.35	0.50	0.35	-	0.03	-	0.010	0.010	Rare Earth

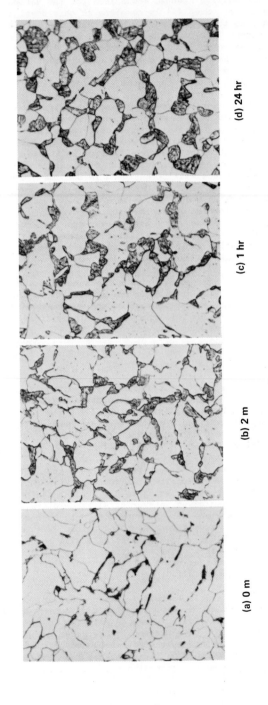

(a) 0 m   (b) 2 m   (c) 1 hr   (d) 24 hr

Fig. 5 - Austenite growth in 0.06C-1.5Mn steel at 740°C (22).

different temperatures in an 0.12C-1.5Mn steel are summarized in Figure 6, with the times and controlling kinetic mechanisms for the various steps indicated at each temperature. Step (1) is complete in this steel in times of less than 15 seconds; at this point the percent austenite is equal to the original pearlite content (~12 percent). Step (2) is complete at high temperatures (carbon-diffusion control) in times of about 1 minute; however, at low temperatures (manganese-diffusion control) the kinetics become very slow and austenite continues to grow over times of the order of hours. Step (3) is extremely slow and is never completed with normal annealing times.

The control of the austenite growth by manganese diffusion in the ferrite at low temperatures implies that manganese enrichment of the austenite phase must occur, and this has been observed by several investigators (22,25-28). A typical example has been shown in Figure 7 (22). Of course, a simple consideration of phase equilibrium in the ternary Fe-Mn-C system (29), as shown in the isothermal section of Figure 8, also indicates that the austenite phase will be enriched in both manganese and carbon if true equilibrium is achieved during intercritical annealing. However, many authors have assumed that because of the short annealing times only carbon redistribution occurs between the two phases, because substitutional manganese diffuses much more slowly than interstitial carbon. This situation is referred to as paraequilibrium (30). The phase boundaries for this situation can be calculated from thermodynamic considerations (31) and are shown as dashed lines in Figure 8.

In principle, neither true equilibrium nor paraequilibrium actually exists during most intercritical-annealing operations but an intermediate situation develops in which some enrichment of the austenite occurs near the ferrite/austenite interface. The thickness of the enriched layer will increase as the intercritical annealing time is increased. The manganese concentration profiles in the austenite phase are complex, however, and can only be obtained by balancing the carbon and manganese fluxes at the interface, as shown by Wycliffe, et al. (25,26).

If it is assumed for simplicity that no manganese redistribution occurs, then vertical sections corresponding to paraequilibrium conditions can be constructed at constant manganese content. The tie line $C^{\alpha}-C_o-C^{\gamma}$ of Figure 8 would then lie in the plane of the vertical section (this would not be the case for the true equilibrium tie line $C^{\alpha\gamma}-C_o-C^{\gamma\alpha}$). Such a diagram is shown in Figure 9.

From the lever rule, it can be seen from Figure 9 that for any given carbon content, the amount of austenite will increase with increasing intercritical temperature, becoming equal to 100 percent at the A_3 temperature. Similarly, for any given intercritical temperature, the amount of austenite will increase with increasing carbon content, becoming equal to 100 percent at a carbon content corresponding to the $\gamma/\alpha+\gamma$ boundary. Below the A_1 temperature (700°C) (32), no austenite will be formed. The data shown in Figure 10 illustrate the variation of the amount of austenite formed at different intercritical temperatures for 1.5Mn steels with different carbon contents (33).

Of course, dual-phase steels may also contain vanadium, columbium, silicon, chromium, phosphorus, or molybdenum, in addition to manganese (Table I). These additional alloying elements may lead to more complex effects on austenite formation than previously discussed. For instance, vanadium and niobium interact strongly with carbon and nitrogen to form a fine dispersion of VCN or NbCN. Upon intercritical annealing this dispersion may be dissolved as the austenite phase grows into the

10

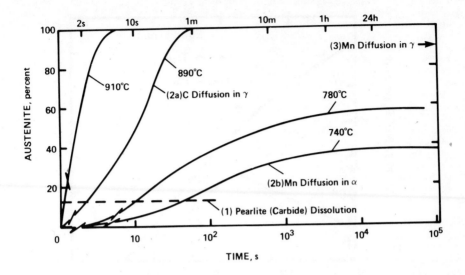

Fig. 6 - Kinetics of austenite formation in 0.12C-1.5Mn steel (22).

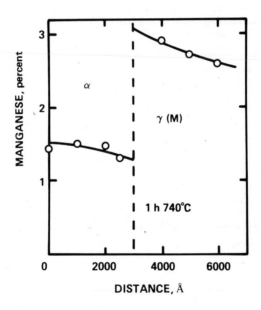

Fig. 7 - Redistribution of manganese between ferrite and austenite phases during intercritical annealing of 0.06C-1.5Mn steel (22).

11

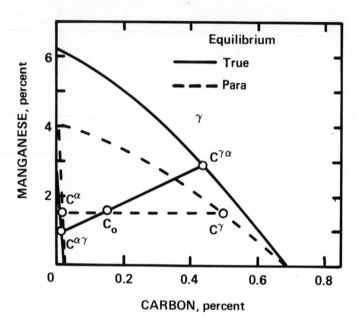

Fig. 8 - Isothermal section at 740°C of Fe-Mn-C ternary phase diagram showing calculated paraequilibrium boundaries (29,31).

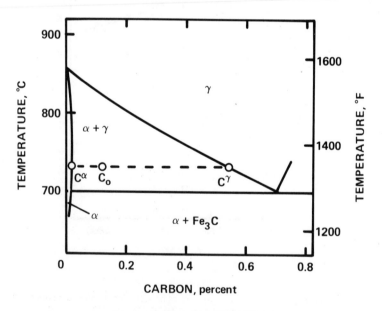

Fig. 9 - Phase diagram for 1.5Mn steel (paraequilibrium conditions).

ferrite (34). Also, silicon widens the α+γ phase field and permits a wider range of intercritical temperatures to be used (35). A complete discussion of all the alloying effects is beyond the scope of this paper.

Transformation of Austenite After Intercritical Annealing

Although the transformation of austenite after intercritical annealing is similar to the transformation of austenite after normal austenitizing, two features make this transformation process unique. First, because the carbon content of the austenite is fixed by the intercritical temperature (see Figure 9), the hardenability of the austenite phase varies with intercritical temperature (20). Thus, at low temperatures where the carbon content of the austenite is high, the hardenability of the austenite is high. Similarly, at high temperatures where the carbon content of the austenite is low, the hardenability of the austenite is low. Second, because ferrite already pre-exists, transformation of the austenite can proceed by epitaxial growth of this "old" ferrite into austenite with no nucleation step required (36,37).

The effect of intercritical temperature and cooling rate on the amount of martensite formed in an 0.12C-1.5Mn steel after intercritical annealing for 1 hour is shown in Figure 11 (33). At the highest cooling rate (240°C/s) essentially all the austenite transforms into martensite (dashed line in Figure 11 taken from Figure 10). At the slower cooling rates, part of the austenite transforms into ferrite or ferrite-carbide aggregates. At the higher intercritical temperatures, the fraction of austenite that transforms into nonmartensitic products is very large, whereas at the lower intercritical temperature this fraction is small. The higher hardenability of the austenite phase at lower intercritical temperatures is, in part, simply a result of its higher carbon content, but may also be a result of manganese redistribution to the austenite phase, as previously discussed.

The transformation products formed after intercritical annealing and cooling at different rates have been studied by several investigators (22,33,34,36-38). Matlock et al. (36) used special etching techniques developed by Lawson et al. (37) to show that epitaxial growth of ferrite into the austenite may occur on cooling after intercritical annealing of 1.5Mn-Nb dual-phase steels for short times at high temperatures (4 min. at 810°C), even when quenched rapidly (1000°C/s). This results in the envelopment of the austenite (martensite) phase with a rim of "new" ferrite, as shown in Figure 12. Both Matlock et al. (36) and Eldis (38) have shown that the epitaxial ferrite front degenerates into acicular ferrite at intermediate cooling rates.

In contrast, Speich et al. (22) have shown that after intercritical annealing of 0.06C-1.5Mn steels at low temperatures (1 hr at 740°C) and slow cooling, the manganese enrichment of the austenite may increase the hardenability near the austenite/ferrite interface, so that a martensite rim formed around the austenite particle, as shown in Figure 13. The center of the austenite particle evidently has transformed to a ferrite-carbide aggregate at higher temperatures. Similar structures have been reported by Wycliffe et al. (25,26). This structure is almost the exact opposite of that shown in Figure 12. Thus, a whole range of morphologies and a whole range of transformation products can be formed from the austenite phase after intercritical annealing, depending on the annealing temperature, time, and cooling rate.

13

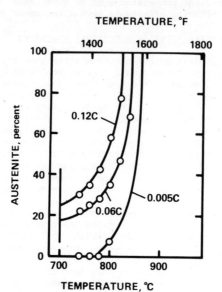

Fig. 10 - Percentage of austenite formed at various intercritical temperatures for 1.5Mn steels containing 0.005 to 0.12C (33).

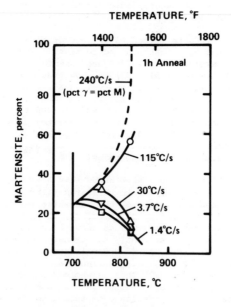

Fig. 11 - Transformation of austenite after intercritical annealing an 0.12C-1.5Mn steel for 1 hr at 820 and 740°C (33).

14

Fig. 12 - Formation of epitaxial ferrite rim around austenite particle in 0.06C-1.5Mn-0.05Nb steel after intercritical annealing 4 m at 810°C and water quenching ["old" ferrite = gray, new ferrite = white, martensite (austenite) = black; (chromic acid etch, light micrograph)] (36).

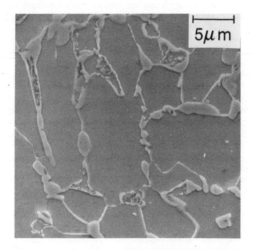

Fig. 13 - Formation of martensite rim about austenite particle in 0.06C-1.5Mn steel after intercritical annealing 1 hr at 740°C and slow cooling [center of austenite particle transforms to ferrite-carbide aggregate upon cooling (2% nital etch, scanning electron micrograph)] (22).

Of course, in addition to the principal effects of carbon and manganese, the hardenability of intercritically formed austenite is affected by
other alloying elements present in dual-phase steels as silicon, molybdenum, vanadium, and chromium, (Table I). Tanaka, et al. (18) have shown
that both chromium and molybdenum increase hardenability. Also,
Repas (12,17) has suggested that vanadium significantly increases hardenability. The role of silicon is less clear. However, a discussion of
these effects is beyond the scope of this paper.

The transformation of the austenite phase into martensite in dual-
phase steels occurs at low temperatures so that the ferrite phase must
plastically deform to accommodate the volume expansion (~2 to 4 percent)
arising from the austenite-to-martensite tranformation (39,40). As a
result, both a high dislocation density and residual stresses are generated in the ferrite phase immediately surrounding the martensite
particle. An example of the increased dislocation density in the ferrite
phase is shown in Figure 14. The residual stress patterns are on too
small a scale to be directly measured, but a theoretical analysis indicates that their maximum value would be of the order of the yield strength
of the ferrite (at the M_s temperature) and decay exponentially away from
the martensite-ferrite interface (41).

The martensite transformation substructure in dual-phase steels can
vary from the lath martensite substructure shown in Figure 14 (20),
typical of low-carbon martensites, to internally twinned substructures
typical of high-carbon martensites (42). These changes in morphology
reflect the effect of intercritical annealing temperature on the carbon
content of the austenite phase and in turn its effect on the M_s temperature (43). Presumably, autotempering of these martensites (42) could
occur at slower cooling rates but such effects have not been reported.

As the transformation of the austenite phase to martensite is not
complete, retained austenite is also generally present in dual-phase
steels in amounts varying from 2 to 9 percent (15,22,38,44). The retained
austenite is generally believed to be contained within the martensite
particles shown in Figure 14, presumably as interlath austenite films, as
reported for freshly quenched low- and medium-carbon martensitic
steels (43). For martensite particles of higher carbon content with an
internally twinned plate microstructure, the situation is less clear. A
number of authors, including Rigsbee and VanderArend (44) and Nakoaka
et al. (16), have presented convincing evidence for the existence of
isolated austenite particles in this case. A typical example of such an
isolated retained-austenite particle is shown in Figure 15. Rigsbee (27)
believes that the small (submicron) size of many of these austenite particles inhibits their transformation to martensite.

The amount of retained austenite appears to be sensitive to a number
of variables, particularly cooling rate. Evidently thermal stabilization
of the austenite phase will occur at slower cooling rates, typical of
effects observed in normal quenched-and-tempered steels (45). Thus,
water-quenched steels have less retained austenite than air-cooled steels,
as shown in Figure 16 (15). Although the retained austenite is generally
stable with respect to subzero cooling, it transforms readily after
straining a few percent, as shown in Figure 16 (15,38,44). Retained-
austenite contents increase slightly with increasing intercritical temperature and with increasing carbon content (44).

Fig. 14 - Increased dislocation density in ferrite phase and martensite substructure of 0.04C-1.5Mn steel intercritically annealed at 726°C and fast cooled (transmission electron micrograph) (20).

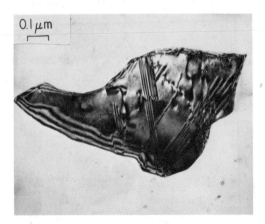

Fig. 15 - Retained austenite particle in 0.072C-1.3Mn-0.08Cb-0.08V steel intercritically annealed 5 m at 899°C and hot-oil quenched (transmission electron micrograph) (44).

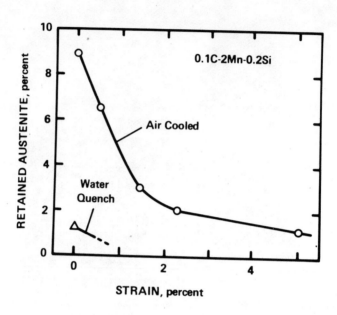

Fig. 16 - Percent retained austenite after air cooling and water quenching, and after plastic straining, in an 0.1C-2Mn-0.2Si steel intercritically annealed 2 m at 750°C (15).

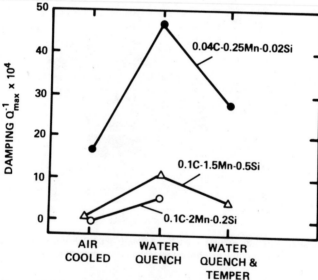

Fig. 17 - Internal friction maxima in several dual-phase steels intercritically annealed 2 m at 750°C and air-cooled, water-quenched, and water-quenched and tempered. (Q^{-1}_{max} x 10^4 ~ ppm C in solution)(15).

Changes in Ferrite Phase During Intercritical Annealing

In cold-rolled steels, recrystallization of the ferrite will occur rapidly and is generally complete before the steel reaches the intercritical annealing temperature, even during the rapid heating encountered on most continuous-annealing lines (23,24). Grain growth of the ferrite phase after recrystallization is generally restricted because of the pinning action of the second-phase austenite particles (23,24).

Changes in the carbon content of the ferrite phase may occur during intercritical annealing, and many investigators believe that this has a critical effect on the ductility (17,35,46,47) These changes may occur because of several reasons. First, the solubility of carbon in the ferrite may be lower at the intercritical temperature than that originally present in the ferrite phase of the as-received material. The solubility of carbon in the ferrite decreases with increasing intercritical temperature, but may also be markedly decreased by increasing the total alloy content of the steel. For instance, under equilibrium conditions, addition of 1.5Mn to a plain carbon steel would lower the solubility of carbon at 760°C from 0.02 to 0.005 percent under equilibrium conditions (see Figure 8). Additions of silicon are also reported to lower the carbon content of the ferrite phase (35,46,47), although Lagneborg (48) has disputed this claim, arguing that just the opposite is true.

Variations in the cooling rate from the intercritical temperature can also affect the carbon content of the ferrite phase. As the cooling rate is lowered, cementite may precipitate in the ferrite resulting in a lower carbon content. Some authors have also suggested that the epitaxial ferrite formed at slower cooling rates will maintain equilibrium with the austenite phase into which it is growing, resulting in a lower carbon content (36).

An example of the effect of alloy content and cooling rate on the internal friction maxima of several dual-phase steels is shown in Figure 17 (15). Increasing the manganese content from 0.25 to 1.5 percent drastically lowered the carbon content of the ferrite ($Q^{-1} \times 10^4 \sim$ ppm C in solution) after either water quenching, air cooling, or tempering. However, for a given steel, water quenching results in a higher carbon content in the ferrite phase than air cooling. The changes in the carbon content of the ferrite in higher alloyed steels for different quenching rates, however, appear to be very small, amounting to only a few ppm carbon. Also, increasing the silicon content from 0.25 to 0.50 percent increased the carbon content of the ferrite, in agreement with a recent suggestion by Lagneborg (48). As shown in Figure 17, tempering after water quenching also lowers the carbon content of the ferrite phase.

Finally, in steels that contain vanadium or niobium, a dispersion of VCN or NbCN precipitates is generally present in the ferrite phase. Upon intercritical annealing, these precipitates may coarsen (34). Also, as discussed earlier, as austenite grows into the ferrite during intercritical annealing, the VCN or NbCN precipitates may dissolve. Upon subsequent cooling, the epitaxial ferrite phase may grown into the austenite without forming a new dispersion of VCN or NbCN precipitates.

Continuous-Cooling Transformations

In addition to the use of intercritical heat treatment, dual-phase steels have been produced in the as-rolled condition by carefully

19

controlling the continuous-cooling transformation characteristics of the steel (49-51). This generally requires the addition of substantial amounts of Si, Cr, and Mo in addition to about 1.0 Mn. A typical as-rolled dual-phase-steel composition is given in Table I.

Although as-rolled dual-phase microstructures have been produced in continuously cooled plate steel for line-pipe applications (40), this is more difficult in sheet steels as a result of the different cooling rates experienced by the steel on a hot-strip mill. After the last roll pass, the strip cools rapidly while traveling on the runout table for about 10 seconds. After the strip is coiled, the cooling rate is reduced approximately to 22°C/h (40°F/h) (50). To produce a dual-phase micro-structure during continuous cooling under these conditions, (1) a large amount of polygonal ferrite, about 80 percent, must form rapidly, but pearlite must be suppressed while the steel cools on the runout table; and (2) the remaining islands of the carbon-enriched austenite must be stable enough to resist transformation to bainitic ferrite during the slow cooling in the coil (50). A desired type of continuous-cooling transfor-mation is shown in Figure 18 (12).

Additions of molybdenum have been found to be particularly beneficial in suppressing the pearlite transformation without preventing the forma-tion of polygonal ferrite during cooling, over a wide range of cooling rates. Silicon and chromium have also been found to be beneficial because silicon accelerates the polygonal ferrite reaction, and because silicon and chromium both increase the hardenability of the remaining austenite. Most as-rolled dual-phase steels therefore contain appreciable amounts of these alloying additions.

A typical microstructure of a Mn-Si-Mo-Cr as-rolled dual-phase steel is shown in Figure 19 (51). Generally, the microstructure contains, in addition to polygonal ferrite and martensite, some bainitic ferrite. The presence of varying amounts of bainitic ferrite has been blamed for the wide variability of the properties obtained in early mill trials of as-hot-rolled dual-phase steels (49-51). However, recent work on developing modified steel compositions appears to have resulted in better reproduci-bility of properties (51).

Production of as-rolled dual-phase steels has the obvious advantage of saving energy costs by eliminating a heat-treatment step. Also, dual-phase steels can be produced when continuous-annealing facilities are not available. However, balanced against these advantages are the disadvan-tages of a higher alloy cost and more variability in properties.

Structure-Property Relationships

Work Hardening and Yielding Behavior

Ferrite-martensite steels, in general, do not show a yield point because the combination of high residual stresses and a high mobile dislo-cation density discussed earlier causes plastic flow to occur easily at low plastic strains (38,40,44), as shown in Figure 3. Because plastic flow begins simultaneously at many sites throughout the specimen, discon-tinuous yielding is suppressed (52).

The work-hardening processes in dual-phase steels are complex but can be separated into three stages (36,44,53). In the first stage (0.1 to 0.5% strain), rapid work hardening is present because of the elimination

20

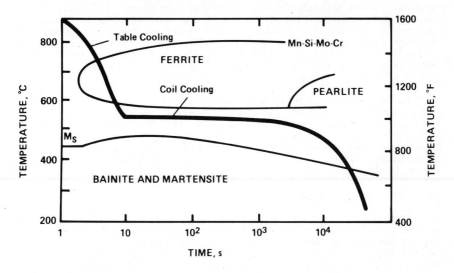

Fig. 18 - Continuous-cooling transformation behavior in Mn-Si-Mo-Cr steel resulting in dual-phase microstructures in as-rolled product (12).

Fig. 19 - Microstructure of as-rolled Mn-Si-Mo-Cr dual-phase steel (51).

21

of residual stresses and the rapid buildup of back stresses in the ferrite caused by the plastic incompatibility of the two phases. In the second stage (0.5 to 4% strain), the work-hardening rate of the ferrite is reduced as the plastic flow of the ferrite is constrained by the hard, undeforming martensite particles. Transformation of retained austenite may also occur during this stage, Figure 16. Finally, in the third stage (4 to 18% strain), dislocation cell structures are formed and further deformation in the ferrite is governed by dynamic recovery and cross slip and by eventual yielding of the martensite phase.

The work-hardening behavior of dual-phase steels is thus very complex, especially in the initial stages. However, the higher initial work-hardening rate is, believed to contribute to the good formability of these steels (53). Also, the lack of a yield point in these steels eliminates Luder band formation and assures a good surface finish after drawing. These topics will be discussed further in the review paper by Duncan and Embury in this volume (54).

Yield and Tensile Strength

From simple composite strengthening theory, it is expected that the strength of dual-phase steels should increase when either the volume fraction or the strength (hardness) of the martensite phase is increased (4,22,55-57). If equal strains are assumed in both phases, Tamura et al. (4,55) have shown that the flow stress of ferrite-martensite mixtures should be given by an equation of the form

$$S = S_\alpha \ (P_\alpha/100) + S_m \ (P_m/100) \tag{1}$$

where S is the nominal flow stress of the composite, S_α and S_m are the nominal stresses in the ferrite and the martensite, respectively, and P_α and P_m are the volume percentages of ferrite and martensite, respectively. Actually, the strains in each phase are far from equal (being concentrated in the softer ferrite phase). Also, the values of S_α and S_m must be dependent on the back stresses generated by the plastic incompatibility of the two phases. In spite of these drawbacks, Equation 1 is a useful expression for discussing the strength of ferrite-martensite mixtures in a semiempirical manner.

For strains corresponding to the yield strength (0.2% offset) or the tensile strength of the composite, Equation 1 becomes

$$S_y = S_{y,\alpha} \ (P_\alpha/100) + S_{y,m} \ (P_m/100) \tag{2}$$

$$S_T = S_{T,\alpha} \ (P_\alpha/100) + S_{T,m} \ (P_m/100) \tag{3}$$

where S_y, $S_{y,\alpha'}$ and $S_{y,m}$ are the yield strength of the composite and the nominal stresses in the ferrite and martensite, respectively, at a plastic strain corresponding to the yield strength (0.2%). Similarly, S_T, $S_{T,\alpha'}$ and $S_{T,m}$ are the tensile strength of the composite and the nominal stresses in the ferrite and martensite, respectively, at a plastic strain corresponding to the tensile strength.

Generally, simplifying assumptions have been made about the values of $S_{y,\alpha}$ and $S_{y,m'}$ and the values of $S_{T,\alpha}$ and $S_{T,m}$ in order to treat the variation of yield and tensile strengths of ferrite-martensite mixtures with volume fraction martensite. One extreme assumption is that the "law of mixtures" is obeyed, in which case $S_{y,\alpha}$ and $S_{y,m}$ become equal to the

22

yield strengths of the 100 percent ferrite and martensite phases respectively, which we shall define as $S^o_{y,\alpha}$ and $S^o_{y,m}$. Similarly, $S_{T,\alpha}$ and $S_{T,m}$ become equal to the tensile strengths of the 100 percent ferrite and martensite phases, respectively, which we shall define as $S^o_{T,\alpha}$ and $S^o_{T,m}$. Thus, we obtain

$$S_y = S^o_{y,\alpha} (P_\alpha/100) + S^o_{y,m} (P_m/100) \qquad (4)$$

$$S_T = S^o_{T,\alpha} (P_\alpha/100) + S^o_{T,m} (P_m/100) \qquad (5)$$

In early work, Tamura et al. (4) found that Equation 4 for the yield strength of ferrite-martensite iron-base alloys was not obeyed except when the values of $S^o_{y,\alpha}$ and $S^o_{y,m}$ were nearly equal, as in well-tempered ferrite-martensite mixtures. In general, the variation of S_y with P_m was more complex and depended on the ratio of the yield strength of the martensite to that of the ferrite, which Tamura et al. (4) called the parameter C, that is,

$$C = S^o_{y,m}/S^o_{y,\alpha} \qquad (6)$$

If C<3, then the law of mixture was nearly obeyed. However, if C>3, the value of S_y first increased linearly with P_m at a rate that was the same as for case for C = 3, but then began to deviate strongly from this relationship at a value of P_m that increased as the value of C increased. These results are summarized in Figure 20.

In subsequent work Davies (56) studied the tensile strengths of ferrite-martensite mixtures in 1.5Mn steels and found that Equation 5 was obeyed over the entire range of volume fractions. Surprisingly, Davies (56) found that even though he varied the carbon content of the martensite phase by varying the intercritical temperature, which resulted in an increase in $S^o_{T,m}$, no effect on S_T was detected.

In later work Speich and Miller (57) studied the dependence of both the yield and tensile strengths of ferrite-martensite mixtures in 1.5Mn steels that had been quenched from different intercritical temperatures, to vary both the volume fraction and strength (carbon content) of the martensite phase, as in the work of Davies (56). Their results are shown in Figure 21, with the appropriate values of C given in each case. (The carbon content of the martensite phase was 0.3% in one case and 0.5% in the other case). The results clearly parallel the findings of Tamura et al. (4), with the yield and tensile strengths varying in a non-linear manner with volume fraction of martensite, and with the strength being dependent on the value of C, in contrast to the work of Davies (56). These data are replotted in Figures 22 and 23, to show that, as suggested by Tamura et al., a linear relationship between strength and volume fraction exists only over a limited range of volume fractions of the martensite phase, when C>3.

Following the analysis of Tamura et al. (4,55), Speich and Miller (57) proposed that the linear relationship between yield and tensile strength and volume fraction could be approximated by the expressions

$$S_{y,m} = S^o_{y,\alpha} + \left(1/3\ S^o_{y,m} - S^o_{y,\alpha}\right) P_m/100 \qquad (7)$$

and

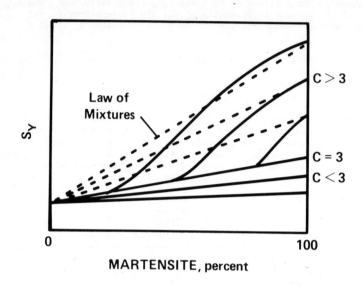

Fig. 20 - Effect of parameter C ($=S^{\circ}_{Y,m}/S^{\circ}_{Y, \alpha}$) on variation of yield strength with percent martensite (4).

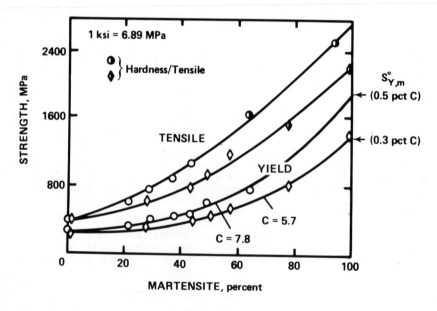

Fig. 21 - Yield and tensile strength of ferrite-martensite mixtures in 1.5Mn steels (57).

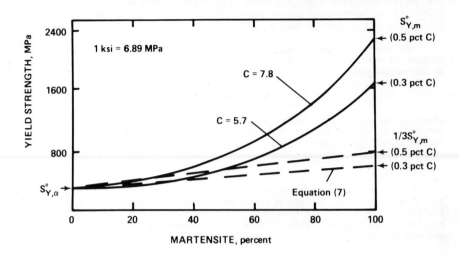

Fig. 22 - Linear approximation for yield strength of ferrite-martensite mixtures (57).

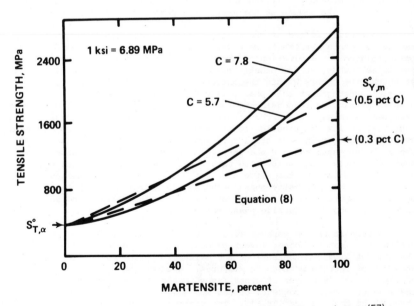

Fig. 23 - Linear approximation for tensile strength of ferrite-martensite mixtures (57).

$$S_{T,m} = S_{T,\alpha}^o + \left(S_{y,m}^o - S_{T,\alpha}^o \right) P_m/100 \qquad (8)$$

Equations 7 and 8 are a "modified law of mixtures" in that they are similar to Equations 4 and 5 but use much lower values for the effective stress in the martensite phase, namely $1/3$ $S_{y,m}^o$ rather than the value of $S_{y,m}^o$, and $S_{y,m}^o$ rather than $S_{T,m}^o$.

The yield strength of the martensite phase is related to the carbon content of the martensite phase, from the work of Leslie and Sober (58), by the expression

$$S_{y,m}^o \text{ (MPa)} = 620 + 2585C_m \qquad (9)$$

where C_m is the carbon content of the martensite phase in weight percent. In simple ferrite-martensite mixtures, the value of C_m is given approximately by the expression

$$C_m = 100 \ (C_o/P_m) \qquad (10)$$

where C_o is the carbon content of the steel, and where it is assumed that the densities of the ferrite and martensite are equal and that the carbon content of the ferrite is negligible. A more complex but more accurate expression which takes these effects into account has been given by Speich and Miller (57).

The straight-line relationships between strength and volume fractions in Figures 22 and 23 are Equations 7 and 8 with the correct value of $S_{y,m}^o$ substituted from Equation 9, corresponding to the carbon content of the martensite phase, given by Equation (10). These equations appear to closely represent the yield and tensile strengths of ferrite-martensite mixtures when the amount of the martensite phase is below 50 percent.

The smaller effect of the martensite phase on yield strength than on tensile strength appears to be a result of the initial yielding behavior being related to elimination of residual stresses about the martensite particles. Thus, at low plastic strains characteristic of the yield strength, the stress within the martensite phase only reaches about 1/3 the yield strength of this phase, resulting in Equation 7. For higher plastic strains corresponding to the tensile strength, residual stresses have been eliminated, and the stresses within the martensite phase increase, so its contribution to the overall strength increases. However, even then the strains are still very small in the martensite phase so that the yield strength and not the tensile strength of the martensite phase must be used in Equation 8. The marked increase in the strength above 50 percent martensite for either the yield or tensile strength appears to result from the martensite phase becoming the matrix phase so that it can begin to support a major fraction of the load with little or no deformation of the occluded ferrite phase occurring.

In support of the work of Tamura et al (4) and Speich and Miller (57), several other investigators (59,60) have found an effect of the strength (carbon content) of the martensite phase on the tensile strength of ferrite-martensite mixtures. However, in support of the work of Davies (56), a number of other investigators have reported no effect of the strength (carbon content) of the martensite phase on the tensile strength of the mixture (38,61). The reasons for these differences are unclear, but appear to be associated with a very large scatter band in the data of these latter workers, so that the effects of carbon content of the martensite phase may be masked.

Although a number of investigators have reported a linear variation of the strength with volume fraction martensite, they have failed to point out that this is really not a "law of mixtures" because the linear relationship does not extrapolate to the strength of the martensite phase. Also, in many cases these linear relationships have been established only over a limited volume fraction range, so that the deviations from linearity at high volume fractions have not been reported (38,59-61).

In addition to changes in the strength of the martensite phase, changes in the strength of the ferrite phase can affect the strength of the ferrite-martensite mixture. Thus, changes in the value of $S^o_{y,\alpha}$ or $S^o_{T,\alpha}$ in Equations 7 and 8 can result from changes in grain size, from solid-solution hardening additions, or from precipitation hardening (62). The effect of ferrite grain size on the strength of dual-phase steels has been analyzed by several investigators (4,59,63,64). As expected, the strength increases with decreasing grain size, but the Petch slope is a function of plastic strain because of the complicated yielding behavior of the steels (63). Also, additions of strong solid-solution-strengthening elements such as silicon or phosphorus increase the strength of dual-phase steels (17,65).

Further treatment of this area would require a discussion of the more sophisticated models that use computer modeling or continuum mechanics (66-71), which is beyond the scope of this paper.

Ductility

The effect of dual-phase microstructure on uniform and total elongation* is probably the most difficult to understand because of the multitude of factors reported to influence the elongation values. Among these are volume fraction of martensite, carbon content of the martensite, plasticity of the martensite, distribution of the martensite, alloy content of the ferrite, carbon content of the ferrite, amount of epitaxial ferrite, and retained austenite. We shall discuss each in that order.

Davies (56) was one of the first to systematically study the effect of the volume fraction of martensite on uniform elongation, using the theory of Mileiko (74). From, the Considere' condition (75).

$$(d\sigma/d\varepsilon) = \sigma \qquad (11)$$

and assuming that a power stress-strain equation

$$\sigma = K\varepsilon^n \qquad (12)$$

is obeyed, it can be shown that (72)

$$n = \varepsilon_u \qquad (13)$$

---

* Total elongation is dependent on gage length, cross-sectional area, and shape of the tension-test specimen. Because the exact method for correcting total elongation to the same test conditions is not clear, comparison of these values from different sources is very difficult. Uniform elongation is believed to be independent of specimen dimensions and shape (72,73).

where σ and ε are true stress and true strain, n is the exponent of the power law, K is a constant, and ε_u is the true uniform strain. If a power stress-strain law is assumed for the mixture and for the ferrite and martensite phases, and in addition the strains are assumed to be equal in each phase, Mileiko (74) showed that the uniform strain of the composite could be calculated from the values of K and n for each of the phases. Davies used this theory to analyze his results for 1.5Mn steels and for V-N HSLA steels, as shown in Figure 24.

Although rather good agreement was obtained between the Mileiko theory and experiment, it has been shown by a number of investigators (57,76) that Equation 12 is not obeyed for dual-phase steels, and also, as previously discussed, that the strain in the two phases is widely different.

Araki et al. (66) also analyzed the uniform strain and work hardening of ferrite-martensite mixtures by modifying the continuum theory of Mori and Tanaka (77). In this theory, the strain energy in each phase and the heat flow between the phases are considered to achieve an overall energy balance. A comparison of observed and calculated n values from this theory is shown in Figure 25.*

Both the work of Davies (56) and that of Araki et al. (66) indicate that the uniform elongation (n value) decreases in a nonlinear manner with increasing percent martensite. Speich and Miller (57) found a similar result, but in addition found that the uniform elongation increased slightly when the carbon content of the martensite phase was decreased. An empirical equation that fit the data for several intercritical temperatures was

$$e_u / e_{u,\alpha}^o = 1 - 2.2\ C_m\ (P_m/100)^{1/2} \qquad (14)$$

where e_u and $e_{u,\alpha}^o$ are the uniform elongation (in percent engineering strain) for the composite and for 100 percent ferrite, respectively. A plot of the data of Speich and Miller according to Equation 14 is shown in Figure 26. (Different intercritical temperatures gave different values of C_m).

Total elongation of ferrite-martensite steels also decreases with increasing percent martensite. Tamura et al. (4) report an exponential decrease in elongation and a higher total elongation for any given percent martensite when the value of C increased. Speich and Miller (57) found a relationship similar to Equation 14 for total elongation:

$$e_t / e_{t,\alpha}^o = 1 - 2.5\ C_m\ (P_m/100)^{1/2} \qquad (15)$$

where e_t and $e_{t,\alpha}^o$ are the total elongations (in percent engineering strain) of the composite and 100 percent ferrite, respectively. Again, total elongation was found to depend not only on the volume fraction martensite but also on the carbon content of the martensite phase, decreasing as the carbon content (strength) of the martensite phase increased. A

---

* The n value of Araki et al.(66) is an average n value over the strain range from 6 to 12 percent. The method of determining the n values in the work of Davies was not reported, but the much higher n values in this work (compare Figures 24 and 25) may result from evaluating n at lower plastic strains.

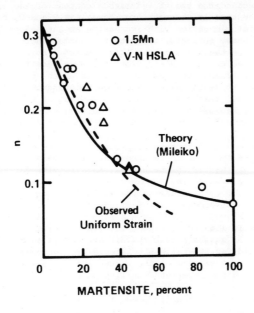

Fig. 24 - A comparison of the calculated and observed n values from theory of
Mileiko (56, 74).

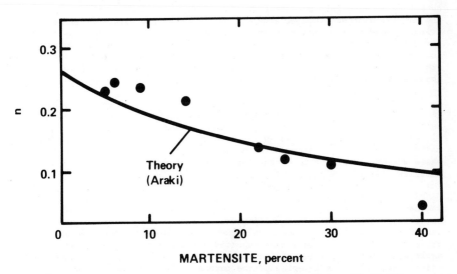

Fig. 25 - Comparison of calculated and observed n values for theory of Araki (66).

plot of the total elongation data of Speich and Miller according to Equation (15) is shown in Figure 26.

Of course, because the amount and carbon content of the martensite phase determine the tensile strength from Equations 8 and 9, the decrease in uniform and total elongation with increasing amount and carbon content of the martensite phase reflects, in part, the normal decrease expected in these values when the tensile strength is increased. However, even when the uniform and total elongation are plotted versus tensile strength, as shown in Figure 27, it appears that lower values of the carbon content (lower strength) of the martensite phase favor higher ductility. Speich and Miller (57) argued that the lower carbon content of the martensite increased ductility because cracking of the martensite particle or decohesion of the martensite/ferrite interface then became more difficult.

Rashid (76) has argued that the improved ductility of ferrite-martensite steels in contrast to ferrite-pearlite steels is caused by the higher plasticity of the martensite phase. As a result, during plastic straining decohesion or cracking of the martensite phase occurs at higher plastic strains than for the pearlite phase, and the ductility is improved. However, uniform-elongation values for ferrite-martensite and ferrite-pearlite mixtures in 1.5Mn steels reported by Davies and Magee (11) are nearly identical, except when the ferrite-pearlite steels are normalized. This suggests that the differences in ductility of ferrite-pearlite and ferrite-martensite steels are more subtle than a simple difference in plasticity of the martensite and pearlite phases, and are more dependent on the inherent ductility of the ferrite phase.

Distribution of the martensite phase must also influence ductility, but very few studies of this variable have been reported, except for the work of Becker and Hornbogen (78). Obviously, for any given percent martensite, a set of widely spaced, small martensite particles is desired. Chains of martensite particles that are linked up may be detrimental to ductility because this may offer an easy crack-propagation path through the matrix, although such effects are sensitive to the nature of the distribution and to the manner in which the sheet is stressed. Because the distribution of the martensite phase is determined by the nucleation of austenite particles at the cementite or pearlite phases present in the starting microstructure, it is important in intercritically annealed dual-phase steels that this microstructure be as uniform and fine as possible.

Because of the reciprocal relation between tensile strength and elongation shown in Figure 27, several investigators have proposed plotting such data as the product of tensile strength and elongation as shown in Figure 28 (47). The decrease in the product of $S_T \times e_t$ at 20 percent martensite has been considered to be significant, indicating that dual-phase steels containing more than this amount of martensite have inferior ductility (47).

Also, as shown in Figure 28, silicon results in a significant increase in this product for a given percent martensite, indicating a significant improvement in ductility even though silicon also increases tensile strength by solid-solution hardening. Repas (17) and several other investigators (35,65) have reported similar effects. The exact explanation for the silicon effect is not known, but it has been suggested that it arises from a lowering of the carbon content of the ferrite phase (35). However, as previously discussed, Lagneborg (48) has challenged this conclusion, arguing that silicon additions would increase the carbon content of the ferrite phase.

30

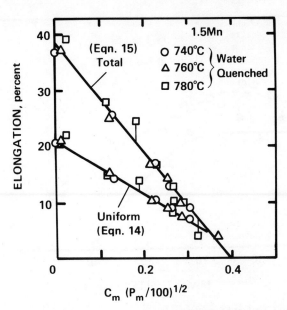

Fig. 26 - Effect of amount and carbon content of martensite phase on uniform and total elongation (57).

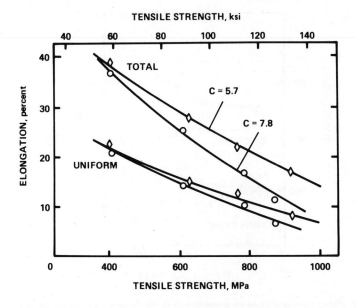

Fig. 27 - Relation between tensile strength and uniform and total elongation for water-quenched 1.5Mn ferrite-martensite steels (57).

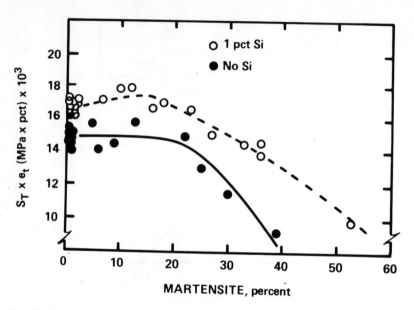

Fig. 28 - Variation of product of tensile strength and elongation with percent martensite for 0.02C-0.3Mn steels with and without silicon (47).

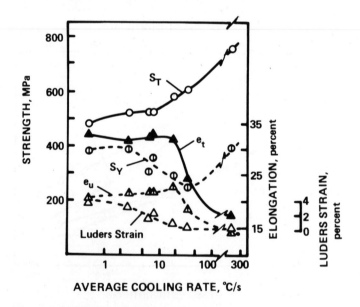

Fig. 29 - Effect of cooling rate on properties of 0.09C-0.4Si-1.5Mn steel after intercritical annealing at 750°C for 2 m (79).

Lowering of the carbon content of the ferrite phase by decreasing the cooling rate after continuous annealing, as discussed earlier, has also been reported by a number of investigators to be critical in obtaining the highest possible ductility in dual-phase steels. In particular Hayami et al. (79) have argued that the ductility obtained after water quenching is much inferior to that obtained with a mild cooling rate, as shown by the results in Figure 29, because of the lower carbon content of the ferrite in the slower cooled material. A decrease in ductility occurs when the cooling rate exceeds 30°C/s in these 0.09C-0.4Si-1.5Mn dual-phase steels. Rashid (76) and Repas (17) have reported a similar effect in V-N dual-phase steels.

Huppi et al. (80) argue that the amount of epitaxial ("new") ferrite formed upon cooling has an important effect on ductility, the ductility improving as the amount of epitaxial ferrite was increased. This improvement in ductility is reported to be caused by a lack of the precipitation of NbCN in the expitaxial ferrite which was formerly present in the retained (old) ferrite. Presumably, this lack of precipitation in the envelope of epitaxial ferrite surrounding the martensite particles results in an improved ductility. A similar argument has been proposed for non-microalloyed dual-phase steels by Krauss and Matlock (81). The higher-ductility in this case is evidently related to the low carbon content of the epitaxial ferrite rim which maintains equilibrium with the austenite phase into which it is growing.

The transformation of retained austenite during plastic straining and the resultant increase in work-hardening rate has also been used to explain the higher ductility of dual-phase steels. This is similar to the well-known transformation-induced-plasticity (TRIP) mechanism used to explain the high ductility of metastable austenitic stainless steels (82). Furukawa et al. (15) showed that the retained-austenite content increased from 2 percent to 9 percent when intercritically annealed dual-phase steels were cooled to successively lower temperatures and then quenched, as shown in Figure 30. Although the strength decreased simultaneously, as shown in Figure 30, and a ductility increase would be expected from this effect alone, Furukawa et al. (15) argued that the ductility increase was primarily due to the increased retained-austenite content. The increase in retained austenite with lower quenching temperatures (or lower cooling rates) is presumably attributed to the stabilization effects previously discussed. Marder (83) and Rigsbee and VandeArend (44) also suggest a close connection between the amount of retained austenite and the ductility of dual-phase steels. Rashid and Rao (84) have also found a relationship between the retained austenite content and the enhanced ductility of dual-phase steels.

In contrast, Eldis (38) found no connection between retained-austenite content and ductility in a number of dual-phase steels. He argued that because the retained austenite transforms during the early stages of plastic straining (see Figure 16), it could not influence the onset of necking in the tension-test specimen, which occurs late in the plastic straining process. Speich and Miller (57) also discounted the effect of retained austenite on ductility because of the small amounts of retained austenite (2 to 4%) present in their specimens.

Arguments about the effect of retained austenite in dual-phase steels are complex and still not resolved. When large amounts of retained austenite are present and when the stability of the austenite is sufficient that transformation of the retained austenite occurs late in the plastic-straining process, then large effects on uniform elongation can be

33

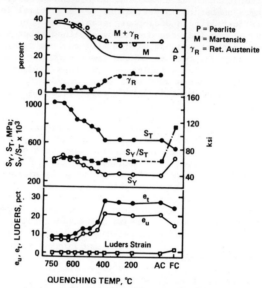

Fig. 30 - Variation in retained austenite content, strength, and ductility of 0.1C-0.2Si-2Mn dual-phase steel cooled to various temperatures from 750°C and quenched (15).

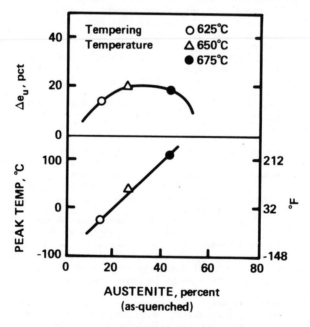

Fig. 31 - Effect of retained austenite on uniform elongation in austenite-martensite steels (85).

expected. An example of the possible magnitude of the effects from the work of Nikura et al. (85) on tempered 5Mn steels with an austenite-martensite structure is shown in Figure 31. The presence of 20 percent retained austenite in these steels increased uniform elongation by 20 percent! (The temperature at which the maximum in uniform elongation occurred varied, depending on the stability of the retained austenite which was fixed by the tempering temperature.) However, when small amounts of retained austenite are present, and especially when the retained austenite transforms early in the straining process, then the effect of retained austenite on ductility may be negligible.

Tempering and Strain Aging

Tempering may be used deliberately in some continuous annealing operations where the dual-phase steels are water quenched from the inter-critical annealing temperature (47). Tempering may also occur during hot-dip galvanizing operations. In addition, tempering of formed parts will occur during paint-baking cycles, although in this case the part has been plastically strained and the tempering is really a strain-aging process. We shall first discuss changes in ferrite-martensite steels that occur upon tempering, followed by a discussion of strain aging.

Tempering

In general, tempering reactions in dual-phase steels are a combination of those effects expected for each of the individual phases. Thus, in the martensite phase we expect to have recovery of the defect structure, precipitation of carbides, and transformation of the retained austenite (42). Similarly, in the ferrite phase we expect to have carbon segregation to dislocations in the ferrite and precipitation of carbides. In addition, there are synergistic effects that can be attributed to the presence of both phases. For instance, the formation of martensite generates residual stresses and a high dislocation density in the ferrite, especially near the martensite/ferrite interface. Carbon segregation to these dislocations, and the elimination of the residual stresses by the volume contraction of the martensite phase during tempering, is an important part of the tempering process (86).

At low temperatures (~200°C), the segregation of carbon to disloca-tions and the elimination of residual stresses results in an increase in the yield strength and a return of discontinuous yielding, but only if the volume fraction of the martensite phase is below 30 percent, as shown in Figure 32 (86). At higher volume fractions of martensite, the initial yielding behavior appears to be less affected by tempering because of higher dislocation density and higher residual stresses. The tensile strength is lowered slightly after tempering at low temperatures because of the decrease in hardness of the martensite phase, for steels with both low and high volume fractions of martensite (86).

At higher temperatures (~650°C), tempering results in a return of discontinuous yielding and a lowering of both the yield and tensile strengths, independent of the amount of martensite (86).

Strain Aging

Strain aging of dual-phase steels can be separated into several problems. In the case of continuously annealed steels which are simply cooled from the intercritical temperature, it is desirable to know how

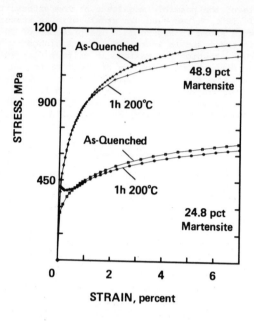

Fig. 32 - Effect of tempering 1 h at 200°C on stress-strain behavior of 0.06-0.20C, 1.5Mn steels containing various percentages of martensite (86).

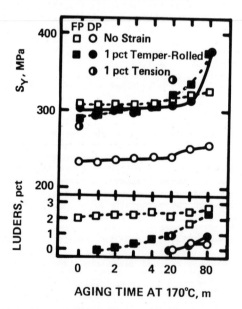

Fig. 33 - Effect of aging at 170°C for various times on yield strength and Luder's strain in 0.08C-1.2Mn-0.5Cr steel in ferrite-pearlite (FP) and dual-phase conditions (DP).(18).

long the steels can be stored at room temperature before discontinuous yielding returns. In the case of continuously annealed steels which are quenched from the intercritical temperature and then tempered, temper rolling may be used to remove discontinuous yielding that returns upon tempering. The relation between the tempering temperature and the subsequent return of discontinuous yielding at room temperature or elevated temperature in these temper-rolled steels is then of interest. Finally, the change in strength and ductility of parts which have been formed from dual-phase steels and subsequently exposed to paint-baking thermal cycles is of interest.

In general, as-annealed dual-phase steels are non-aging at room temperature (87), and exhibit sluggish aging behavior at elevated temperatures up to 260°C. However, deformation either by cold rolling or by tensile straining accelerates the aging process.

Tanaka et al. (18) have studied the return of the yield point in 0.06C-1.2Mn-0.5Cr dual-phase steels after intercritical annealing, after temper rolling 1 percent, and after straining in tension, for various aging times at 100 and 170°C. After intercritical annealing, the yield point did not return after aging for 125 hours at 100°C, but did return after aging 40 minutes at 170°C. This was also true after 1 percent temper rolling. However, after straining 1 percent in tension, a return of yield point was observed after aging only 2 hours at 100°C, and after aging 20 minutes at 170°C. Aging data for dual-phase steels, along with comparable data for the same steels in the ferrite-pearlite condition, are shown in Figure 33.

Tanaka et al. (18) explained the more rapid strain aging after straining 1 percent in tension by changes in dislocation distribution. In the as-quenched condition, the dislocation distribution in dual-phase steels is markedly nonuniform, and in the region of high dislocation density the excess carbon is drained locally before sufficient pinning of dislocations has been achieved to cause a return of the yield point. To get additional pinning, carbon must diffuse from the grain interior to the region of high dislocation density near the martensite/ferrite interface. However, this requires a long aging time. If the specimen is slightly strained in tension, the dislocation density is made much more uniform than that which was present in the as-quenched or temper-rolled conditions. As a result carbon diffusion only needs to occur over short distances, and the yield point returns quickly. Tanaka et al. also studied the effect on the yield strength of prestraining various amounts and then aging at 100 and 170°C. The maximum change in yield strength appeared to occur at 1 percent strain as shown in Figure 34.

Nakoaka et al. (47) have studied increases in yield strength of dual-phase steels that were water quenched from different temperatures, and then tempered to vary the solute carbon content. They found that the subsequent increase in yield strength upon tempering for 20 minutes at 170°C (referred to as bake hardenability) was directly related to the solute carbon content, as shown in Figure 35.

Rashid (88) has studied the change in strength and ductility of a number of different dual-phase steels after prestraining up to 25 percent and aging for 1 hour at 205°C. In all cases the yield strength increased and the ductility decreased. The change in tensile strength was more complex, increasing for all degrees of prestrain in microalloyed steels but decreasing for low degrees of prestrain in steels without microalloying additions. Rashid explains these differences by a reversal from a net

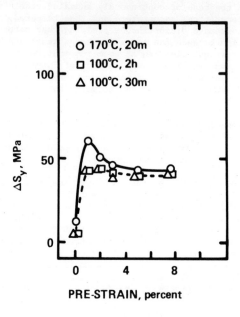

Fig. 34 - Effect of prestrain on change in yield strength after aging at 100 and 170°C in ⋅ 0.08C, 1.2Mn, 0.5Cr dual-phase steel.(18).

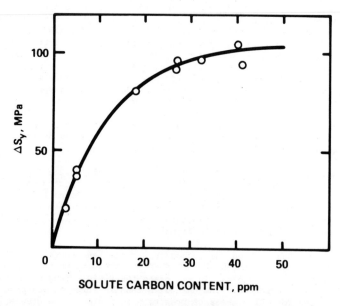

Fig. 35 - Change in yield strength of 0.07C-0.38Mn dual-phase steel intercritically annealed, water quenched and tempered, followed by aging 20 m at 170°C (47).

softening when the tempering of the martensite phase is important to a net strengthening when ferrite strengthening resulting from the interaction between dislocations and NbCN or VCN precipitates is important.

Davies (89) has also proposed that the strain-aging of microalloyed steels is influenced by the interaction of carbon atoms with clusters of vanadium atoms in the ferrite matrix.

Summary

Intercritical heat treatment, continuous cooling transformations, structure-property relationships, and tempering and strain aging reactions in dual-phase steels have been reviewed. Dual-phase steels can be produced by intercritical heat treatment or directly after hot rolling. The formation of austenite during intercritical annealing and the transformation of this austenite after intercritical annealing are now reasonably well understood and form a basis for understanding the production of dual-phase steels by continuous annealing or box annealing. Continuous cooling transformation behavior has been studied in less detail but the essential features for producing dual-phase steels in the as-rolled condition have been defined.

Structure-property relationships in these steels have been extensively studied. In particular, the effect of the volume fraction martensite, the carbon content of the martensite, the grain size of the ferrite, and the solute content of the ferrite on the yield and tensile strength are now well documented and reasonably well understood. The ductility of these steels is less well understood because it is influenced not only by the above factors but also by more subtle factors such as carbon content of the ferrite, amount of retained austenite, and the amount of epitaxial ferrite.

Tempering of dual-phase steels at low temperatures (~200°C) increases the yield strength, causes a return of the yield point, and lowers the tensile strength. Dual-phase steels are nonaging at room temperature in the as-annealed condition, but straining in tension accelerates the return of the yield point. The increase in yield strength after pre-straining and aging at low temperatures (~200°C) is primarily related to the total carbon content of the ferrite, but interactions of carbon with microalloying elements may also be important.

References

1. Grange, R. A., "Fibrous Microstructures Developed in Steel by Thermo-chemical Processing," pp. 861-76 in Proceedings 2nd Int. Conf. on Strength of Metals and Alloys, ASM, Cleveland, 1970.

2. Cairns, R. L. and Charles, J. A., "Production of Controlled Martensite-Ferrite Microstructures in Steel," Journal Iron and Steel Inst., 205 (1967) pp. 1044-50.

3. Cairns, R. L. and Charles, J. A., "Mechanical Properties of Steels with Controlled Martensite-Ferrite Microstructures," ibid, pp. 1051-65.

4. Tamura, I., Tomata, Y., Akao, A., Yamaoha, Y., Ozawa, M., and Kanotoni, S., "On the Strength and Ductility of Two-Phase Iron Alloys," Trans. Iron and Steel Inst. Japan, 13 (1973) pp. 283-92.

5. Hayami, S. and Furukawa, T., "A Family of High Strength, Cold-Rolled Steels," in Microalloying 75, Session 2A, Vanitec, London, 1975, pp. 78-87.

6. Rashid, M. S., "A Unique High-Strength Sheet Steel with Superior Formability," SAE Preprint 760206, 1976.

7. Rashid, M. S., "GM980X-Potential Applications and Review," SAE Preprint 770211, 1977.

8. Formable HSLA and Dual-Phase Steels, A. T. Davenport, ed.; AIME, New York, NY, 1979.

9. Dual-Phase and Cold Pressing Vanadium Steels in the Automobile Industry, Vanitec, Berlin, 1978.

10. Structure and Properties of Dual-Phase Steels, R. A. Kot and J. W. Morris, ed.; AIME, New York, NY, 1979.

11. Davies, R. G. and Magee, C. L., "Physical Metallurgy of Automotive High-Strength Steels," pp. 1-19, ibid.

12. Repas, P. E., "Physical Metallurgy of Dual-Phase Steels," pp. 277-305, in Mech. Working and Steel Processing Conf. XVII, AIME, New York, NY, 1979.

13. Owen, W. S., "Can a Simple Heat Treatment Help to Save Detroit" Metals Technology, 7 (1980) pp. 1-13.

14. Bucher, J. H., and Hamburg, E. G., "High-Strength Formable Sheet Steel," SAE Reprint 770164, 1977.

15. Furukawa, T., Morikawa, H., Takechi, H., and Koyama, K., "Process Factors for Highly Ductile Dual-Phase Steel Sheet," pp. 281-303 in Structure and Properties of Dual-Phase Steels, R. A. Kot and J. W. Morris, ed.; AIME, New York, NY, 1979.

16. Nakaoka, K., Hosoya, Y., Ohmura, M., and Nishimoto, A., "Reassessment of the Water-Quench Process as a Means of Producing Dual-Phase Formable Steel Sheets," pp. 330-45, ibid.

17. Repas, P. E., "Metallurgy, Production, Technology, and Properties of Dual-Phase Steels," pp. 13-22 in Dual-Phase and Cold-Forming Vanadium Steels in the Automobile Industry, Vanitec, Berlin, 1978.

18. Tanaka, T., Nishida, M., and Hashiguchi, K., and Kato, T., "Formation and Properties of Ferrite Plus Martensite Dual-Phase Structures," pp. 221-41 in Structure and Properties of Dual-Phase Steels, R. A. Kot and J. W. Morris, ed.; AIME, New York, NY, 1979.

19. Matsuoka, T., and Yamomori, K., "Metallurgical Aspects of Cold-Rolled High Strength Steel Sheets," Metallurgical Transactions, 6A (1975), pp. 1613-22.

20. Mould, P. R. and Skena, C. C., "Structure and Properties of Cold-Rolled Ferrite-Martensite (Dual-Phase) Steel Sheets," pp. 183-205 in Formable HSLA and Dual-Phase Steels, R. A. Kot and J. W. Morris, ed.; AIME, New York, NY, 1979.

21. Coldren, A. P., Eldis, G. T., Buck, R. M., Tither, G., Boussel, P., and Chihara, T., Journal of Molybdenum Technology, 14 (1980) No. 3, pp. 3-12.

22. Speich, G. R., Demarest, V. A., and Miller, R. L., "Formation of Austenite During Intercritical Annealing of Dual-Phase Steels," submitted to Metallurgical Transactions, 1980.

23. Garcia, C. I. and DeArdo, A. J., "The Formation of Austenite in Low Alloy Steels," pp. 40-61 in Structure and Properties of Dual-Phase Steels, R. A. Kot and J. W. Morris, ed.; AIME, New York, NY, 1979.

24. Garcia, C. I. and DeArdo, A. J., "Formation of Austenite in 1.5 Mn Steels," submitted to Metallurgical Transactions, 1980.

25. Wycliffe, P. A., Purdy, G. R., and Embury, J. D., "Growth of Austenite in the Intercritical Annealing of Fe-C-Mn Dual-Phase Steels," submitted for publication, 1980.

26. Wycliffe, P. A., Purdy, G. R., and Embury, J. D., "Austenite Growth in the Intercritical Annealing of Fe-C-Mn and Fe-C-Mn-Si Dual-Phase Steels," this book.

27. Rigsbee, J. M., "Inhibition of Martensitic Transformation of Small Austenite Particles in Low-Alloy Steels," pp. 381-5 in Proc. Int. Conf. Martensitic Transformations, Boston, 1979.

28. Matsuoka, T., Takahashi, M., and Okamato, A., "Production of Cold-Rolled Dual-Phase Steel Sheet for Outer Panel," pp. 62-3 in Dual-Phase and Cold Pressing Vanadium Steels in the Automobile Industry, Vanitec, Berlin, 1978.

29. Hillert, M. and Waldenstrom, M., "Isothermal Sections of the Fe-Mn-C System in the Temperature Range 873 k-1373K," Calphad, 1 (1977) pp. 97-132.

30. Hultgren, A., "Isothermal Transformation of Austenite," Trans ASM, 39 (1947) pp. 915-86.

31. Gilmour, J. B., Purdy, G. R., and Kirhaldy, J. S., "Thermodynamics Controlling the Pro-Eutectoid Ferrite Transformations in Fe-Mn-C Alloys," Met. Trans., 3 (1972) pp. 1455-64.

32. Andrews, K. W., "Empirical Formula for the Calculation of Some Transformation Temperatures," J. Iron and Steel Inst., 203 (1965) pp. 721-27.

33. Speich, G. R. and Miller, R. L., private communication, Research Laboratory, U. S. Steel Corporation, Monroeville, Pa., 1980.

34. Geib, M. D., Matlock, D. K., and Krauss, G., "The Effect of Intercritical Temperature on the Structure of Niobium Microalloyed Dual-Phase Steel," Met. Trans., 11A (1980) pp. 1683-89.

35. Thomas, G. and Koo, J. Y., "Developments in Strong, Duplex, Ferritic-Martensitic Steels, pp,. 183-201 in Structure and Properties of Dual Phase Steels, R. A. Kot and J. W. Morris, ed.; AIME, New York, NY, 1979.

36. Matlock, D. K., Krauss, G., Ramos, L. F., and Huppi, G. S., "A Correlation of Processing Variables with Deformation Behavior of Dual-Phase Steels," pp. 62-90 in Structure and Properties of Dual-Phase Steels, R. A. Kot and J. W. Morris, ed.; AIME, New York, NY, 1979.

37. Lawson, R. D., Matlock, D. K., and Krauss, G., "An Etching Technique for Microalloyed Dual-Phase Steels," Metallography, 13 (1980), pp. 71-87.

38. Eldis, G. T., "The Influence of Microstructure and Testing Procedure on the Measured Mechanical Properties of Heat-Treated Dual-Phase Steels," pp. 202-20, in Structure and Properties of Dual-Phase Steels, R. A. Kot and J. W. Morris, ed.; AIME, New York, NY, 1979.

39. Moyer, J. M. and Ansell, G. S., "The Volume Expansion Accompanying the Martensite Transformation in Iron-Carbon Alloys," Met. Trans., 6A (1975) pp. 178-91.

40. Dabkowski, D. S. and Speich, G. R., "Transformation Products and the Stress-Strain Behavior of Control-Rolled Mn-Mo-Cb Line-Pipe Steels," pp. 284-312 in Proc. Mechanical Working and Steel Processing Conf. XV, AIME, New York, NY, 1977.

41. Hillert, M., "Pressure Induced Diffusion and Deformation During Precipitation, Especially Graphitization," Jerkontorets Annaler, 141 (1957) pp. 67-89.

42. Speich, G. R. and Leslie, W. C., "Tempering of Steel," Met. Trans., 3 (1972) pp. 1043-54.

43. Rao, Narashima B. V., and Thomas, G., "Transmission Electron Characterization of Dislocated Lath Martensite," pp. 12-21 in Proc. Int. Conf on Martensite Transformations, Cambridge, Mass., 1979.

44. Rigsbee, J. M. and VanderArend, P. J., "Laboratory Studies of Microstructures and Structure-Property Relationships in Dual-Phase Steels," pp. 56-86 in Formable HSLA and Dual-Phase Steels, R. A. Kot and J. W. Morris, ed.; AIME, New York, NY, 1979.

45. Z. Nishiyama, _Martensitic Transformation_, Academic Press, New York, NY, 1978.

46. Davies, R. G., "The Deformation Behavior of a Vanadium Strengthened Dual-Phase Steel," _Met. Trans._, 9A (1978), pp. 41-52.

47. Nakoaka, K., Araki, K., and Kurihara, K., "Strength, Ductility, and Aging Properties of Continuously Annealed Dual-Phase High-Strength Sheet Steels," pp. 126-41 in _Formable HSLA and Dual-Phase Steels_, R. A. Kot and J. W. Morris, ed.; AIME, New York, NY, 1979.

48. Lagneborg, R., "Structure Property Relationships in Dual-Phase Steels," pp. 43-50 in _Dual-Phase and Cold-Pressing Vanadium Steels in the Autombile Industry_, Vanitec, Berlin, 1978.

49. Coldren, A. P., and Tither, G., "Development of a Mn-Si-Cr-Mo As-Rolled Dual-Phase Steel," _J. Metals_, 30 (4) (1978) pp. 6-19.

50. Coldren, A. P., Tither, G., Cornford, A., and Hiam, J. R., "Development and Mill Trial in As-Rolled Dual-Phase Steel," pp. 207-28 in _Formable HSLA and Dual-Phase Steels_, R. A. Kot and J. W. Morris, ed.; AIME, New York, NY, 1979.

51. Eldis, G. T. and Coldren, A. P., "Using CCT Diagrams to Optimize the Composition of an As-Rolled Dual-Phase Steel," _J. Metals_, 32 (No. 3) (1980), pp. 41-48.

52. Baird, J. D., "Strain Aging of Steel--A Critical Review," Parts I and II, _Iron and Steel_, 63 (1963) pp. 186-91, pp. 326-334, pp. 368-74.

53. Gerbase, J., Embury, J. D., and Hobbs, R. M., "The Mechanical Behavior of Some Dual-Phase Steels--With Emphasis on the Initial Work Hardening Rate," pp. 118-44 in _Structure and Properties of Dual-Phase Steels_, R. A. Kot and J. W. Morris, ed.; New York, NY, 1979.

54. Duncan, J., and Embury, J. D., "Mechanical Metallurgy of Dual-Phase Steels," this book.

55. Tamura, I., Tomata, Y., and Ozawa, H., "Strength and Ductility of Fe-Ni-C Alloys Composed of Austenite and Martensite with Various Strengths," pp. 611-15 in _Proc. 3rd Int. Conf. on Strength of Metals and Alloys_, Vol. 1 Cambridge, 1973.

56. Davies, R. G., "Influence of Martensite Composition and Content on the Properties of Dual-Phase Steels," _Met. Trans._, 9A (1978) pp. 671-79.

57. Speich, G. R., and Miller, R. L., "Mechanical Properties of Ferrite Martensite Steels," pp. 145-82 in _Structure and Properties of Dual-Phase Steels_, R. A. Kot and J. W. Morris, ed.; AIME, New York, NY, 1979.

58. Leslie, W. C. and Sober, R. J., "The Strength of Ferrite and of Martensite as Functions of Composition, Temperature, and Strain Rate," _Trans ASM_, 60 (1967) pp. 459-84.

59. Koo, J. Y., Young, M. J., and Thomas, G., "On the Law of Mixtures in Dual-Phase Steels," _Met. Trans. A._, 11A (1980) pp. 852-54.

60. Ramos, L. F., Matlock, D. K., and Kraus, G., "On the Deformation Behavior of Dual-Phase Steels," Met. Trans. A., 10A (1979) pp. 259-61.

61. Marder, A. R. and Bramfitt, B. L., "Processing of a Molybdenum-Bearing Steel," pp. 242-59 in Structure and Properties of Dual-Phase Steel, R. A. Kot and J. W. Morris, ed.; AIME, New York, NY, 1979.

62. Pickering, F. B., "Physical Metallurgy and the Design of Steels," Applied Science Publishers Ltd., London, 1978.

63. Ramos, L. F., Matlock, D. K., and Krauss, G., "The Effects of Ferrite Grain Size on the Deformation Behavior of Dual-Phase Steels and Quenched Armco Iron," submitted to Met. Trans. A., 1980.

64. Repas, P. E., "Microstructural Effects in Dual-Phase Sheet Steels," AIME Meeting, Las Vegas, Nev., 1980.

65. Davies, R. G., "Effect of Silicon and Phosphorus on the Mechanical Properties of Both Ferrite and Dual-Phase Steels," Met. Trans. A., 10A (1979) pp. 113-18.

66. Araki, K., Takada, Y., and Nakoaka, K., "Work Hardening of Continuously Annealed Dual-Phase Steels," Trans. Iron and Steel Inst. Japan, 17 (1977) pp. 710-17.

67. Fishmeister, H., and Karlson, B., "Plastizitätseigenschaften Grobzweiphasiger Werkstoffe," Z. Metallkunde, 68 (1977) pp. 311-27.

68. Karlson, B. and Sundstrom, B. O., "Inhomogeneity in Plastic Deformation of Two-Phase Steels," Mat. Sci. Eng., 16 (1974) pp. 161-68.

69. Bhadeshia, H. K. D. H., and Edmonds, D. V., "Analysis of the Mechanical Properties and Microstructure of High Silicon Dual-Phase Steel," Metal Science Journal, 14 (1980) pp. 41-49.

70. Korzewa, D. A., Lawson, R. D., Matlock, D. K., and Krauss, G., "A Consideration of Models Describing the Strength and Ductility of Dual-Phase Steels," Scripta Met., 14 (1980) pp. 1023-28.

71. Tomata, Y., Kuroki, K., Mori, T., and Tamura, I., "Tensile Deformation of Two-Ductile-Phase Alloys: Flow Curves of α-γ Fe-Cr-Ni Alloys," Mat. Sci. Eng., 24, (1976) pp. 85-94.

72. Dieter, G., Mechanical Metallurgy, McGraw Hill Inc., New York, NY, 1961.

73. Holt, J. M., private communication, Research Laboratory, U. S. Steel Corporation, Monroeville, Pa., 1980.

74. Mileiko, S. T., "The Tensile Strength and Ductility of Continuous Fiber Composites," J. Materials Science, 4 (1969) pp. 974-77.

75. Considere', A., Memoire sur l'emploi du fer et de l' acier dans les Constructions, Ann. Ponts de Chauss, 9 (1885) pp. 574-75.

76. Rashid, M. S., "Relationship Between Steel Microstructure and Formability," pp. 1-24 in Formable HSLA and Dual-Phase Steels, R. A. Kot and J. W. Morris, ed.; AIME, New York, NY, 1979,

77. Mori, T. and Tanaka, K., "Average Stress in Matrix and Average Elastic Energy of Materials with Misfitting Inclusions," Acta Met., 21 (1973) pp. 571-74.

78. Becker, J., and Hornbogen, E., "Microscopic Analysis of the Formation of Dual-Phase Steels," pp. 20-39 in Structure and Properties of Dual-Phase Steels, R. A. Kot and J. W. Morris, ed.; AIME, New York, NY, 1979.

79. Hayami, S., Furakawa, T., Gondoh, H., and Takechi, H., pp. 169-182 in "Recent Developments in Formable Hot and Cold Rolled HSLA Including Dual-Phase Sheet Steels," Formable HSLA and Dual-Phase Steels, R. A. Kot and J. W. Morris, ed.; AIME, New York, NY, 1979.

80. Huppi, G. S., Matlock, P. K., and Krauss, G., "An Evaluation of the Importance of Epitaxial Ferrite in Dual-Phase Microstructures," Scripta Met., 14 (1980) pp. 1239-43.

81. Krauss, G., and Matlock, D. K., "An Evaluation of the Strain-Hardening Behavior of Dual-Phase Steels at Both Low and High Plastic Strains," this book.

82. Zackay, V. F., Parker, E. R., Fahr, D., and Burch, R., "The Enhancement of Ductility in High Strength Steels," Trans ASM, 60 (1967) pp. 252-59.

83. Marder, A. R., "Factors Affecting the Ductility of 'Dual-Phase' Alloys," pp. 89-100 in Formable HSLA and Dual-Phase Steels, R. A. Kot and J. W. Morris, ed.; AIME, New York, NY, 1979.

84. Rashid, M. S., and Rao, B. V. N., "Tempering Characteristics of a Vanadium Containing Dual-Phase Steel," this volume.

85. Niikura, M., Yamada, M., Tanaka, J., and Ichinose, H., "Transformation-Induced-Plasticity Observed in Low-Alloy Steels With Mixed Structure of Austenite and Martensite," pp. 321-6 in Proc. Int. Conf. on New Aspects of Martensitic Transformation, Kobe, 1976.

86. Speich, G. R. and Miller, R. L., "Tempering of Ferrite-Martensite Steels," this book.

87. Krupitzer, R. P., "Strain-Aging Behavior in a Continuous-Annealed Dual-Phase Steel," this book.

88. Rashid, M. S., "Strain Aging Characteristics of Some Dual-Phase Steels," to be published in Met. Trans., 1980.

89. Davies, R. G., "Early Stages of Yielding and Strain Aging of a Vanadium Containing Dual-Phase Steel," Met. Trans., 10A (1979), pp. 1549-55.

MICROSTRUCTURE AND PROPERTIES OF DUAL—PHASE

STEELS CONTAINING FINE PRECIPITATES[+]

J. S. Gau*, J. Y. Koo**, A. Nakagawa* and G. Thomas*

* Department of Materials Science and Mineral Engineering,
University of California, Berkeley, CA 94720

**Corporate Research Science Laboratory, Exxon Research and
Engineering Company, Linden, NJ   07036

Very fine particles (carbides or carbonitrides) of the order of 20 Å
were extensively examined in the ferrite regions of dual-phase steels subjected to intercritical annealing followed by fast quenching to room
temperature. These particles are probably formed during quenching after
intercritical annealing. The driving force for the precipitation reaction
may arise from the supersaturation of carbon (or nitrogen) in the ferrite
phase. These precipitates in certain alloy compositions cause a deviation
from the generally observed two phase mixture rule in that the strength of
the dual-phase steels having a higher volume fraction of martensite is <u>lower</u>
than that having a lower volume fraction of martensite. Thus, the influence
of such precipitates must be considered in the structure-property relations
of dual-phase steels when fast quenching is employed after intercritical
annealing.

+ This work was supported by the Director, Office of Energy Research, Office
of Basic Energy Sciences, Materials Science Division of the U.S. Department
of Energy under Contract No. W-7405-ENG-48.

Introduction

Significant advances have been made in the past several years in our understanding of structure-property relations in intercritically annealed dual-phase steels. Many of the results are published in two recent conference proceedings (1,2). Among the variables which may be important in influencing these relationships are the volume fraction of martensite, size and distribution of martensite, properties of the constituent phases, retained austenite, and fine precipitation in the ferrite.

It is only very recently that very fine carbides (or carbonitrides) of the order of 20 Å have been observed in the ferrite region of as-quenched dual-phase steels (3,4). Important consequences of this precipitation on the mechanical behavior of dual-phase steels was first reported by Thomas and Koo (3). This paper presents recent developments on the microstructure and mechanical properties of various dual-phase steels containing fine precipitates. Emphasis will be placed on the origin and characteristics of the particles, and the influence of alloying elements on the formation and morphology of the particles.

Formation of Fine Precipitates

By holding in the ($\alpha+\gamma$) region, the steel will consist of low carbon ferrite (α) and higher carbon austenite (γ). Although the carbon content in the ferrite is negligably small compared to that in the austenite, it could be significant enough to cause precipitation in the ferrite. This can be seen from the Fe-rich portion of the Fe-C phase diagram (5), which clearly indicates that the solubility of carbon in ferrite at the intercritical annealing temperature is many times ($\sim$100 times at about 50% volume fraction of ferrite) greater than that at room temperature. This situation is enhanced when a ferrite stabilizing element, e.g., Si is present in the steel as an alloying element.

Thus, quenching from the intercritical annealing temperature creates a large supersaturation of interstitial carbon which can provide a driving force for carbide precipitation in the ferrite. The quenching rate is critical in the formation of the precipitates since, if slowly cooled from the ($\alpha+\gamma$) region, the supersaturation of carbon in the ferrite will be reduced or eliminated due to the diffusion of carbon into the shrinking austenite.

It is also important to note from the Fe-C phase diagram that, in the ($\alpha+\gamma$) field, the solubility of carbon in ferrite decreases as the intercritical annealing temperature increases. This means that the carbon supersaturation level at room temperature increases as the intercritical annealing temperature decreases, or alternatively as the volume fraction of martensite decreases in the final quenched product. As a result, it should be expected that the higher the volume fraction of martensite, the lower the density of precipitates in the ferrite region.

The mechanism for this precipitate formation will be discussed in a later section.

Alloy Systems Studied

The microstructures of dual-phase steels which were extensively investigated with respect to the precipitation phenomena are described in this section. The essential part of the dual-phase heat treatment is that after intercritical annealing all the alloys were subjected to either a water quench (WQ) or an iced brine quench (IBQ) unless otherwise specified.

Fe/Nb/0.1C

The chemical composition of the steel containing Nb was Fe/0.4 Si/0.04 Nb/0.1 C. The specimens were intercritically annealed at 800 to 850°C which correspond to a volume fraction of 20% and 40% austenite (martensite after quenching), respectively. At each temperature, the specimens were held for 10 minutes in a vertical tube furnace under an argon atmosphere.

The optical microstructures observed in the Nb containing dual-phase steels are shown in Figure 1.

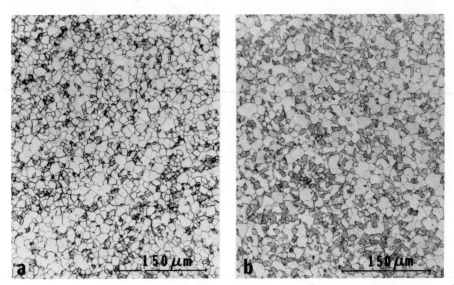

Fig. 1 - Optical micrographs showing dual-phase microstructures in the Fe/0.4 Si/0.04 Nb/0.1 C steel. (a) annealed at 800°C for 10 minutes and water quenched, (b) annealed at 850°C for 10 minutes and water quenched. 2% nital etch.

Transmission electron microscope examination of the structure revealed extensive precipitation in the ferrite region, as shown in Figure 2. Figure 2(a) is a bright field image showing extremely fine particles of the order of ~10 Å wide and ~50 Å long in the ferrite region of Figure 1(a) having 20% martensite. These particles are visible by strain contrast only when favorable diffracting conditions are met. Another ferrite region of interest is shown in Figure 2(b). It can be noted from the figures that the particles show dislocation loop contrast which is typical of coherent precipitates, and that there are precipitate free zones near dislocations and the α/α grain boundary.

A slow cooling experiment was conducted in an attempt to determine if the particles are formed during intercritical annealing or if they are undissolved particles. After annealing at 800°C, the specimens were forced air cooled to room temperature. The resultant TEM micrograph is illustrated in Figure 3, where the ferrite regions are relatively clean and are not associated with extensive precipitation. Thus, this result rules out the possibility of undissolved particles or of particle formation during

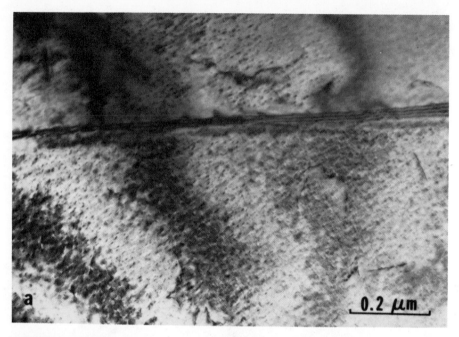

Fig. 2 - Transmission electron micrographs showing very fine
precipitates in the ferrite regions of Figure 1(a).

intercritical annealing, and suggests that they may be formed during quenching to room temperature. A further discussion will follow in a later section.

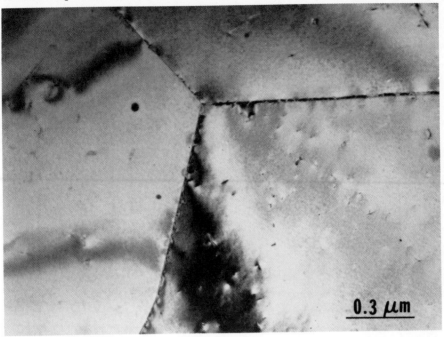

0.3 μm

Fig. 3 - Transmission electron micrograph showing the ferrite regions in the Nb containing steel. Intercritically annealed at 800°C and subsequently forced air cooled.

The morphology of the particles in the ferrite region of Figure 1(b) having 40% martensite was very similar to that shown in Figures 2(a) and 2(b), except for a definite decrease in the density of precipitation as expected. Quantitative measurements of the particle density in the two specimens, together with habit plane determinations of the particles, have been described by Hoel and Thomas (13). At present, the crystal structure and chemical composition of the precipitates are not clearly known. However, all the experimental evidence to date seems to indicate that they are ε carbide, cementite, or carbonitrides.

The room temperature tensile properties of the two specimens (Figure 1(a) and 1(b)) are summarized in Table 1 and the stress-strain curves are plotted in Figure 4. It is interesting to note from the data that the strength of the 850°C WQ specimen having 40% martensite is lower than that of the 800°C WQ specimen having 20% martensite. The apparent relaxation of the law of mixtures observed in the Nb containing steels can be explained based on the varying density of precipitation in the ferrite as the volume fraction of martensite varies (13).

Fe/N/0.1C

The alloying addition of V was studied to determine the role of V on the

Table 1. Summary of Tensile Data

Heat Treatment	Volume Pct Martensite	σ_y	σ_{uts}	Uniform Elong.(%)	Total Elong.(%)
800°C annealing + water quench	20	70	106	10	15
850°C annealing + water quench	40	64	101	11	17
800°C annealing + forced air cool	0*	48	61.5	28.5	36

Fe/0.4 Si/0.04 Nb/0.1C Steel

*The second phase consisted of ∿10% pearlite.

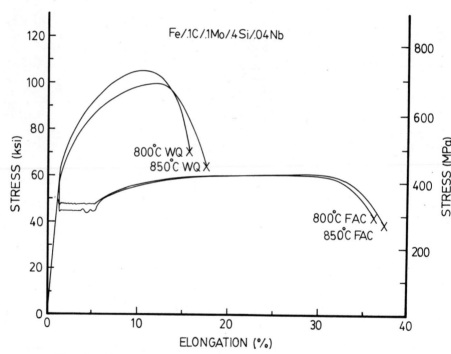

Fig. 4 - Stress-strain curves for the Nb dual-phase steels. Forced air cooled (FAC) specimens are also included in the plot.

structure-property relationship in dual-phase steels. The chemical compositions of the base steel and the V modified steel were Fe/1.15 Mn/ 0.6 Si/0.12C and Fe/1.22 Mn/0.6 Si/0.12 V/0.1 C, respectively. The specimens were intercritically annealed at 800°C for 10 minutes followed by an IBQ. The TEM micrographs taken from the base steel after the 800°C/ IBQ treatment are shown in Figure 5. Figure 5(a) illustrates the morphology of the martensite/ferrite interface and its vicinity. The martensite consists of predominantly dislocated laths and the ferrite region near the interface is heavily dislocated. These features represent the typical microstructure of dual-phase steels as reported by many investigators (1,2). The other transformation products of austenite, e.g., bainite and pearlite were not observed due to the fast quenching rate. The "retained ferrite", the ferrite regions retained during the intercritical annealing treatment (6), showed a relatively low density of fine precipitation (Figure 5(b)).

The microstructures observed in the V modified steel after the 800°C IBQ treatment were similar to those of the base steel as described above, except for discontinuous precipitation in the "new" ferrite near the martensite/ferrite boundaries, and for an increased density of fine precipitation in the retained ferrite. The morphology of the discontinuous carbide precipitation is shown in Figure 6(a). The precise identification of the carbides was not attempted in this study, but they are presumably cementite (6,7), or a vanadium carbide (8,9). This type of structure is similar to the new eutectoid transformation products observed in the isothermal decomposition of austenite in steels containing strong carbide forming elements.

In the retained ferrite regions, very fine precipitates were extensively observed, as shown in Figure 6(b). The density was higher in this alloy than that in the base steel, and this indicates that V enhances the formation of such precipitates. The contrast characteristics of the particles appear to be identical to those observed in the Nb containing steel.

As described above, V modifies the morphology of the carbides in the base steel. Nevertheless, the modified microstructures did not result in improved mechanical properties for the steel composition and the heat treatment considered. Further experiments have shown that the V addition to the base steel was not beneficial even for air cooled or forced air cooled material after intercritical annealing.

Other Steels

Similar fine precipitation morphologies were observed in the ferrite regions of dual-phase steels containing Mo and Al (4). This precipitation is not limited to steels bearing carbide or nitride forming elements. 1010 steels (10) and Fe/1.5 Si/0.1C steel (11) also revealed extensive precipitation. In all cases, the characteristic features of the particles were identical. The only experimentally observable effect of the alloying elements was the increase in the density of precipitation with alloying additions, especially with Nb and V.

Discussion

It is clear from the preceding experimental results that the precipitation phenomenon found in the ferrite of steels fast quenched after intercritical annealing is a general observation rather than a random occurrence. The presence of carbide forming elements in dual-phase steels can enhance precipitation, but it is not required to form the precipitates. A question arises as to when are the precipitates formed. The possibility that they form during intercritical annealing or that they are undissolved particles present in the starting microstructure is ruled out by the following experimental evidence. The slowly cooled specimens after intercritical annealing do not have the fine particles. Secondly, a

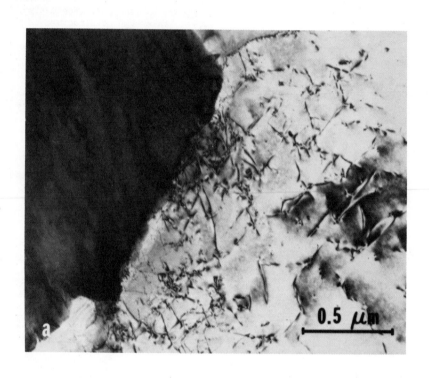

Fig. 5 - Transmission electron micrographs showing:
(a) martensite and ferrite regions

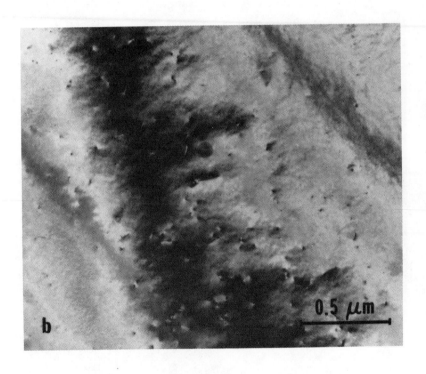

Fig. 5 - Transmission electron micrographs showing:
(b) fine particles in the retained ferrite

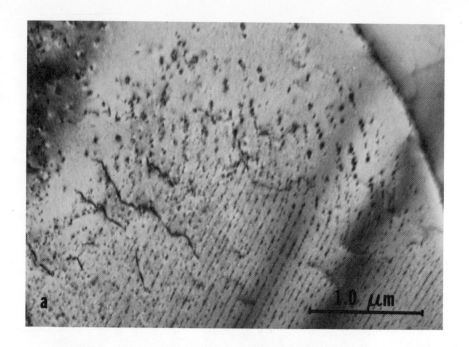

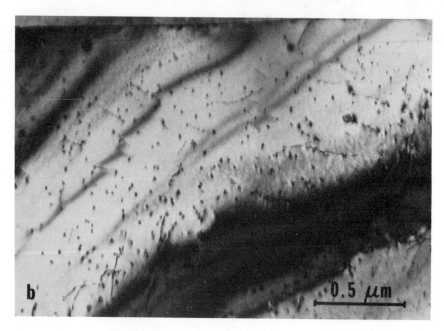

Fig. 6 - Transmission electron micrographs showing (a) aligned, discontinuous carbides in the vicinity of martensite/ferrite interfaces, and (b) the fine precipitation in the retained ferrite.

56

dual-phase 1010 steel which does not contain carbide forming elements revealed extensive precipitation (10).

A possibility exists that natural aging or thin foil surface preparation effects may account for the precipitation. To check for natural aging, the microstructures were examined immediately after heat treatment, and after several prolonged natural aging treatments. The results show the same size and morphology of the particles in all cases, which excludes the possibility of room temperature precipitaion. Likewise, thin foil surface effects were eliminated by sputtering the thin foil surface and also using different electropolishing solutions. At present, therefore, the most plausible explanation is the formation of the particles during quenching to room temperature. This proces may be similar in nature autotempering observed in steels with high M_S temperatures.

The presence of precipitates does not always give rise to a violation of the two phase mixture rule. Violation did occur in the alloys containing Nb, Mo and Al but did not occur in 1010 steel (10), Fe/Si/C (12), and Fe/Cr/C steels (12). The results suggest that, although the image characteristics of the particles were identical in all the alloys studied, the elements Nb, Mo and Al apparently influence the particle properties in such a way that they act as more effective barriers to dislocation motion. In this case, dispersion strengthening becomes significant. Thus, the strength of ferrite with a higher volume fraction of martensite is lower than that with a lower volume fraction of martensite because of the change in precipitate density. This situation is responsible for the relaxation of generally observed law of mixtures.

Conclusions

A very fine dispersion of coherent particles (carbides or carbonitrides) of the order of 20 Å were observed in the ferrite region of dual-phase steels subjected to intercritical annealing followed by fast quenching to room temperature. These particles probably form during quenching after intercritical annealing. The driving force for the precipitation reaction may arise from the supersaturation of carbon (or nitrogen) in the ferrite phase.

The particles were observed in all the steels we have thus far investigated regardless of alloy compositions. These precipitates in certain alloy compositions, especially those containing Nb caused a relaxation of the two phase mixture rule, i.e., the strength of the dual-phase steels having a higher volume fraction of martensite was <u>lower</u> than that of a lower volume fraction of martensite.

Thus, the influence of such precipitates must be considered in the structure-property relations of dual-phase steels when the fast quenching is employed after intercritical annealing.

Acknowledgements

We are grateful to Climax Molybdenum Company, Foote Mineral Company, and Republic Steel Company who supplied the alloys used in this investigation. This work was supported jointly by the Materials and Molecular Research Division of the U.S. Department of Energy under contract W-7405-Eng-48, and Exxon Research and Engineering Company.

References

1. A. T. Davenport, ed., _Formable HSLA and Dual-Phase Steels_, AIME, New York, N.Y., 1979.

2. R. A. Kot and J. W. Morris, eds., _Structure and Properties of Dual-Phase Steels_, AIME, Warrendale, PA, 1979.

3. G. Thomas and J. Y. Koo, "Developments in Strong, Ductile Duplex Ferritic-Martensitic Steels", pp. 183-201 in _Structure and Properties of Dual-Phase Steels_, R. A. Kot and J. W. Morris, eds., AIME, Warrendale, PA, 1979.

4. T. O'Neill, M.S. Thesis, University of California, Berkeley, CA, 1979, (LBL #9047).

5. D. T. Hawkins and R. Hultgren, "Constitution of Binary Alloys", p. 276, in _Metals Handbook_, Vol. 8, ASM Publication, 1973.

6. M. D. Geib, D. K. Matlock, and G. Krauss, "The Effect of Intercritical Annealing Temperature on the Structure of Niobium Microalloyed Dual-Phase Steel, _Metallurgical Transactions A_, $\underline{11A}$ (1980) pp. 1683-1689.

7. J. Y. Koo and G. Thomas, "Thermal Cycling Treatments and Microstructures for Improved Properties of Fe-0.12%C-0.5% Mn Steels," _Materials Science and Engineering_, $\underline{24}$, (1976) pp. 187-198.

8. R. W. K. Honeycombe, "Transformation from Austenite in Alloy Steels," _Metallurgical Transactions_, $\underline{7A}$, (1976) pp. 915-936.

9. A. T. Davenport and P. C. Becker, "Interphase Precipitation of Cementite in a Continuously-Cooled Plain-Carbon Steel," _Metallurgical Transactions_, $\underline{2}$, (1971) pp. 2962-2964.

10. M. J. Young, M.S. Thesis, University of California, Berkeley, CA, 1977, (LBL #6620).

11. A. Pelton and K. Westmacott, University of California, Berkeley, CA, unpublished results.

12. J. Y. Koo and G. Thomas, "Design of Duplec Fe/X/.1C Steels for Improved Mechanical Properties," _Metallurgical Transactions_, 8A, (1977) pp. 525-528.

13. R. H. Hoel and G. Thomas, "Ferrite Structure and Mechanical Properties of Low Alloy Duplex Steels", Scripta Met., in press (1981).

AUSTENITE GROWTH IN THE INTERCRITICAL ANNEALING

OF TERNARY AND QUATERNARY DUAL-PHASE STEELS

P. Wycliffe, G.R. Purdy and J.D. Embury
Metallurgy and Materials Science
McMaster University
Hamilton, Ontario, Canada

Austenite growth during the intercritical annealing of a dual-phase steel is studied. A local equilibrium model assuming a diffusion controlled reaction is developed to estimate the volume fraction and composition of austenite as a function of time during the intercritical annealing period. Models are developed for both ternary, e.g. Fe-C-Mn and quaternary, e.g. Fe-C-Mn-Si steels.

In both cases a rapid carbon diffusion controlled period of growth goes to completion in times of the order of a second followed by much slower growth controlled by diffusion of alloying element(s) in ferrite. The rate and extent of austenite growth is related to the alloy chemistry, the scale of the microstructure and the distribution coefficient(s) of the alloying element(s). Predicted growth rates are compared with published data and predicted concentration profiles are compared with profiles observed using the Scanning Transmission Electron Microscope.

Introduction

Dual-phase steels typically consist of fine scale mixtures of martensite and ferrite. This structure may be produced in either an as-hot-
rolled product or by intercritically annealing in the $\alpha + \gamma$ range followed
by rapid cooling. This latter process is investigated here.

It is well established that the mechanical properties of a dual-phase
steel depend on the composition and volume fraction of martensite produced
in the steel (1,2,3). Thus, the growth rate and composition of austenite
in the intercritical annealing period is of interest. In the binary Fe-C
case, equilibrium compositions of ferrite (α) and austenite (γ) are fixed
for a given temperature and carbon is the only diffusing species that need
be considered.

When additional alloying elements are present the situation is more
complex. The equilibrium compositions of α and γ are not fixed for a given
temperature, and diffusion of more than one alloy species must be considered. Since typical alloying elements, e.g., Si, Mn, Mo, have diffusion
rates three or more orders of magnitude lower than carbon, their effect on
the reaction rate is of interest.

The objective of the current work is to estimate the growth rate of
γ and the compositions of γ and α as a function of time during the intercritical annealing of a ternary, e.g., (Fe-C-Mn) alloy and a quaternary,
e.g., (Fe-C-Mn-Si) alloy. The model assumes that local equilibrium exists
at the α/γ interface and that growth occurs by diffusion control. In
addition, the reaction is assumed to occur at a plane front so that the diffusion process can be treated as a one-dimensional problem.

Calculation for Isothermal Growth

Ternary Case Fe-C-Mn

Initial Conditions. The following calculations are concerned with the
annealing of a ferrite-pearlite mixture which upon heating into the intercritical range becomes a mixture of α and γ. The carbon content of the α
is assumed to be negligible and the carbon content of the γ is assumed uniform and of eutectoid composition C_1^E, i.e., the time to nucleate γ from
pearlite and to remove any inhomogeneity in the γ is neglected. See
Appendix I for additional notation explanation. (The time to form γ from
pearlite may be of some importance when considering continuous annealing
cycles, or situations where the pearlite initially has Mn partitioned
between ferrite and carbide.)

The initial concentration of manganese is here assumed to be uniform,

at C_2^0, i.e., the same both in the α and γ which leads to concentration profiles as illustrated in Figures 1b and 1c and for a structure such as that
depicted schematically in Figure 1a.

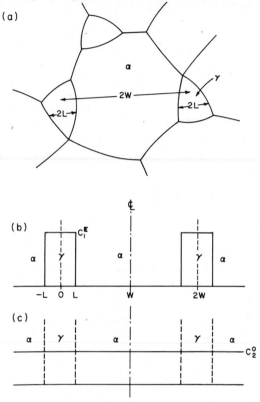

Fig. 1 - Initial Conditions. A ferrite-pearlite mixture is heated into
the intercritical range to produce austenite particles of width, 2L,
separated by distance, 2G, as illustrated schematically in (a). The
austenite is initially of eutectoid composition C_1^E with a carbon concentration profile as in (b). The manganese concentration is uniform
as shown in (c).

As noted above it is assumed that the reaction is volume diffusion controlled; interface compositions are therefore given by equilibrium tie
lines on the phase diagram. If the distribution coefficients k_0, k_1, k_2 are
known (for Fe, C and Mn, respectively), then the other interface concentrations may be expressed in terms of the interface carbon concentration in
the γ, C_1 (4), i.e.:

$$C_2^\gamma = a + bC_1 \tag{1a}$$

$$C_2^\alpha = k_2(a + bC_1) \tag{1b}$$

where

$$a = \frac{1-k_0}{k_2-k_0} \quad , \quad b = \frac{k_0-k_1}{k_2-k_0}$$

To simplify subsequent calculations, the carbon concentration in the is considered as zero, and distribution coefficients are assumed invariant with composition. This results in linear phase boundaries on the isotherm, Figure 2c. They are estimated using data compiled by Urhenius (5) and Gilmour et al. (6).

The process may be considered in two stages: a carbon diffusion controlled growth stage accompanied by negligible partitioning of manganese into the austenite, followed by a stage where manganese is partitioned from the α and growth is controlled by manganese diffusion. (The initial carbon diffusion controlled stage may be further subdivided into parabolic and logarithmic growth stages).

Initial Growth. During the carbon diffusion controlled growth the equilibrium tie line describing the interface composition changes by only a negligible amount. Its position is uniquely defined at this point by equating velocities based on both manganese diffusion and carbon diffusion. This is treated quantitatively in reference (7) but may be explained qualitatively by reference to Figure 2, which shows manganese and carbon concentration profiles after a small amount of growth in the initial stage. For both manganese and carbon the interface velocity, $\frac{dX_0}{dt}$ is related to the flux at the interface, J, and the concentration difference at the interface ΔC, by:

$$\left|\frac{dX_0}{dt}\right| = \left|\frac{J_2}{\Delta C_2}\right| \tag{2a}$$

$$= \left|\frac{J_1}{\Delta C_1}\right| \tag{2b}$$

while the flux is given by Fick's first law:

$$J = D\frac{dC}{dX} \tag{3}$$

D_C is roughly 10^4 greater than D_{Mn}, and in order that the two values for interface velocity be equal the low diffusion coefficient for Mn must be compensated by a steep gradient for Mn in the ferrite. Considering a mass balance for manganese (i.e., the area of its negative "spike" in the α in Figure 2b must equal the area of the rectangle in the γ) leads to negligible manganese partitioning in the initial stage ($C_2^0 \approx C_2^\gamma$). This corresponds to the tie line position "s" in Figure 2c.

62

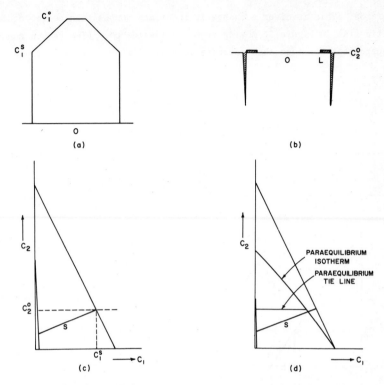

Fig. 2 - Carbon Diffusion Controlled Growth. In order to have the same value for interface velocity from both the flux of carbon and the flux of manganese, the manganese gradient in (b) must be $\sim 10^4$ x steeper than the carbon gradient (a). Associated with the steep manganese gradient is a minute amount of manganese partitioning in the γ, $C_2^{\gamma} \approx C_2^0$. The equilibrium tie line is compared with the para-equilibrium tie line corresponding to the same manganese content in (d).

The local equilibrium tie line "s" is compared in Figure 2d with a para-equilibrium tie line. Para-equilibrium, as discussed by Gilmour et al. (6) and Hillert (8) is a constrained form of equilibrium where the energy is minimized subject to the constraint $C_2^{\gamma} = C_2^{\alpha} = C_2^0$, i.e., no manganese partitioning occurs. Para-equilibrium thus involves no manganese depletion in the ferrite and (generally) a lower carbon concentration in the austenite.

Growth with Manganese Partitioning. After carbon has reached uniform activity throughout the material, as depicted in Figure 3a, the reaction will proceed under manganese diffusion control. Further growth of γ is not possible unless the carbon concentration in the γ is reduced from C_1^S as in

Figure 3a. This involves a change in interface composition corresponding to the tie line in Figure 3c moving from "s" toward "p". (The carbon content of austenite, C_1, must then decrease.) Since diffusion in γ is slower than

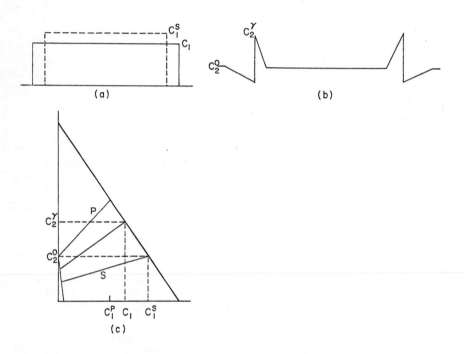

(a)

(b)

(c)

Fig. 3 - Growth with Manganese Partitioning. Further γ growth occurs by a reduction of γ carbon content to C_1 below the C_1^S value as in (a). This results in a shift of the tie line from s toward p in (c). The tie line shift corresponds to an increase in γ manganese content with the equilibrium interface manganese compositions being "frozen" in the γ as in (b).

in α, it is assumed that the moving interface leaves behind it a manganese profile which is a locus of equilibrium compositions of γ as a function of distance (as shown in Figure 3).

A carbon mass balance relates C_1 to the interface position, X_0:

$$C_1 = \frac{C_1^E L}{X_0}$$

(4)

which, combined with equation 1 gives C_2^γ as a function of position $C_2^\gamma [X]$:

64

$$c_2^{\gamma}[X] = a + b\frac{c_1^E L}{X} \tag{5}$$

Approximating the Mn diffusion profile in the α to a triangle with base "f" allows the f to be calculated for a given interface position using a mass balance for Mn, i.e.:

$$\int_0^{X_0} (c_2^{\gamma}[X] - c_2^0)\,dX = \frac{1}{2}f(c_2^0 - c_2^{\alpha}) \tag{6}$$

where f is then used to determine the magnitude of the concentration gradient in the α. This quantity, combined with equations 3 and 2a gives the component of interface velocity as determined by a flux in the ferrite:

$$D_2^{\alpha}\frac{c_2^0 - c_2^{\alpha}}{f} = (c_2^{\gamma} - c_2^{\alpha})(\frac{dX_0}{dt})^{\alpha} \tag{7}$$

Combining equations 6 and 7 allow $(\frac{dX_0}{dt})^{\alpha}$ to be determined as a function of position. The gradient of Mn in the γ leads to a flux in the γ which tends to reduce the interface velocity. The net interface velocity $(\frac{dX_0}{dt})$ is the sum of the component due to flux in the α and flux in γ:

$$\frac{dX_0}{dt} = (\frac{dX_0}{dt})^{\alpha} + (\frac{dX_0}{dt})^{\gamma} \tag{8}$$

where

$$(\frac{dX_0}{dt})^{\gamma} = \frac{J^{\gamma}}{c_2^{\gamma} - c_2^{\alpha}} \tag{9}$$

J^{γ} is derived from the differential of equation 5 using Fick's first law:

$$J^{\gamma} = D_2^{\gamma}(\frac{dc_2^{\gamma}[X]}{dt}) \tag{10}$$

The flux in the austenite only makes a significant contribution as austenite growth nears completion. Using equations 5 - 10 allows the interface velocity and concentrations to be determined as a function of interface position. Separating variables and integrating allows calculation of the time associated with each interface position.

Austenite growth ends when the flux of manganese in the austenite equals that in the ferrite:

$$|J^{\gamma}| = |J^{\alpha}| \tag{11}$$

This corresponds to a very flat gradient in the α since $D_2^{\gamma} << D_2^{\alpha}$. In most cases this flat gradient is achieved by the impingement of diffusion fields in the α so that the α is almost uniformly depleted in manganese.

Impingement of the diffusion fields leads to a series of concentration profiles as depected in Figure 4. The growth rate at this stage may be estimated by modifying the manganese mass balance (cf. equation 5). If the profile in the ferrite is approximated as linear then:

$$\int_0^{X_0} (C_2^\gamma [X] - C_2^0)\, dX = \frac{1}{2}(G-X_0)(C_2^\alpha + C_2^{\alpha I}) \qquad (12)$$

γ growth ends when a concentration profile as depicted in Figure 4c is produced.

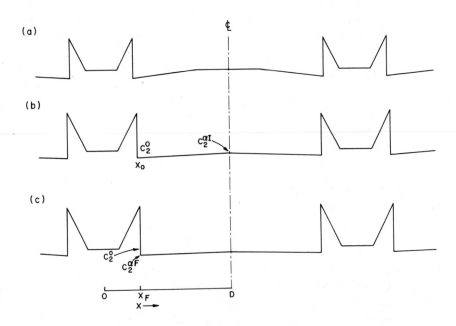

Fig. 4 - Manganese Concentration Profiles for Late Stages of γ Growth.
In the late stages of manganese-diffusion-controlled growth, manganese diffusion fields in the ferrite impinge so that all the ferrite becomes depleted in manganese. (a) shows concentration profiles before impingement. (b) shows concentration profiles shortly after impingement and defines $C_2^{\alpha I}$, the ferrite manganese concentration midway between austenite grains. (c) shows the concentration profile at the end of austenite growth where a slight gradient in the ferrite remains.

Neglecting the slight gradient in the α, makes it possible to estimate the volume fraction of austenite at the end of growth without reference to the growth kinetics. A mass balance for the Mn based on the profile in Figure 4c gives:

$$\int_0^{X_F} (C_2^\gamma[X] - C_2^0)\,dX = (G - X_F)(C_2^0 - C_2^{\alpha F}) \tag{13}$$

A mass balance for carbon gives:

$$C_1^E L = C_1^0 G = C_1^F X_F \tag{14}$$

while volume fraction, V_f, is given by:*

$$V_f = \frac{C_1^0}{C_1^F} = \frac{X_F}{G} \tag{15}$$

which with equations 1, 5 and 13 lead to an expression relating volume fraction V_f to bulk carbon content, C_1^0:

$$(a-C_2^0)V_f + bC_1^0 + bC_1^0 \ln \frac{V_f}{bC_1^0}(C_2^0-a)$$

$$= (C_2^0-k_2 a)(1-V_f) - k_2 bC_1^0(\frac{1}{V_f}) - 1) \tag{16}$$

Since L and D are related by equation 14, the above expression for volume fraction is independent of the scale of the structure and depends only on chemistry and distribution coefficients.

However, the time to reach this volume fraction will be greater for coarser structures (and varies as L^2). The volume fraction derived from equation 16 is within 1% of that calculated using the kinetic equation.

Quaternary Case: Fe-C-Si-Mn

Initial Conditions. The initial conditions are assumed to be the same as for the ternary case with both alloying elements Si and Mn of uniform concentration throughout the material at C_3^0 and C_2^0, respectively.

The addition of Si contributes another degree of freedom so that the isotherm may be represented as a tetrahedron while the $\alpha/\alpha + \gamma$ and $\gamma/\alpha + \gamma$ boundaries are surfaces as in Figure 5a. Once again interface

---

*Strictly speaking, equation 15 is true when V_f is mass fraction of γ. However, since martensite and ferrite have approximately equal densities, V_f will be a good approximation of volume fraction of martensite in the room temperature structure (for rapidly quenched materials).

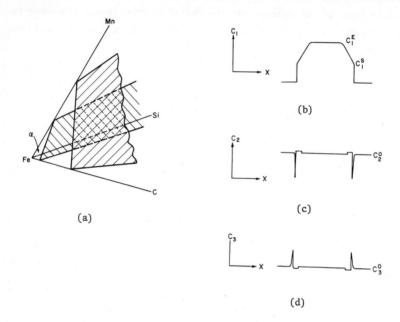

(a)

(b)

(c)

(d)

Fig. 5 - The quaternary isotherm (a) is a tetrahedron where the $\alpha/\alpha + \gamma$ $\gamma/\alpha + \gamma$ boundaries are surfaces. These surfaces are connected by tie lines so that specifying two interface concentrations fixes all other interface concentrations (cf. equation 17). The initial growth stage in the quaternary case is carbon-diffusion-controlled. The finite gradient in the carbon concentration profile (b) is accompanied by very steep gradients in the manganese and silicon concentration profiles: (c) and (d) respectively. A mass balance associates only a minute amount of partitioning with the carbon-diffusion-controlled growth: $C_2^{\gamma} \sim C_2^{0}$, $C_3^{\gamma} \sim C_3^{0}$.

compositions are given by equilibrium tie lines on the phase diagram. However, now two independent compositions must be specified to determine the tie line. If the interface silicon concentration in the γ, C_3^{γ} is used as an independent variable then the other interface concentrations are given by:

$$C_2^{\gamma} = a + bC_1 + cC_3^{\gamma} \tag{17a}$$

$$C_2^{\alpha} = k_2 C_2^{\gamma} \tag{17b}$$

$$C_3^{\alpha} = k_3 C_3^{\gamma} \tag{17c}$$

68

where $a = \dfrac{1-k_0}{k_2-k_0}$, $b = \dfrac{k_0-k_1}{k_2-k_0}$ and $c = \dfrac{k_0-k_3}{k_2-k_0}$.

The process may again be considered in two stages although the second stage is more complex. The first stage involves carbon diffusion control and no partitioning of alloying elements. The second stage is controlled by diffusion of alloying elements and involves partitioning of Si and Mn.

Initial Growth. In the first stage the tie line is fixed. Although there is an additional degree of freedom providing three unknowns: C_1, C_2^{γ}, and C_3^{γ} (as compared with two unknowns in the ternary case). There are also three expressions for interface velocity so that the tie line is specified.

The diffusion coefficients for Si, D_3^{α}, is still three orders of magnitude less than that for carbon, D_1^{γ}, so a similar argument to that for the ternary case applies.

The concentration profiles associated with the carbon diffusion controlled stage are shown in Figures 5b, 5c and 5d. To arrive at the same velocity based on diffusion of all three additions, steep concentration profiles are necessary in the α for both Si and Mn so that considering a mass balance for these elements gives:

$$C_2^{\gamma s} \simeq C_2^0 \tag{18a}$$

$$C_3^{\gamma s} \simeq C_3^0 \tag{18b}$$

which specifies the tie line, i.e.:

$$C_1^s = (C_2^{\alpha}-a-cC_3^{\alpha})/b \tag{19}$$

Alloying Element Partitioning. Once the carbon has reached uniform activity throughout the material further growth of γ requires reduction of the γ carbon content from C_1^s. As in the ternary case a carbon mass balance gives C_1 as a function of interface position. This is not, however, sufficient to specify the tie-line for a given interface position.

A trajectory on the $\alpha + \gamma/\gamma$ surface of the isotherm must be determined by equating interface velocities based on manganese and silicon diffusion.

Figure 6 shows concentration profiles for manganese and silicon after some growth with alloying elements partitioning. The diffusion profile in the ferrite is again approximated as linear leading to a mass balance for each element:

$$\text{for Mn:} \quad \int_{X_s}^{X_0} (C_2[X] - C_2^0)dX = I_2 = \tfrac{1}{2} g(C_2^0-C_2^{\alpha}) \tag{20}$$

69

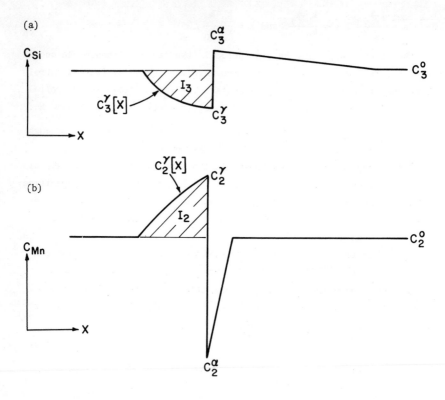

(a)

C_{Si}

C_3^α

C_3^0

$C_3^\gamma[X]$

I_3

C_3^γ

X

(b)

$C_2^\gamma[X]$

C_2^γ

I_2

C_{Mn}

C_2^0

X

C_2^α

Fig. 6 - Concentration profiles for Si (a) and Mn (b) during growth with alloying-element-control. Fluxes of both Si and Mn are used to determine separate expressions for interface velocity for a given interface position. $C_2^\gamma[X]$ and $C_3^\gamma[X]$ are determined iteratively for increments of interface position such that both velocities are equal.

for Si: $\quad \int_{X_s}^{X_0} (C_3^\gamma[X] - C_3^0) \, dX = I_3 = \frac{1}{2} h(C_3^0 - C_3^\alpha)$ (21)

Proceeding in a manner similar to the ternary, results in two expressions for the components of interface velocity due to flux in the γ:

for Mn: $\quad (\dfrac{dX_0}{dt})_2^\alpha = \dfrac{D_2^\alpha}{C_2^\gamma - C_2^\alpha} \dfrac{(C_2^0 - C_2^\alpha)^2}{I_2}$ (22)

for Si: $\quad (\dfrac{dX_0}{dt})_3^\alpha = \dfrac{D_3^\alpha}{C_3^\gamma - C_3^\alpha} \dfrac{(C_2^0 - C_2^\alpha)^2}{I_3}$ (23)

70

For both the Mn and the Si the contribution of flux in the γ are
derived and give rise to complete expressions for net interface velocity:

$$(\frac{dX_0}{dt})_2^\gamma = \frac{D_2^\gamma}{C_2^\gamma - C_2^\alpha} (\frac{dC_2^\gamma [X]}{dX})_{X=X\emptyset} \qquad (24a)$$

$$(\frac{dX_0}{dt})_2 = (\frac{dX_0}{dt})_2^\gamma + (\frac{dX_0}{dt})_2^\alpha \qquad (24b)$$

$$(\frac{dX_0}{dt})_3^\gamma = \frac{D_3^\gamma}{C_3^\gamma - C_3^\alpha} (\frac{dC_3^\gamma [X]}{dX})_{X=X\emptyset} \qquad (25a)$$

$$(\frac{dX_0}{dt})_3 = (\frac{dX_0}{dt})_3^\gamma + (\frac{dX_0}{dt})_3^\alpha \qquad (25b)$$

In the quaternary case, as distinct from the ternary, $C_2^\gamma [X]$ and
$C_2^\gamma [X]$ may not be expressed analytically. Rather, the system of equations
must be solved for increments of interface position. At each interface
position C_2^γ and C_3^γ are determined iteratively such that the interface
velocity based on the flux of manganese equals the interface velocity
based on the flux of silicon, i.e.:

$$(\frac{dX_0}{dt})_2 = (\frac{dX_0}{dt})_3 \qquad (26)$$

As for the ternary, the mass balance equations 20 and 21 may be modified
to account for impingement of diffusion fields in the α.

Computed Results

Figures 7 and 8 show calculated concentration profiles for Fe-.08%C-
1%Mn and Fe-.08%C-1%Mn-1%Si alloys annealed at 750°C for different times.
Figures 9 to 12 show computed austenite volume fraction vs. annealing time
for these alloys, and for a number of other ternary alloys. Figure 9
compares our calculated results with experimental data published by
Speich (9). Diffusion data came from References 10, 11 and 12.

It is interesting to note the effect of various parameters on the
calculated results.

Volume fractions for a given time and at the completion of growth are
extremely sensitive to the input thermodynamic data. Consider for example
the computed growth curve for Fe-0.12%C-1.5% Mn in Figure 9. Changing k_0
from 1.0339 to 1.0349 results in a change of final volume fraction
austenite from 0.395 to 0.376. Uncertainties in the thermodynamic data are
probably the largest single source of error in the calculation of volume
fraction.

71

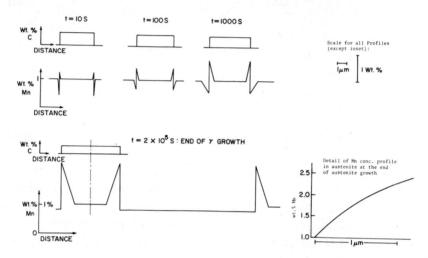

Fig. 7   Concentration profiles for carbon and manganese for different times during the annealing of an Fe-.08%C-1%Mn alloy at 750°C. Thermodynamic data from Reference 5: k_0 = 1.0297, k_1 = .027, k_2 = .336.

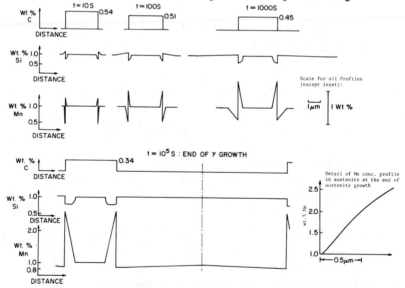

Fig. 8   Concentration profiles for carbon and manganese for different times during the annealing of an Fe-.08%C-1%Mn-1%Si alloy at 750°C. Same data as in Figure 7 with k_3 - 1.3.

The reaction rate is faster for finer structures: the time to reach a given set of interface concentrations (tie line position) varies as L^2 (the scale of the concentration profiles varies as L).

The effect of composition is shown in Figure 10 where calculated results for a range of compositions in the Fe-C-Mn system are plotted. A greater concentration of alloying element leads to a greater amount of growth by alloying element diffusion. Reducing carbon content (for a given Mn content) also leads to a greater amount of growth by alloying element diffusion since the ratio of particle spacing to particle size is increased. This results in impingement of Mn diffusion fields in ferrite at a later stage of growth. If reduction of carbon content increases particle spacing with particle size constant a longer time to completion of γ growth will result.

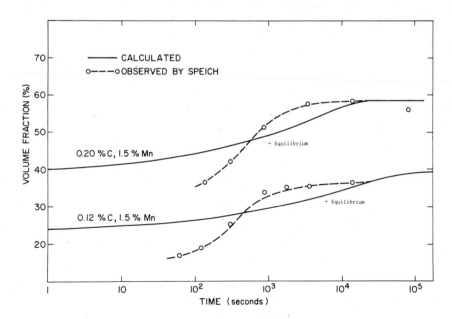

Fig. 9 - Calculated γ growth rate curves (volume fraction vs. log time) compared with observed growth rate curves from Speich(9) for Fe-C-Mn steel annealed at 740°C. Volume fractions associated with complete equilibrium (i.e. uniform Mn in γ) are indicated. Thermodynamic data from Reference 6: k_0 = 1.0339, k_1 = 0.026, k_2 = .407.

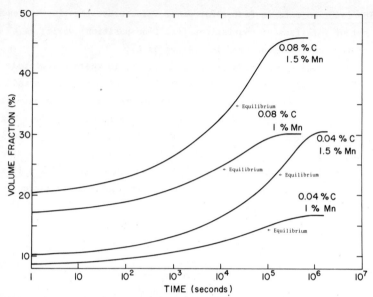

Fig. 10 - Calculated γ growth rate curves (volume fraction vs. log time) for Fe-C-Mn alloys of different composition annealed at 750°C. Volume fractions associated with complete equilibrium (i.e. uniform Mn in γ) are indicated. Thermodynamic data from Reference 5: k_0 = 1.0297, k_1 = .027, k_2 = .336.

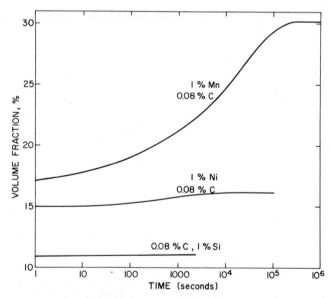

Fig. 11 - Calculated γ growth rate curves (volume fraction vs. log time) for different ternary alloys annealed at 750°C. Thermodynamic data from Reference 5: k_0 = 1.0297, k_1 = .027, k_{Mn} = .336, k_{Si} = 1.3, k_{Ni} = .57.

Both the rate and the extent of growth by alloying element diffusion are very sensitive to the alloying element distribution coefficient, k_2, and to the bulk composition of the alloying element C_2^0. As might be expected the rate is also affected by the alloying element diffusion coefficient. Elements with distribution coefficients close to unity cause less austenite growth. Further, for these elements, growth goes to completion sooner than for elements whose distribution coefficients indicate large equilibr-um concentration differences between ferrite and austenite. The combination of these effects is best illustrated by comparing the calculated results for Fe-.08%C-1%Mn and Fe-.08%C-1%Ni. Here the 1% Ni alloy (k = .57, D = 1.0 x 10^{-13} cm^2/sec.) alloy attains

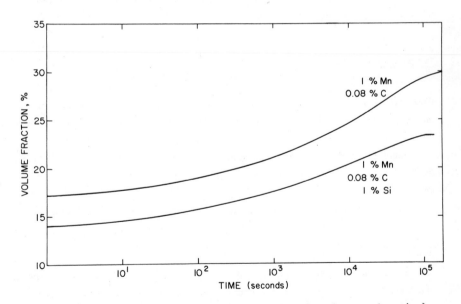

Fig. 12 - Calculated γ growth rate curves (volume fraction vs. log time) for Fe-.08%C-1%Mn and Fe-.08%C-1%Mn-1%Si alloys annealed at 750°C. Data is the same as for Figures 7 and 8.

completion of austenite growth in roughly the same time as the 1% Mn alloy ($k = .34$, $D = 2.0 \times 10^{-12}$ cm^2/sec) despite the fact that its diffusion coefficient is lower.

In the Fe-C-Si case (Figure 11) only a small amount of austenite growth occurs, and this nears completion in times of the order of 1 second since here, a distribution coefficient close to 1 is combined with a high rate of diffusion.

In moving from the Fe-C-Mn ternary to the Fe-.08C-1%Mn-1%Si quaternary alloy no large rate effects are noted. The main effect of the silicon additions is a constitutional one which reduces the volume fraction for a given annealing time(and reduces the terminal volume fraction).

Because Si has a higher rate of diffusion in ferrite, impingement of silicon diffusion fields in ferrite occurs sooner than for Mn. This leads to an interesting variation in interface concentration of Si as γ growth proceeds. In the ternary case the interface concentration of Si in the austenite decreases continually as γ growth continues. However, in the quaternary case, after impingement of diffusion fields in the ferrite have made the concentration gradients quite flat, the interface concentration of Si begins to increase (albeit by only a minute amount) in order to maintain a positive interface velocity consistent with that associated with the steeper (unimpinged) Mn diffusion profiles.

Experimental Study

Procedure

Samples of a ternary Fe-.08%C-1%Mn alloy were homogenized for two weeks at 1200°C, then quenched, followed by five cycles of reaustenitizing and quenching to refine the structure. Further reaustenitizing was followed by air cooling to produce a fine pearlite-ferrite mixture. Small specimens (1 cm cubes) were then annealed in the intercritical region for different times and water quenched. Thin foils were prepared and examined in the Phillips 400 TEM/STEM and the V.G. HB5 STEM. These instruments are capable of X-ray microanalysis using probe sizes of the order of 10 - 100 Å (in the absence of beam spreading in the specimen) (13).

Results

Experimental concentration profiles are shown in Figures 13, 14 and 15. The peaks in the profiles are broader than those predicted if a section perpendicular to the α/γ interface is considered.

The apparent spreading of the peak is most likely due to sectioning effects. For example, if the profile in Figure 14 was for a planar

76

interface of width W, inclination of the interface at angle θ to the foil normal with foil thickness, t, would expand the apparent width to $(\frac{W}{\cos\theta} + \frac{t}{2}\tan\theta)$, i.e., from .34 μm to .62 μm, for θ = 45° and t = 3000 Å. Further, such a sectioning effect decreases the observed peak height (even in the absence of beam spreading) since for a finite thickness regions of different concentration will be sampled.

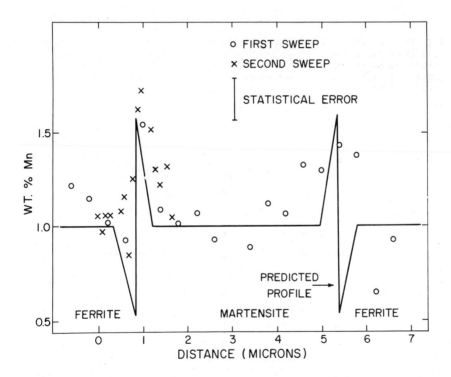

Fig. 13 - Phillips 400 TEM/STEM results. An observed manganese concentration profile is shown along with a predicted profile. The sample is an Fe-.08%C-1%Mn steel annealed for 500 seconds at 750°C. The predicted profile neglects effects due to sectioning so that the interface is not normal to the foil.

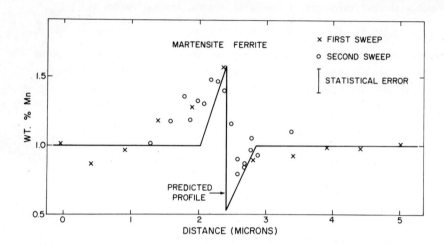

Fig. 14 - HB5 STEM results. An observed manganese concentration profile is shown along with a predicted profile. The sample is an Fe-.08%C-1%Mn steel annealed for 500 seconds at 750°C. The predicted profile neglects effects due to sectioning so that the interface is not normal to the foil.

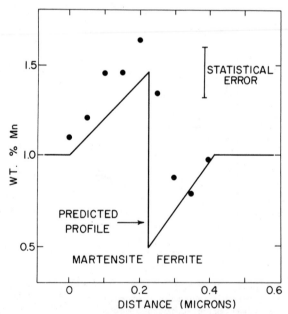

Fig. 15 -   HB5   STEM results. An observed manganese concentration profile is shown along with a predicted profile. The sample is an Fe-.08%C-1%Mn steel annealed for 100 seconds at 750°C. The predicted profile neglects effects due to sectioning so that the interface is not normal to the foil.

Discussion

Comparison of calculated growth data with published growth data reveals a reasonable correspondence. The lower volume fraction for early times in the published data is probably due to growth limited by austenite nucleation.

The difference in volume fraction at the end of growth is most likely due to inaccuracies in the input thermodynamic data (the sensitivity of calculated results to variations in such data has been noted).

A comparison of the experimental and calculated profiles for the ternary Fe-C-Mn alloys, taking cognizance of sectioning effects further lends credence in the model. We therefore suggest that the local equilibrium analysis given here is adequate for the prediction and understanding of the main features of austenite growth during intercritical annealing.

The times for attainment of uniform carbon activity from pearlite-ferrite mixtures initially homogeneous with respect to alloying elements, are of the order of seconds. Subsequent growth of austenite is then determined by the diffusion (and the consequent partitioning) of the alloying elements. Thus, we expect a central region for each austenite particle that is essentially unpartitioned, and a rim that is enriched or depleted in each of the alloying elements, depending on the values of the individual alloying element distribution coefficients.

Since the volume fraction of martensite/austenite is vital in determining the work-hardening response and strength of the material, the prediction of the rate of austenite growth is a central objective of this work. The ternary case can be understood with the aid of Figures 3 and 4 which demonstrate that the slow alloying-element-controlled growth stage is accompanied by the steady variation of the interfacial equilibrium compositions. This variation is of course possible using a ternary analysis but does not arise using a pseudo-binary approximation.

An analogy may be drawn between γ growth by alloying element partitioning in the ternary case and classical non-equilibrium freezing for a binary system. The ternary Fe-C-Mn isotherm may be drawn as in Figure 16 if chemical potential of carbon is substituted for carbon concentration. Carbon chemical potential then becomes analogous to temperature in the binary case. In each case, the rate of transport in the growing phase is much less than that in the consumed phase. In each case, the concentration profile "frozen" into the growing phase is a locus of interface equilibrium concentrations. In the binary freezing case the change in interface

79

composition, C_L, is related to mass fraction solidified $(1 - M_L)$ by considering a mass balance with the well-stirred liquid:

$$M_L dC_L = - C_L (1 - K) dM_L \tag{27}$$

In the ternary austenite growth case the interface composition of carbon is related to the austenite mass fraction, M_f, via a mass balance for carbon:

$$C_1 = \frac{C_1^0}{M_f} \tag{28}$$

Differentiating gives:

$$(M_f)^2 \, dC_1 = - C_1^0 dM_f \tag{29}$$

Manganese concentration is then related to carbon concentration by the phase diagram:

$$C_2 = a + bC_1 \tag{30}$$

Equations 28 and 30 allow equation 13 to be re-written:

$$\int_0^{V_f} (C_2^\gamma [M_f^\gamma] - C_2^0) \, dM_f^\gamma = (1 - V_f)(C_2^0 - C_2^{\alpha F}) \tag{31}$$

This means that the volume fraction of γ at the end of growth (given by equation 16) is independent of particle shapes as well as scale.

In the freezing case growth of a pro-eutectic phase ends, e.g., when the liquid reaches eutectic composition; in the ternary analogue considered here, austenite growth ends when the ferrite approaches uniform composition. In both cases non-equilibrium fractions of the two phases result. At the completion of γ growth in the ternary case the volume fraction of γ is generally greater than the equilibrium volume fraction.

Consideration of the behaviour of a binary Fe-C alloy suggests that, upon cooling from the intercritical range, ferrite growth should occur as the interfacial carbon content of austenite increases. It has been noted (4) that addition of a slow diffusing alloying element leads to the requirement of a finite degree of undercooling before rapid carbon-diffusion-controlled austenite shrinkage can occur. Manganese enrichment of the perimeter of γ particles tends to stabilize that region with respect to ferrite growth; the non-enriched centres become preferred sites for ferrite growth on cooling. The possibility of γ shrinkage must be considered before attempting to apply predicted volume fractions to material which is not rapidly quenched.

The quaternary case is similar to the ternary in that the interfacial equilibrium tie line is expected to change steadily during the period of alloying-element-controlled growth. One interesting result of the numerical calculations is that in moving from Fe-C-1%Mn to Fe-C-1%Mn-1%Si there is little change in growth rate despite the fact that Si has a diffusion rate ten times greater than Mn.

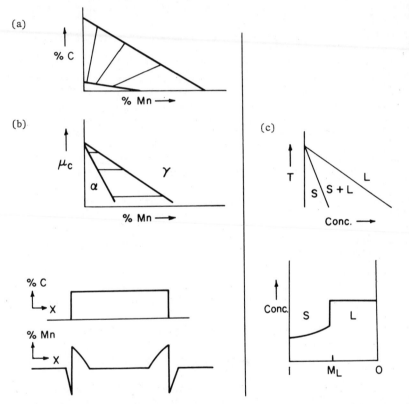

Fig. 16 - An analogy may be drawn between γ growth controlled by alloying element diffusion and classical non-equilibrium freezing. If the ternary isotherm, (a), is drawn with chemical potential of carbon substituted for carbon concentration, (b), then carbon chemical potential becomes analogous to temperature in the freezing case, (c). In both cases a locus of equilibrium concentrations is left behind by the advancing interface with this equilibrium concnetration being related to the fraction transformed by a mass balance.

81

Acknowledgements

Financial support was provided by Energy, Mines and Resources Canada (CANMET) and the Natural Sciences and Engineering Research Council Canada. Experimental material was provided by G. Speich of U.S. Steel. The scanning transmission electron microscopy was carried out at the University of Cambridge, U.K. and the author, P. Wycliffe, is indebted to Drs. A.J. Bourdillon, L.M. Brown and D. McMullen for their assistance.

References

(1) A.E. Cornford, J.R. Hiam and R.M. Hobbs, "Properties of As-Rolled Dual Phase Steels", S.A.E. Tech. Paper Series 79007, Detroit, 1979.

(2) M.S. Rashid, "GM-980 X - Potential Applications and Review", S.A.E. Paper 770211, February, 1977.

(3) J.M. Rigsbee and P.J. Van der Arend, "Laboratory Studies of Structure Property Relationships in Dual Phase HSLA Steels", pp. 58-88 in Formable HSLA and Dual Phase Steels, AIME, New York, N.Y., 1979.

(4) M. Home, S.V. Subramanian and G.R. Purdy, "Estimating Terminal Phase Equilibria in Ternary Systems", Can. Met. Quart., $\underline{8}$, 251, 1969.

(5) B. Uhrenius, "A Compendium of Ternary Iron-Base Phase Diagrams", pp. 28-81 in Hardenability Concepts with Applications to Steel, D.V. Doane, J.S. Kirkaldy, Editors, (Met. Soc. of AIME, Warrendale), p. 28, 1978.

(6) J.B. Gilmour, G.R. Purdy and J.S. Kirkaldy, "Thermodynamics Controlling the Proeutectoid Ferrite Transformation in Fe-C-Mn Alloys", Met. Trans., $\underline{3}$, 1455, 1972.

(7) P. Wycliffe, G.R. Purdy and J.D. Embury, "Growth of Austenite in the Intercritical Annealing of Fe-C-Mn Dual Phase Steels", Can. Met. Quart., in press.

(8) M. Hillert, "Paraequilibrium", Internal Report of Swedish Institute for Metals Research, Stockholm, 1953.

(9) G. Speich, private communication.

(10) K. Nohara and K. Hirano, "Diffusion of Mn in Iron and Iron-Manganese Alloys", Proc. Int. Conf. Sci. & Tech., Iron & Steel Suppl., Trans. ISIJ, Vol. 79, p. 1267, 1971.

(11) W. Batz, H.W. Mead and C.E. Birchenall, "Diffusion of Silicon in Iron", J. of Metals, p. 1070, 1952.

(12) Y. Adda and J. Philibert, "La Diffusion dan les Solides, Vol. II", pp. 1168-1170, Press Univ. de France, Saclay, 1966.

(13) J.I. Goldstein, D.B. Williams and A.D. Romig, Jr., "Spatial Resolution in STEM - X-ray Microanalysis", Proc. of a Specialist Workshop in Analytical Electron Microscopy", Cornell Univ., July, 1978, ed. P.L. Fejes.

(14) M.C. Flemings and R. Mehrabian, "Segregation in Casting and Ingots", pp. 311-340 in Solidification, ASM, Metals Park, 1971.

Appendix I. Explanation of Notation

C, D and J refer to concentration, diffusion coefficient and flux, respectively. Subscripts 2 and 3 indicate manganese and silicon, respectively which combine with superscripts α to indicate ferrite and γ to indicate austenite. C_1 indicates carbon content of austenite and C_1^E indicates eutectoid carbon concentration. C with only a Greek superscript refers to interface concentration with the exception of expressions such as C_2^γ (X) which indicate concentration (of manganese in austenite) varying as a function of distance. Superscript 0 indicates bulk concentration, while other Roman superscripts are as indicated.

ELECTRON MICROSCOPIC STUDY OF DEFORMED DUAL-PHASE STEELS

A. M. Sherman, R. G. Davies and W. T. Donlon

Ford Motor Company
Dearborn, MI 48121

ABSTRACT

Dual-phase steels having martensite contents ranging from 16 to 60% were examined using transmission electron microscopy after having been subjected to various amounts of monotonic and/or cyclic deformation. The dislocation density of the ferrite in the as-formed steels increased with martensite content. Monotonic deformation of 7% resulted in an increase in the dislocation density by a factor of 25 to 40. Cyclic deformation resulted in the formation of a well-defined cell structure with an average cell size of about 0.5 μm, independent of martensite content. However, differences were noted in cell distributions depending on the morphology of the martensite/ferrite structure, which varied with martensite content. These observations are used to suggest an explanation for the previously observed fatigue properties of these steels.

Introduction

The mechanical properties of dual-phase steels, under both monotonic and cyclic loading, have been previously studied (1,2). It has been found that monotonic strength is essentially proportional to the martensite content of the steel; this result has been explained by a rule-of-mixtures hypothesis (3). On the other hand cyclic properties were found to improve with increasing martensite content only up to about 30% (2); no satisfactory explanation for this behavior has been proposed. Due to the high strain hardening rate of dual-phase steels, prior deformation is an important factor in their monotonic strength properties. Conversely, the effect of pre-straining on cyclic stress-strain response has been shown to be relatively modest (1).

While property measurements of the type cited above are necessary for the practical application of dual-phase steels, detailed observations of their submicrostructures are needed to develop a fundamental understanding of their mechanical behavior. This paper reports on transmission electron microscopic studies on dual-phase steels which had been monotonically and cyclically deformed. Among the variables investigated were the percent martensite in the structure and the amount of prior deformation.

Experimental Procedures

A vanadium bearing dual-phase steel (0.11 C, 1.40 Mn, 0.48 Si, 0.08 V, bal Fe), supplied in the form of 3.9mm (0.154 in) thick sheets, was used for these observations. The original martensite content was 16.5% and samples having other martensite contents ranging from 24 to 100% were produced by reheating to and quenching from an appropriate temperature (2). Samples of these steels were subjected to various combinations of montonic and/or cyclic deformation. The results of these mechanical property tests have been previously reported (1,2).

Samples for election microscopic examination were cut from the gage sections of the mechanical property test specimens. Samples were also cut from the ends of these specimens in order to ascertain the structure of the undeformed steel and thus characterize the changes which occurred as a result of mechanical deformation. The 0.35mm thick samples were polished to a thickness of 0.15mm using 1 μm diamond paste. Disk shaped foils were prepared by electropolishing in a jet polishing apparatus using a solution consisting of 20% perchloric acid and 80% methanol. The foils were then examined in a Hitachi HU-650 high voltage electron microscope and in a Seimens 102 microscope operated at 125 kV.

The dislocation densities in the ferrite were estimated by measuring the number of intersections made by a series of concentric circles with the dislocations (4), and determining the foil thickness by the number of thickness fringes associated with a g=400 reflection. These results were in good agreement with those obtained using another method in which the number of dislocations intersecting the foil surfaces are counted (4).

Results and Discussion

As-Quenched Structures

The microstructure of "as-quenched" dual-phase steels containing 24 and 60% martensite are shown in Figure 1. The structures, in general, consisted of fine grained ferrite (~ 5μm diameter) containing a random array of dislocations; there was no evidence of any dislocation pile-up at the ferrite-martensite interface. However, it was noted that the dislocation density in the ferrite increases with the martensite content of the steel. This is clearly shown in Figure 2 where the ferrite dislocation density is plotted as a function of the martensite content. The increase in dislocation density with martensite content no doubt arises from the increasing amount of strain that has to be accommodated when the austenite transforms to martensite. The above observations are important when considering the strength of dual-phase steels as a function of percent martensite. The inherent strength of the ferrite has been assumed constant (3,5), however, it is now clear that the higher the martensite content of the material the more "cold-worked," i.e. stronger due to the higher dislocation density is the ferrite. At the moment it is not possible to quantify this increase in ferrite strength; this will have to await a study of the dislocation density as a function of strain in a fine grained ferrite of similar composition.

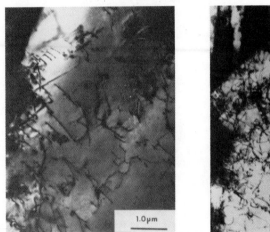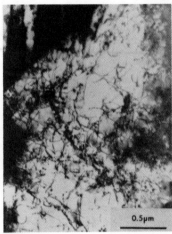

Figure 1   Dislocation substructure in the ferrite phase of as quenched dual phase steels (a) specimen quenched from 760°C containing 24% martensite and (b) specimen quenched from 820°C containing 60% martensite. 125 kV.

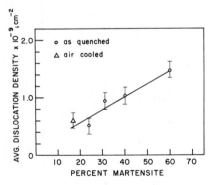

Figure 2   The dislocation density, measured in the ferrite phase, of dual phase steels versus martensite content.

The structure of the martensite was also not constant throughout the series of dual-phase steels. Specimens quenched from the higher temperatures (lower martensite carbon content) contained a mixture of plate and lathe martensite. Figure 3 shows lath martensite typical of that formed in a low-carbon steel. Micrographs of specimens quenched from the lower temperatures (highest martensite carbon content) consisted of ferrite and predominantly plate martensite, which as shown in Figure 4 contained some retained austenite. The role that these microstructural features play in influencing the monotonic and cyclic properties of dual phase steels, with the exception of the effect of carbon content (6), remain to be determined.

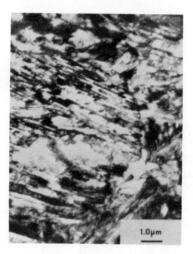

Figure 3   Microstructure in a specimen which is nearly 100% martensite, produced by quenching from the austenite single phase field (880°C); 650 kV.

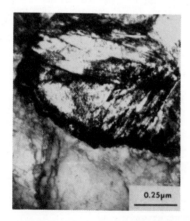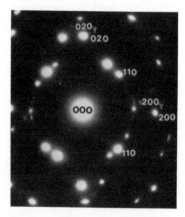

Figure 4   Plate martensite in a specimen quenched from 820°C containing 60% martensite (a) bright field micrograph and (b) selected area defraction slowing some retained austenite; 125 kV.

Monotonically Deformed Structures

For this initial study of the dislocation structure of monotonically deformed dual-phase steels, the specimens containing from 24 to 60% martensite were uniformly strained 7 percent; the figure of 7% strain was chosen as it was the maximum uniform strain obtainable in the specimens containing 60% martensite. As shown in Figure 5 the dislocations in the ferrite after this deformation are in fairly random arrays with no indication of well developed cells. The ferrite dislocation density as a function of percent martensite is shown in Figure 6; since individual dislocations could not be resolved in some areas of the ferrite the measured dislocation densities probably tend to be lower limit values. From a comparison of Figures 2 and 6 it can be seen that 7% monotonic strain has resulted in an increase in the ferrite dislocation density by a factor of between 25 and 40, however, the ferrite dislocation density of the strained specimens still increases with increasing martensite content. This variation in dislocation density with martensite content is most probably due to the partitioning of strain between the ferrite and martensite; the higher the martensite content the more of the total strain will be carried by the ferrite and the higher will be its dislocation density.

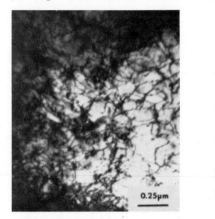

Figure 5   Dislocation substructure in the ferrite phase of monotonically deformed dual phase steels uniformly strained 7 percent (a) bright field micrograph of a specimen containg 24% martensite and (b) bright field micrograph of a specimen containing 60% martensite; 125 kV.

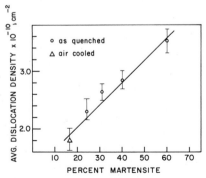

Figure 6   Measured dislocation density in the ferrite versus martensite content for specimens monotonically deformed 7 percent.

Cyclically Deformed Specimens

The cyclic and monotonic stress strain curves for dual-phase steels as a function of martensite content are compared in Figure 7. It can be seen that while the monotonic curves increase approximately linearly with percent martensite, the cyclic curves for dual-phase steels show an increase with up to 31% martensite but then remain essentially constant to at least 60%.

The dislocation structure in the ferrite after cyclic deformation at a strain amplitude of 0.01 is shown in Figure 8. For the ferrite, the dislocations are arranged in well defined cells independent of the percentage of martensite in the structure. No dislocation pile-ups were observed at either grain boundaries or ferrite-martensite interfaces; the dislocation cells simply terminated whenever they encountered a boundry or interface. The dislocation structures of cyclically deformed specimens which had been monotonically prestrained by various amounts were identical to those discussed above. Thus, the way in which the material accommodates different types of deformation by changes in its submicrostructure explains the relatively slight influence of prior deformation on cyclic properties as opposed to the large effect on monotonic strength (1).

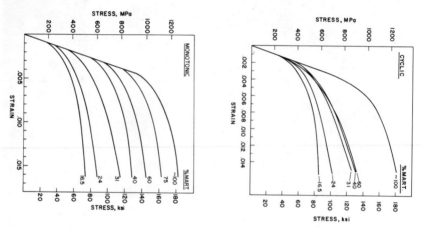

Figure 7   Monotonic (a) and cyclic (b) stress-strain curves. Note cyclic hardening in alloys up to 31% martensite and softening at higher martensite contents (2).

For each of the four dual-phase steels containing from 24 to 60% martensite a histogram of the cell size distribution similar to Figure 9 was obtained. The mean dislocation cell size and the standard deviation are, as shown in Table I, essentially independent of the martensite content of the dual-phase steel. The dislocation cell sizes observed in dual-phase steels are the same as those seen in conventional HSLA steels having the same grain size (7). Dislocation cell sizes have been found to vary inversely to the cyclic strain amplitude, although such variations are not large (25% for a factor of 10 in strain amplitude). The constancy of cell size in dual-phase steel as martensite content is increased may indicate that the ferrite is subjected to roughly the same plastic strain independent of martensite content. Thus, to accommodate the imposed strain amplitude, the martensite phase must also undergo cyclic deformation, although the structure of the martensite phase is apparently too

Figure 8 Dislocation substructures in cyclicly deformed $(\frac{\Delta \epsilon}{2} = 0.01)$ dual-phase steels; 650 kV,
(a) specimen quenced from 760°C containing 24% martensite (b) specimen quenched from 780°C containing 30% martensite (c) specimen quenched from 800°C containing 40% martensite and (d) specimen quenched from 820°C containing 60% martensite.

complex to allow evidence of this to be observed. This suggests that the partitioning of strains between the martensite and ferrite may be different under cyclic loading than under monotonic loading.

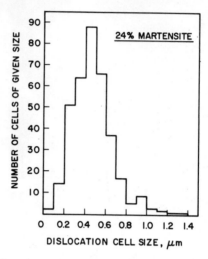

Figure 9   Histogram showing the distribution of dislocation cell sizes in a dual-phase steel containing 24% martensite cyclicly deformed ($\frac{\Delta \varepsilon}{2}$ = 0.01) (total number of disloction cells measured was approximately 200).

It was noted however, that in the higher martensite content dual-phase steels there were areas of the ferrite totally surrounded by martensite in which the dislocations remained in random arrays and did not form a cell structure. Figure 10 shows such an area in the dual-phase steel containing 60% martensite. The volume fraction of these areas increased as the martensite content increased and the lack of a dislocation cell structure indicates that these areas have not been as highly strained as the other ferrite regions. The intimate presence of the martensite appears to be shielding this ferrite from plastic deformation of an amount sufficient to form cells, which suggests that the martensite around these less deformed ferrite patches must be undergoing some cyclic plastic deformation. Thus, as well as the partitioning of strain mentioned above, this non-uniformity of strain distribution which increases with increasing martensite content, may also be part of the reason for the constancy of cyclic stress-strain behavior in steels containing between 31 and 60% martensite.

Conclusions

This investigation has shown that the dislocation substructure of dual-phase steel depends on the proportion of martensite. With higher martensite contents the martensite is increasingly of the lath variety and the greater is the dislocation density of the ferrite. Observations of dislocation densities due to monotonic deformation indicates that the partitioning of plastic strain to the ferrite increases with martensite content. Cyclic deformation results in the formation of a dislocation cell structure in the ferrite, with a cell size averaging 0.5 μm independent of martensite content. However, as martensite content increased small patches of less deformed ferrite shielded by the martensite were observed. These observations imply that cyclic plastic strains are partitioned differently from monotonic strains and thus suggest an explanation for the differences in the effect of martensite content on monotonic and cyclic behavior.

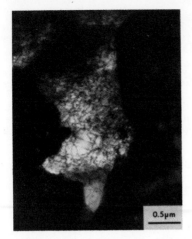

Figure 10   Dislocation substructure in a specimen containing 60% martensite cyclicly
deformed ($\frac{\Delta \epsilon}{2}$ = 0.01). Areas of ferrite surrounded by martensite; 125 kV.

Acknowledgements

The authors wish to express their thanks to Messers. B. T. Crandell and W. S. Stewart for help in sample preparation.

TABLE I

Martensite Content	Mean Dislocation Cell Size	Standard Deviation
24%	0.47 μm	.22
31%	0.58 μm	.21
40%	0.49 μm	.20
60%	0.48 μm	.20

References

1.   A. M. Sherman and R. G. Davies, "Fatigue of a Dual-Phase Steel," Met Trans A, 10A, (7) (1979), pp. 929-933.

2.   A. M. Sherman and R. G. Davies, "The Effect of martensite content on the fatigue of a dual-phase steel," International Journal of Fatigue, (1) (19981) pp. 36-40.

3.   R. G. Davies, "Deformation Behavior of a Vanadium-Strengthened Dual Phase Steel," Met Trans. A, 9A, (1) (1978) pp. 41-51.

4.   Hirsh, P. B., Howie, A., Nicholson, R. B., Pashley, D. W., Whelan, M. J., Electron Microscopy of Thin Crystals, 1st ed. reprinted and revised, Butterworth & Co. LTD., London, 1965.

5.   J. Y. Koo and G. Thomas, "Thermal Cycling Treatments and Microstructures for Improved Properties of Fe-0.12% C-0.5% Mn Steels," Materials Science and Engineering 24, (1976) pp. 187-198.

6.   A. M. Sherman and R. G. Davies, "Influence of Martensite Carbon Content on the Cyclic Properties of Dual Phase Steel," TMS-AIME Fall Meeting, Pittsburgh, Pa, Oct., 7, 1980.

7.   R. W. Landgraf, A. M. Sherman and J. W. Sprys, "Fatigue Behavior of Low Carbon Steels as Influenced by Microstructure," Proceedings of the Second International Conference on Mechanical Behavior of Materials, Aug., 16-10, 1976, Boston, Mass. pp. 513-517.

EFFECT OF DEFORMATION ON THE $\gamma \to \alpha$ TRANSFORMATION

IN TWO HIGH SILICON DUAL-PHASE STEELS

J.J. Jonas and R. A. do Nascimento
Dept. of Metallurgical Engineering, McGill University,
3450 University St., Montreal H3A 2A7, Canada

I. Weiss
Department of Engineering, Wright State University, Dayton, Ohio, 45435

A.B. Rothwell
Nova Corporation, P.O. Box 2511, Calgary, Alberta, T2P 2M7, Canada

Rolling simulations were carried out on two steels with the following compositions: steel A - .07 C, 1.03 Mn, 1.11 Si, 0.54 Cr and 0.42 Mo; steel B - .06 C, 0.92 Mn, 1.46 Si, 0.45 Cr and 0.40 Mo. The deformation-time-temperature profiles involved in rolling and in cooling on the run-out table were simulated using a computerized, servo-controlled hot torsion machine. The samples were austenitized at 1100°C for 5 min. and deformed in two passes of 30 and 40% (steel A) and 60 and 80% (steel B) to simulate roughing. After cooling into the temperature range 1000 to 900°C, two further strains of 30 and 40% (steel A) and 60 and 80% (steel B) were applied to simulate finishing. The specimens were cooled at 10°C/s after finishing and then water quenched from temperatures in the range 800 to 650°C.

The progress of ferrite transformation during cooling was followed by means of optical metallography. The results obtained in this way are compared with the CCT diagrams obtained from undeformed samples using dilatometry. It is shown that the principal effect of deformation is to raise the polygonal ferrite (PF_s) temperature by about 50°C. Deformation also increases the rate of ferrite formation by up to one order of magnitude in time. The acceleration of the transformation increases with the extent to which the finish rolling temperature is depressed below the Ac_3 temperature. The shifting of the deformed CCT diagram to shorter times is judged to be of secondary importance compared to the increase in the PF_s temperature because of the relatively slow rate of cooling during finish rolling. Thus, as long as deformation occurs substantially below the Ac_3 temperature, the transformation takes place in the vicinity of the plateau region of the 'strained' CCT diagram, and not near the accelerated 'nose'.

Introduction

Since the first commercial mill trials in 1977 (1), several authors have been concerned with establishing the basic metallurgy of the production of dual-phase steels directly by hot rolling (2-7). Much of this research has involved the determination of the CCT diagrams for the undeformed material by dilatometry. References 2, 5 and 7 fall into this category. However, one shortcoming of this approach is that the CCT curves obtained apply to recrystallized austenite while, under rolling conditions, the transformation to ferrite takes place in strained austenite. The deformed nature of the austenite, in turn, can be expected to increase the rapidity of both the ferrite and pearlite reactions (7,8).

In order to 'correct' the CCT curves obtained by dilatometry for the absence of strain in the austenite, several approaches have been used. For example, Eldis et al. (7), make use of the ratio of the pearlite start time (P_s) to the time for the formation of 75% polygonal ferrite (PF_{75}) in the undeformed material. According to these authors, a ratio of 10 or more indicates that pearlite formation will be avoided, and ferrite formation favored, under appropriate rolling conditions. Although the CCT diagram for deformed austenite is not actually known, it has also been estimated (5) that severe straining shifts the CCT diagram to the left by 'two orders of magnitude in time'.

It is evident from the above that an accurate diagram for austenite in the strained condition would be of considerable use. It could, in principle, enable as-rolled dual-phase steels to be produced with smaller alloying additions. It could, also, reduce the amount of off-grade material made under adverse conditions of finishing and/or coiling temperature. Thus the aim of the present investigation was to determine the transformation diagram for deformed austenite by means of torsional simulation of the rolling process, coupled with quantitative metallography for measurement of the volume fractions of the transformed phases. It was also an objective of this work to assess the effect of different finish rolling temperatures and to gain an insight into how these influence the amount of ferrite formed under run-out table conditions of cooling.

Materials and Experimental Procedure

Two dual-phase steels were selected for these experiments; their compositions are given in Table I. Steel A was received in the form of 13 mm thick hot-rolled plates and steel B as hot-rolled bars of 15 mm dia. Torsion specimens were machined from the plates and bars with their axes in the rolling direction. Gage lengths of 26 and 15 mm were used for steels A and B, with diameters of 6.4 and 5.0 mm, respectively. The torsion experiments were performed on a servo-controlled, hydraulic machine operated in the range of 2 to 25 revolutions per second with a maximum applicable torque of 100 N-m. Steels A and B were deformed at 2.24 and 22.4 revolutions per second, which correspond to equivalent strain rates of 1 and 10 s^{-1}, respectively. The torsion machine was driven by an MTS hydraulic power supply and the data acquisition and control functions were performed by a PDP 1104 minicomputer. Prior to testing, the test pieces were austenitized at 1100°C for 5 min. directly in the torsion machine by means of an RI dual elliptical radiant heater mounted on the bed of the torsion machine. This heat treatment produced austenite grain sizes of 50 and 45 μm in steels A and B, respectively.

The computerized torsion machine has the capability of applying up to twenty-four simulated 'passes' of rolling separated by intervals of load-free

Table I. Chemical Composition of the Steels Tested - WT%

	Steel A	Steel B
C	.07	.063
Mn	1.03	.92
Si	1.11	1.46
Cr	.54	.45
Mo	.42	.40
P	.006	.010
S	.010	.006
Al	.03	.028
N	.006	.0046
Ac_3	970°C	985°C
Ac_1	707°C	735°C

holding and cooling. In this way, plate rolling schedules can be readily
simulated and, by quenching or air cooling samples after 'passes' or holding
intervals of interest, the evolution of both the austenite and the transformed ferrite structures can be followed in detail (10,11). In the present
study, only four deformation passes were applied: two at 1100°C, to represent the roughing process, and two, in the interval from 1000 to 900°C, to
represent the effects of the finishing process. The deformation-time-
temperature sequences employed for the two steels are listed in Table II and
displayed in schematic form in Figures 1 and 2. It should be noted that

Table II. Testing Conditions in the Rolling Simulations

	Steel A	Steel B
Austenitization - temperature	1100°C	1100°C
- time	5 min.	5 min.
Austenite grain size	50 μm	45 μm
Testing strain rate	1 s^{-1}	10 s^{-1}
Roughing - ε_1	30%	60%
interpass time	150 s.	5 s.
ε_2	40%	80%
isothermal hold time	150 s.	-
Cooling time to finish rolling (approximate)	(i) 120 s. (to 980°C)	(i) 15 s. (to 950°C)
	(ii) 180 s. (to 920°C)	(ii) 20 s. (to 900°C)
Cooling rate during finish rolling	1°C/s.	10°C/s.
Finishing - ε_1	30%	60%
interpass time	5 s.	5 s.
ε_2	40%	80%
Finishing temperatures	(i) 980°C (ii) 920°C	(i) 950°C (ii) 900°C
Cooling rate after finish rolling	10°C/s.	10°C/s.

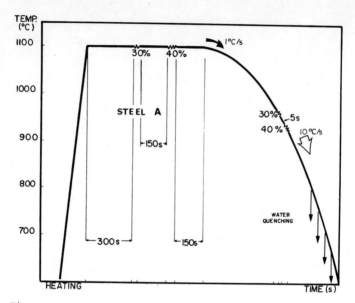

Fig. 1 - Deformation-time-temperature profile for steel A.
Note that the cooling rate after roughing is 1°C/s.
and is 10°C/s. after finishing.

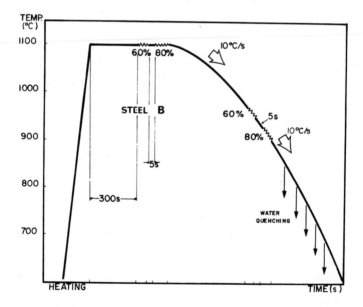

Fig. 2 - Deformation-time-temperature profile for steel B.
Note that the cooling rate after both roughing and
finishing is 10°C/s.

steel A was held at temperature after roughing for 150 s. before the cooling program of 1°C/s. was applied. This was done to permit full recrystallization. In steel B, the holding period was omitted as it was considered unnecessary, and cooling at 10°C/s. was initiated directly after roughing. The lengths of the cooling periods were then adjusted so that the first finishing pass was applied about 10 and 50°C above the desired finishing temperature for steels A and B, respectively.

Each deformation sequence was carried out under the control of the mini-computer, which applied the required strain per pass, unloaded the sample for the appropriate delay times and then continued the straining sequence for the total of four passes. After roughing, the temperature was decreased by shutting off the furnace and cooling each sample with a regulated stream of argon. After finishing, all the samples were argon cooled at 10°C/s. to simulate water cooling on the run-out table, and then water quenched. The specimen temperature was measured using a flexible thermocouple mechanically fixed to the center of the gage length and free to rotate with the twisted specimen. Since the last two passes were carried out while the sample was cooling, there was a tendency for the measured temperatures to be 'low' (i.e. less than the average for the specimen as a whole) because of the higher cooling rate associated with the exterior of the specimen. To minimize this effect, 15 turns of nichrome wire were wrapped around the center of the gage length to provide a measure of thermal shielding.

After testing, the samples were quenched directly in the torsion machine, using a pressurized water spray. Quench rates of about 400°C/s. were achieved. Prior to metallographic examination, the side of the specimen in the direct path of the spray was identified on each deformed sample, which was then sectioned longitudinally so as to contain this direction. The polished samples were etched in 2% nital and the volume fraction of polygonal ferrite was determined by point counting, which was carried out on layers adjacent to the most rapidly quenched part of the test-piece surface. The equivalent stresses and strains were calculated from the torques and twists by the method of Fields and Backofen (9-11), using average values of 0.17 for the rate sensitivity of the torque and 0.13 for the torque-based coefficient of work hardening.

Results.

A typical stress-strain diagram for steel B is displayed in Figure 3, together with the relevant pass temperatures. It is evident that the flow curve for the second roughing pass can be superimposed on that for the first. This indicates that full static recrystallization took place during the 5 s. unloading interval at 1100°C. The two finishing passes were carried out during cooling after roughing, so that, unlike the first and second curves, the third and fourth flow curves were determined at different temperatures. As a result of the continuous cooling, the third curve is at a higher general stress level than the second and, in a similar way, the fourth curve is at a higher level than the third. It is evident that the 'yield' stress at 950°C is higher than at 1100°C and that the rate of work hardening is also greater. On reloading at 900°C, 5 s. after deforming at 950°C, it can be seen that the 'yield' strength has increased still further and the shape of the flow curve has also changed. As the temperature decrease between passes 3 and 4 is relatively small, these features indicate that only static recovery (and no static recrystallization) occurred between the third and fourth passes. The flow curves for steel A had similar features, and a larger selection of such curves is presented in References 10 and 11 for a plain C and two microalloyed steels.

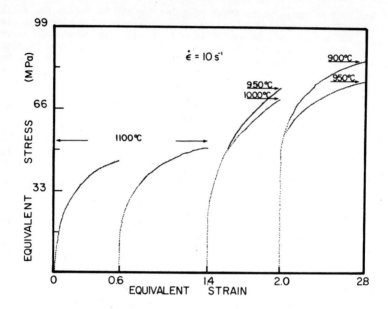

Fig. 3 - Examples of flow stress-strain curves for steel B produced by
torsion testing. The two flow curves produced at 1100°C are
similar due to the occurrence of static recrystallization between 'passes'. The third pass differs from the second principally because of the decrease in temperature. The fourth pass
differs from the third principally because of the absence of
recrystallization during the holding interval.

Ferrite Formation in Undeformed Austenite

The CCT diagram for steel A was determined at the Noranda Research
Centre by means of standard dilatometric methods. That for steel B was not
established directly; instead an undeformed CCT diagram published as Figure
4 in Reference 7 for a similar steel was used for comparison purposes.*
Examples of the ferrite structures that developed in the dilatometer in
steel A are given in Figure 4. In the sample which was cooled to room
temperature at about 25°C/s. (Figure 4a), about 5% polygonal ferrite can be
seen, located primarily at the prior austenite grain boundaries. When the
cooling rate was decreased to 10°C/s., about 35% ferrite was formed (Figure
4b). Finally, when the cooling rate was reduced to 1°C/s. or less, the
amount of polygonal ferrite attained its maximum value of 92%.

_____

*
Steel B and the steel of Reference 7 had similar C, Cr, Mn and Mo
contents (see Table I). However, the Reference 7 steel contained only 1.20%
Si vs. 1.46% Si in the present material. Thus, in the comparisons between
the undeformed and deformed conditions to be made below (Figures 11 and 13),
the 'true' undeformed behavior of steel B can be expected to be slightly more
'rapid' than indicated by the undeformed CCT diagram pertaining to the Reference 7 steel.

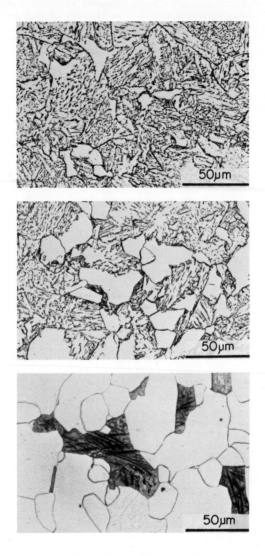

Fig. 4 - Examples of polygonal ferrite structures produced
in steel A in the dilatometer (undeformed).

(a) cooled to R.T. at 25°C/s. (5% ferrite)
(b) cooled to R.T. at 10°C/s. (35% ferrite)
(c) cooled to R.T. at 1°C/s. (92% ferrite)

Similar microstructures were produced in steel B when it was cooled at 10°C/s. in the torsion machine in the absence of deformation. Some representative structures are presented in Figure 5, from which it can be seen that the proportion of polygonal ferrite increases from about 0 to about 10% as the quenching temperature decreases from about 900 to 750°C.

Ferrite Formation in Deformed Austenite

The ferrite proportions produced by deformation according to the schedules of Figures 1 and 2 are displayed in Figures 6 and 7 for steels A and B, respectively. The structures of specimens 'finish rolled' at 980°C and quenched from 810 and 660°C are presented in Figures 6a and 6b, whereas Figures 6c and 6d represent samples finish rolled at 920°C and quenched from 800 and 650°C. It can be seen that the polygonal ferrite is nucleated principally at the prior austenite boundaries, as before, but that the grain size is finer than in the undeformed case. It is also of interest that more ferrite is formed at a given temperature after rolling at 920°C than after rolling at 980°C (compare the ferrite proportions of 35 and 50% pertaining to Figures 6a and 6b with those of 65 and 82% pertaining to Figures 6c and 6d).

Qualitatively similar results were obtained with steel B, see Figure 7, despite the higher Si concentration (1.46% vs. 1.11% for steel A). In addition, the Mn and Cr concentrations were lower in the B material and the strains and strain rates applied to the B specimens were higher. Thus ferrite formation was expected to be favored in the B steel, although the influences described above may have been counteracted by the occurrence of partial recrystallization after finishing in steel B due to the higher pass strains. This is a point to which we will return below.

Discussion

Ferrite Transformation in Steel A

The effect of finish rolling temperature on the amount of ferrite formed during continuous cooling is illustrated in Figure 8 for steel A. The two deformation temperatures were chosen to be 10° above (980°C) and 50° below (920°C) the Ac_3 temperature of 970°C. It is readily apparent that more ferrite has been formed by the time a given quench temperature is attained when final straining is performed at the lower temperature. In the present case, the magnitude of the effect, as well as a general knowledge of recrystallization rates in such steels (12), suggests that some partial recrystallization might have occurred after rolling at the higher temperature. The small difference in dislocation density produced by rolling at 920 as opposed to 980°C (deduced from the small difference in the flow stresses at these two temperatures) is judged not to be responsible for the discrepancy in transformation rate. The role of the dislocations retained in the austenite after rolling at 920°C is possibly to permit pipe diffusion, and in this way to enable more substitutional solute partitioning to take place during the $\gamma \rightarrow \alpha$ transformation.* The presence of the dislocations also increases the energy of the austenite, making it more unstable below 970°C. According to this view, the effect of rolling below the Ac_3 temperature is simply to decrease

_____

*Although C diffusion is rapid enough to permit partitioning under run-out table cooling conditions, the bulk diffusion rates for the substitutional solutes are too low to allow the substantial partitioning of these elements in the short times available.

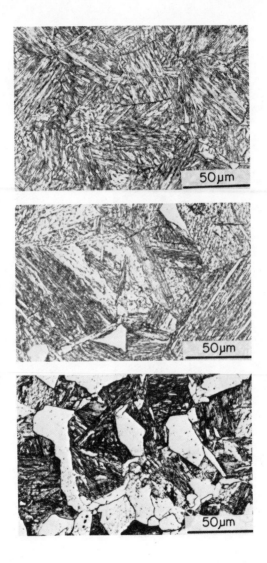

Fig. 5 - Examples of polygonal ferrite structures produced in
undeformed samples of steel B during cooling in the
torsion machine.

(a) quenched at 900°C (100% martensite)
(b) quenched at 800°C   (1% ferrite)
(c) quenched at 750°C (10% ferrite)

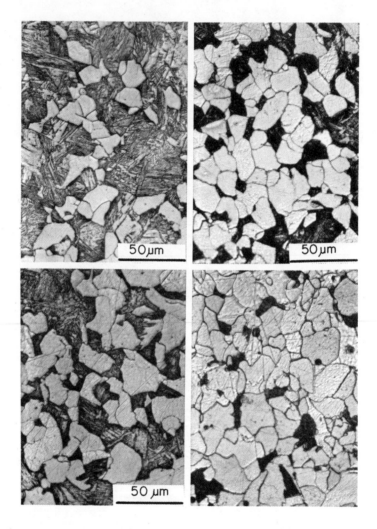

Fig. 6 - Examples of the ferrite structures produced in deformed samples of steel A.

    (a)   'finish rolled' at 980°C; quenched at 810°C (35% ferrite)
    (b)   'finish rolled' at 980°C; quenched at 660°C (50% ferrite)
    (c)   'finish rolled' at 920°C; quenched at 800°C (65% ferrite)
    (d)   'finish rolled' at 920°C; quenched at 650°C (82% ferrite)

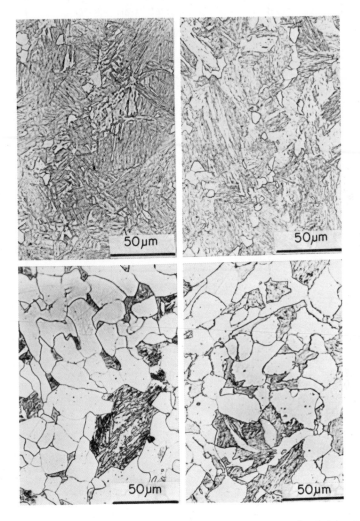

Fig. 7 - Examples of the ferrite structures produced in deformed
samples of steel B.

(a) final 'pass' at 950°C; quenched at 800°C (2% ferrite)
(b) final 'pass' at 950°C; quenched at 650°C (75% ferrite)
(c) final 'pass' at 900°C; quenched at 760°C (14% ferrite)
(d) final 'pass' at 900°C; quenched at 670°C (83% ferrite)

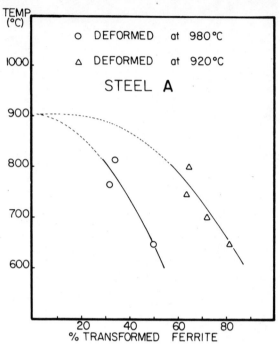

Fig. 8 - Dependence of the amount of polygonal ferrite formed at a
given quench temperature on the temperature of final deformation (steel A).

the rate of recrystallization after straining, thereby promoting the retention of the dislocations, and increasing the transformation rate in this way.
Although rolling below the Ac_3 could also accelerate ferrite formation more
directly, the mechanism for doing so is not evident, bearing in mind that no
shear transformation is involved, and that the volume change on transformation is, in fact, positive.*

For ease of comparison with the transformation data for the undeformed
austenite, the present results relating to steel A in the deformed state are
displayed in Figure 9 in the form of two cooling curves, one for each of the
finish rolling temperatures. The zero time base for the cooling curves was
taken here as the instant when deformation ceased. It is readily apparent
that, in addition to making the phase change more rapid, as already discussed
above, deformation increases the PF_s temperature. A measure of the former
effect is indicated directly on the diagram, where the time required to produce 72% of ferrite on cooling to 700°C is identified for each of the two
conditions. Although the two sets of data are not completely compatible
(because the austenitization conditions and grain sizes were not identical),
the data suggest that deformed austenite transforms about three times more
quickly than the unstrained material. The extent of recrystallization can
also be judged in the 980°C finish-rolled material in that it attains a

---
*Rolling below the recrystallization temperature also increases the
austenite grain boundary area per unit volume and contributes to increasing
the rate of transformation in this way.

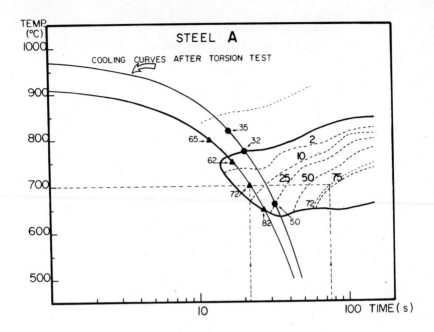

Fig. 9 - The data of Figure 8 for the deformed austenite superimposed on
the CCT diagram for undeformed austenite determined by dilatometry.
Here the zero time base for the two cooling curves is taken as the
instant that the final deformation pass is completed. The final
deformation temperatures were 980°C for the upper curve and 920°C
for the lower curve. Note the increase in the PF_s temperature
indicated by the present results.

transformation level of 50% at about 30 s., whereas the undeformed material
requires 45 s., i.e. it takes $1\frac{1}{2}$ (and not 3) times as long. The two cooling
curves thus support the view that the 'state' of the austenite (i.e. deformed,
partly recrystallized, or undeformed) affects the amount of ferrite formed
to a greater degree than either the temperature or the time.

Ferrite Transformation in Steel B

The progress of the ferrite reaction in the higher silicon material is
illustrated in Figure 10. Here there is no significant difference in the
amount of ferrite formed after deformation at the two temperatures. Also,
it appears that the ferrite transformation does not begin until the steel
has cooled to about 800°C, whereas PF_s temperatures of at least 850°C are
suggested by the results for steel A. The cooling curves for the two deformation temperatures are presented in Figure 11 where, as in Figure 9, the
curves are considered to begin when straining ceases. The ferrite portion
of the CCT diagram determined by Eldis et al. (7) for austenite with a 60
μm grain size is included for reference purposes. Once again it can be seen
that deformation increases the PF_s temperature and accelerates the transformation by a factor of about three in time.

107

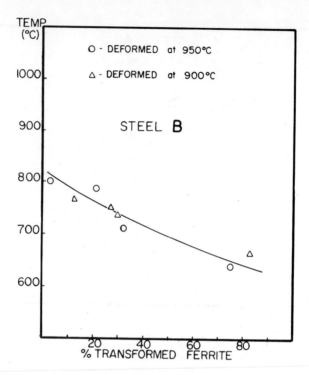

Fig. 10 - Dependence of the amount of polygonal ferrite formed in steel B
on the quench temperature after finish rolling at 950 and 900°C.

Effect of Accumulated Time below the Ac_3 Temperature

Before concluding this discussion, it must be noted that the time zero
for the cooling curve could also be taken as the moment when the material
passes through the Ac_3 (or Ae_3) temperature. The effect of such an interpretation is depicted for Steel A in Figure 12, where the cooling rate during 'rolling' is taken as 1°C/s. (see Figure 11). From Figure 12 it is
evident that, when finish rolling takes place below the Ac_3 temperature, the
time to cool down to the final deformation temperature may exceed the minimum
PF_s time determined from the dilatometer experiments. Thus, when rolling is
completed below the Ac_3 temperature, the transformation diagram will normally
be approached from above, and the principal effect of deformation will be to
increase the PF_s temperature.

Under these conditions, the concept of the 'acceleration' of the transformation must be modified. In Figure 12, for example, the PF_{32} time for
the material rolled at 980°C (i.e. above the Ac_3) is about 20 s., whereas
the undeformed PF_{32} time is about 100 s., i.e. a factor of 5 greater. When
the sample rolled at 920°C (i.e. below the Ac_3) is considered, an even
greater acceleration is obtained. In this case, for the deformed material,
PF_{65} is about 60 s. at 800°C, and PF_{65} cannot be attained at all at 800°C
in the undeformed condition.

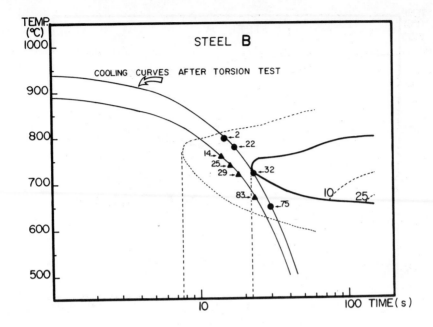

Fig. 11 - The data of Figure 10 for the deformed austenite superimposed
          on the CCT diagram for undeformed austenite (7). As in Figure
          9, the zero for the time base is taken as the moment when
          final straining ceases. The last pass temperatures were 950°C
          (upper curve) and 900°C (lower curve). As in Figure 9, de-
          formation increases the PF_s temperature.

In the steel B experiments depicted in Figure 13, the cooling rate
during simulated rolling was 10°C/s. down to both 950 as well as to 900°C
(see Figure 2). Thus the two cooling curves are superimposed, and the
transformation is not approached from above, as would generally be expected,
and as shown in Figure 12 for the 920°C finish temperature. Nevertheless,
it can be seen that the PF_{80} time for the deformed material is about 30 s.,
which is more than an order of magnitude faster than the PF_{80} time of 350 s.
indicated by the dilatometer results. Similarly, the deformed PF_{32} time
is about 25 s. compared to about 540 s. in the dilatometer. In the region
of the 'nose' of the PF_s curve, the increase in the PF_s temperature is
about 50°C. As in the case of steel A, it is probably the raising of the
PF_s temperature that is principally responsible for the increase in the
volume fraction of ferrite formed after rolling above that predicted from
the conventional CCT diagram.

Conclusions

1. When austenite is deformed, the PF_s temperature is <u>increased by about
    50°C</u> by comparison with the CCT diagram for the undeformed material.
    This is attributed to the effect of dislocations on increasing
    (a) the energy of the deformed austenite; and (b) the rate of dif-
    fusion during solute partitioning. The increase in the specific
    grain boundary area is also likely to make a contribution.

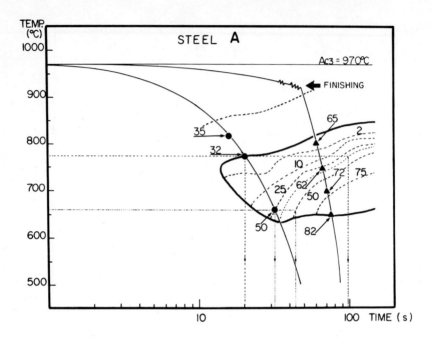

Fig. 12 - The data of Figure 9 replotted on the basis of the accumulated
time elapsing since cooling through the Ac_3 temperature. The
upper curve is representative of finishing below the Ac_3 temperature ($920°C$ in this case). When run-out table cooling rates
are applied, transformation occurs in the plateau region of the
deformed CCT curve. The lower curve is representative of finishing above the Ac_3 temperature ($980°C$ in this case). On applying
the run-out table cooling rate, transformation occurs near the
nose region of the deformed CCT curve. Note that deformation
increases the PF_s temperature.

2. The ferrite transformation in deformed austenite is about three to
    twenty times faster than in undeformed austenite of approximately the
    same grain size. As a result of these two effects, under cooling
    conditions which produce 25% polygonal ferrite in undeformed austenite,
    over 80% polygonal ferrite is formed.

3. Finish rolling in the vicinity of Ac_3 leads to the formation of less
    polygonal ferrite than when lower rolling temperatures are used (e.g.
    (Ac_3 - 80)$°C$). This may result from the partial recrystallization of
    austenite after rolling in the vicinity of $1000°C$. The principal
    factor, however, is likely to be the difference in the accumulated
    time below the Ac_3 temperature between the two deformation-time-
    temperature profiles. Thus, rolling below the Ac_3 leads to more
    complete transformation because rapid cooling (on the run-out table)
    takes place in the plateau region of the deformed CCT diagram. By
    contrast, rolling above the Ac_3 (even in the absence of recrystalliza-
    tion) leads to less transformation because the rapid cooling profile
    passes through the nose region of the deformed diagram instead.

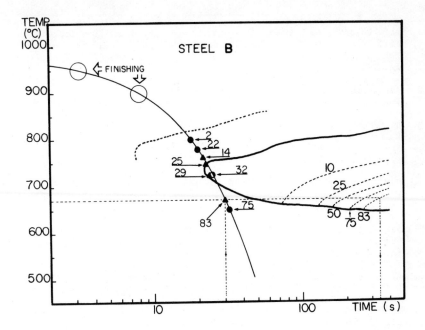

Fig. 13 - The data of Figure 11 replotted on the basis of the time elapsed
since cooling through the Ac_3 temperature. Note that the cooling
curves are identical for finishing at both 950 and 900°C. The
unrealistically high cooling rates applied during finish rolling
(10°C/s. - see Table II) have the effect of taking the cooling
curve through an earlier portion of the deformed CCT curve. Under
industrial conditions, the temperature-time profile is expected
to be closer to the upper curve of Figure 12.

Acknowledgements

        This research project was initiated by the Noranda Research Centre,
which also provided one of the experimental materials (steel A). The
authors are grateful to the Climax Molybdenum Co. for supplying the second
alloy (steel B). They also acknowledge with appreciation financial support
received from the following sources: Canadian International Development
Agency; Energy, Mines and Resources, Canada; Natural Sciences and Engineering
Research Council of Canada; and the Ministry of Education of Quebec (FCAC
program). One of the authors (R.N.) is thankful to the Companhia Siderúrgica
Nacional of Brazil for granting a period of study leave.

References

1. A.P. Coldren, G. Tither, A. Cornford, and J.R. Hiam, "Development and Mill Trial of As-Rolled Dual-Phase Steels," pp. 205-226 in Formable HSLA and Dual-Phase Steels, A.T. Davenport, ed.; TMS-AIME, Warrendale, PA, 1979.

2. A.P. Coldren and G. Tither, "Development of a Mn-Si-Cr-Mo As-Rolled Dual-Phase Steel," Journal of Metals, 30 (4) (1978) pp. 6-9.

3. T. Greday, H. Mathy, and P. Messien, "About Different Ways to Obtain Multiphase Steels," pp. 260-280 in Structure and Properties of Dual-Phase Steels, R.A. Kot and J.W. Morris, Jr., eds.; TMS-AIME, Warrendale, PA, 1980.

4. J.M. Rigsbee, J.K. Abraham, A.T. Davenport, J.E. Franklin and J.W. Pickens, "Structure-Processing and Structure-Property Relationships in Commercially Processed Dual-Phase Steels," Reference 3, pp. 304-329.

5. A.P. Coldren and G.T. Eldis, "Using CCT Diagrams to Optimize the Composition of an As-Rolled Dual-Phase Steel," Journal of Metals, 32 (3) (1980) pp. 41-48.

6. A.E. Cornford, J.R. Hiam, and R.M. Hobbs, "Properties of As-Rolled Dual-Phase Steels," pp. 1-10 in Technical Paper Series, Society of Automotive Engineers, Inc., Warrendale, PA, 1979.

7. G.T. Eldis, A.P. Coldren, and F.B. Fletcher, "Alloying and Transformation Control in Mn-Si-Cr-Mo As-Rolled Dual-Phase Steels," pp. 37-57 in Alloys for the 80s, Climax Molybdenum Company, Ann Arbor, Michigan, June 1980.

8. Y.E. Smith and C.A. Siebert, "Continuous Cooling Transformation Kinetics of Thermomechanically Worked Low-Carbon Austenite," Metallurgical Transactions, 2A (1971) pp. 1171-1725.

9. S. Fulop, K.C. Cadien, M.J. Luton, and H.J. McQueen, "A Servo-Controlled Hydraulic Hot-Torsion Machine for Hot Working Studies," Journal of Testing and Evaluation, 5 (1977) pp. 419-426.

10. I. Weiss, J.J. Jonas, P.J. Hunt, and G.E. Ruddle, "Simulation of Plate Rolling on a Computerized Hot Torsion Machine and Comparison With Mill Results," pp. 1225-1236 in International Conference on Steel Rolling, Iron and Steel Institute of Japan, Tokyo, September 1980.

11. I. Weiss, J.J. Jonas, and G.E. Ruddle, "Hot Strength and Structure in Plain C and Microalloyed Steels During the Simulation of Plate Rolling by Torsion Testing," in Process Modelling Tools, American Society for Metals, Cleveland, Ohio, October 1980, in press.

12. I. Weiss and J.J. Jonas, "Dynamic Precipitation and Coarsening of Niobium Carbonitrides During the Hot Compression of HSLA Steels," Metallurgical Transactions, 10A (1979) pp. 831-840.

STRUCTURE/PROPERTY RELATIONSHIPS AND CONTINUOUS YIELDING

BEHAVIOR IN DUAL-PHASE STEELS

S.S. Hansen and R.R. Pradhan
Research Department
Bethlehem Steel Corporation
Bethlehem, Pennsylvania

A laboratory study involving various annealing temperatures and cooling rates was carried out on a series of C-Mn-Si alloys to determine the specific effects of the amount and nature of the second phase on the properties of dual-phase steels. Linear regression equations were developed that related mechanical properties to the microstructural constituents in ferrite-martensite and ferrite-martensite-pearlite microstructures. The addition of pearlite to ferrite-martensite structures had an additive but less significant effect on the mechanical properties beyond the separate effects of martensite. A clear transition from discontinuous to continuous yielding was observed in these steels as the cooling rate from the annealing temperature was increased. This transition was unaffected by changes in annealing temperature or microstructure. Interrupted cooling experiments identified mechanisms responsible for continuous yielding.

Introduction

Since the introduction of dual-phase steels in the mid-1970's (1,2) much research (3,4) has been aimed at defining the microstructural parameters that contribute to their improved strength-ductility balance. The microstructure of dual-phase steels generally consists of a dispersed martensitic second-phase* in a relatively soft ferritic matrix, although an acicular ferrite-martensite aggregate has also been observed in as-hot-rolled, dual-phase steels (6). The mechanical properties of these steels are characterized by:

- Continuous yielding behavior with a 310-380 MPa yield strength.

- A high work-hardening rate and a tensile strength in excess of 550 MPa.

- Superior uniform and total elongations compared with conventional HSLA steels of similar tensile strength.

Several investigations have related the tensile strength of dual-phase steels to the amount of martensite present (7-10). Still other studies (11,12) provided evidence that the carbon content of the martensite, and hence the strength of the second phase, must also be considered in assessing the overall strength of these ferrite-martensite composites. The effects reported by Ramos et al (11) are relatively small, e.g., in a ferrite-20% martensite mixture an increase in martensite carbon content from 0.30% to 0.65% increases the overall tensile strength by only about 55 MPa. In contrast, Speich and Miller's (12) equation would predict a tensile strength increase of about 185 MPa over the same carbon range. Along with these effects of martensite on tensile strength there is a clear trade-off between tensile strength and total elongation in dual-phase steels, and several correlations of this type have also been developed (7,9,10,13). Generally, increasing amounts of martensite result in reduced ductilities, but other factors such as ferrite strength or retained austenite have also been proposed as significant (7,14).

In contrast to the fairly extensive work on the tensile strength-total elongation balance in dual-phase steels, the factors that control the continuous yielding behavior or yield strength level in dual-phase steels have received little attention. Some studies (8,10,15-20) have touched on this phenomenon, and various explanations have been proposed to account for continuous yielding. Before examining these suggestions, let us first consider the fundamentals of discontinuous yielding. The three criteria necessary for discontinuous yielding behavior are now well understood (21-23). For a material to yield discontinuously, there must be:

- A low density of <u>mobile</u> dislocations prior to deformation

- Rapid dislocation multiplication during deformation

- Significant dependence of dislocation velocity on the applied stress.

Since the latter two conditions are satisfied in b.c.c. ferrite, the important factor governing the yielding behavior in steels is the presence

---

*In actual fact, the second phase also contains small amounts of retained austenite and has been called a martensite-austenite aggregate (5). For brevity, this aggregate will be referred to simply as martensite.

of a sufficient number of mobile dislocations. Hahn's model for discontinuous yielding (23) predicts that mobile dislocation densities of 10^2-10^4/cm^2 will lead to discontinuous yielding, while mobile dislocation densities of 10^6-10^8/cm^2 will result in continuous flow. Although dislocation densities of 10^6-10^8/cm^2 are typical of hot-rolled or annealed steels, the majority of these dislocations are immobile, the result of pinning by interstitial atoms (21).

Rashid, among others, has advanced the possibility that dual-phase steels develop continuous yielding as a result of mobile dislocations introduced during the austenite-to-martensite transformation (2). While this appears to be a reasonable proposal, the factors responsible for introducing a sufficient number of mobile dislocations have not been clearly defined. As several authors have suggested, if discontiuous yielding is to be suppressed there must be a minimum amount of martensite (15,20,24,25). These authors reported the critical amount of martensite to be between 5 and 15%; however, it should be noted that the alloy content and heat treatment parameters vary considerable between studies. It has also been suggested that a heterogeneous distribution of mobile dislocations is necessary for discontinuous yielding (17), and variations in austenitic grain size, annealing temperature and cooling rate after annealing have all been shown to influence the yield behavior of dual-phase steels. Finally, the presence of pearlite in a ferrite-martensite microstructure has been linked to discontinuous yielding (10,18), but no mechanism has been proposed.

As this survey of the literature indicates, several features of the mechanical behavior of dual-phase steels require clarification. The present study was initiated to specifically evaluate the influence of the following factors: the volume fraction of the second phase, the nature of the second phase, and the heat treatment parameters on the mechanical properties of dual-phase steels. To this end a series of steels was designed at constant levels of Mn and Si with a systematic variation in carbon from 0.04 to 0.20% C. It was anticipated that such a series would permit separation of the effects of varying martensite volume fraction and strength via martensite carbon content.

Experimental Procedure

The steels used in this investigation (Table I) were C-Mn-Si alloys containing about 1.5% Mn and 1.0% Si. Five carbon levels were specified, and all steels were produced as 68-kg air-induction heats. The ingots were first slabbed to 19 mm, and then reduced to a 2.54 mm hot band in a four-pass rolling sequence. The hot-band microstructures were typically fine-grained ferrite-pearlite mixtures, although occasional martensite islands were also observed. The ferrite grain sizes of the hot-rolled steels ranged from 3 μm (steel 5) to 7 μm (steel 1).

Standard sheet tensile samples of 203 mm gage length were cut from the hot bands. Duplicate tensile specimens were annealed for 1 minute at 760, 815 and 870 C and cooled continuously to ambient at rates ranging from air cooling, 5°C/sec, to water quenching, 833°C/sec. (Cooling rates are based on the cooling time between the annealing temperature and 425 C). Following heat treatment, mechanical properties were measured and samples were taken from the grip section for metallographic examination. All tensile tests were carried out at a strain rate of 0.05/min to about 3.5% plastic strain and then at 0.5/min to failure. Yield strength was measured at either the lower yield point or at 0.2% offset plastic strain. During metallographic evaluation the ferrite grain size and the volume fraction of martensite and

pearlite were determined for all samples. A picral etch was applied prior
to point counting for pearlite, and sodium metabisulfite was applied prior
to determination of the martensite volume fraction.

Table I. Composition of the C-Mn-Si Steels

Steel	C	Mn	P	S	Si	Al	N
1	.037	1.55	.006	.016	1.16	.021	.0069
2	.070	1.52	.003	.015	1.06	.041	.0056
3	.11	1.50	.007	.015	1.17	.035	.0069
4	.145	1.46	.008	.016	1.12	.048	.0078
5	.20	1.50	.008	.014	1.14	.037	.0079

    As an extension of the structure-property investigation, we carried out
experiments aimed at identifying parameters that affect the continuous
yielding behavior of the ferrite-martensite aggregates. In these experiments various intercritical annealing treatments were followed by continuous
as well as discontinuous cooling to room temperature. All samples were
evaluated for the amount of martensite and its carbon content, the presence
and amount of pearlite, and the ferrite grain size. Some samples were analyzed for retained austenite, ferrite substructure and ferrite strength.
Retained austenite measurements were made using an x-ray technique (26) in
the rolling plane of samples ground to half thickness. The ferrite
substructure was studied by transmission electron microscopy of thin foils
taken from the grips section of various broken tensiles. To qualitatively
follow changes in ferrite strength, ferrite microhardness was measured using
a Reichert microhardness tester under a 5-g load. Finally, the martensite
start (M_s) temperature of the austenite pools formed during intercritical
annealing was determined by dilatometry.

Results and Discussion

    The mechanical properties and microstructures of all steels were
thoroughly investigated, and clear effects of carbon and heat treatment
parameters observed.

Mechanical Properties

    The effect of cooling rate on the mechanical properties of the steels
is illustrated in Figure 1, which shows the stress-strain curves obtained
for the 0.11% C steel annealed at 815 C. These curves show the typical pattern of tensile behavior versus cooling rate, the levels of strength and
ductility varying significantly with steel and heat treatment. For example,
the variation in strength and ductility for all steels is shown in Figures 2
and 3 for an annealing temperature of 760 C. On the basis of Figures 1

116

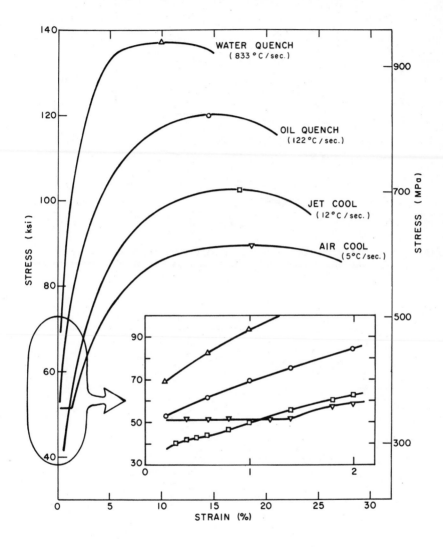

Figure 1. Stress-strain curves for Steel 3 (0.11% C) annealed at 815 C and cooled at indicated rates.

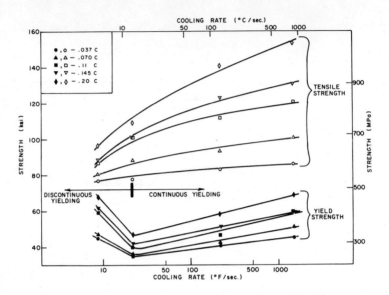

Figure 2.   Variation in yield and tensile strength with cooling rate for
various carbon additions to a 1.5% Mn, 1.0% Si steel (annealing
temperature = 760 C).

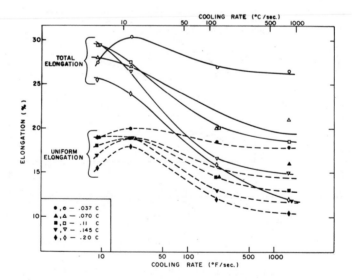

Figure 3.   Variation in uniform and total elongation with cooling rate for
various carbon additions to a 1.5% Mn, 1.0% Si steel (annealing
temperature = 760 C).

to 3, the following observations can be made:

- For all steels, tensile strength increases and total elongation decreases as the cooling rate increases. The rate of increase in strength or decrease in elongation increases with increasing carbon content.

- The yield strength goes through a minimum at a cooling rate of about 12°C/sec, independent of alloy or annealing temperature. At this cooling rate a transition from discontinuous to continuous yielding is observed, with the stress-strain curves exhibiting an inflection at yield point. This minimum defines the critical cooling rate which must be exceeded to ensure continuous yielding.

- A maximum in the uniform elongation is observed at this cooling rate. Our observations that at a given tensile strength this transitional yield behavior provides the highest uniform elongations are in agreement with those of Matlock et al (17).

The effects of annealing temperature and carbon content on tensile strength are illustrated in Figure 4 for the extremes in cooling rate considered. The effects of annealing temperature are much more pronounced with water quenching than with air cooling. The air-cooled samples show a plateau in tensile strength between 0.10 and 0.15% C, whereas in the water-quenched condition the tensile strength increases continuously with carbon. In the case of water quenching, as seen in Figure 4:

- At all annealing temperatures, the tensile strength is very sensitive to carbon content of the steel. For example, the rate of increase in tensile strength with carbon is about 310 MPa/0.1% C at 760 C and about 480 MPa/0.1% C at 870 C.

- Sensitivity of tensile strength to annealing temperature increases markedly with carbon content. For example, in a 0.037% C steel, increasing the annealing temperature from 760 to 870 C produces a tensile strength increase of about 69 MPa; whereas at 0.145% C a similar change in annealing temperature produces a strength increase of about 345 MPa.

Table II. Ferrite Microhardness of 0.037% C Steel Annealed at 760 C and Cooled at Indicated Rates

Cooling Rate		Ferrite Microhardness (MPa)
Air Cool	(5°C/sec)	1410
Jet Cool	(12°C/sec)	1568
Oil Quench	(122°C/sec)	1600
Water Quench	(833°C/sec)	1672

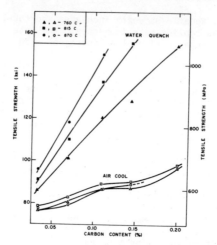

Figure 4. Effect of annealing temperature and carbon content on the tensile strength of water-quenched and air-cooled samples.

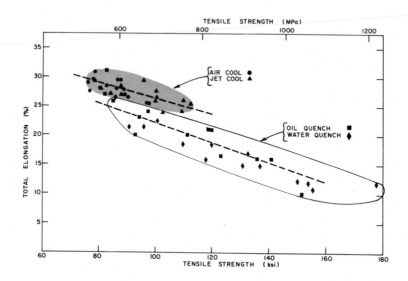

Figure 5. Total elongation-tensile strength relationship for C-Mn-Si steels.

120

The mechanical properties for all steels and heat treatments considered are consolidated in the form of a total elongation-tensile strength plot in Figure 5. The data can be grouped by cooling rate, with air- and jet-cooled samples displaying higher elongations at a given tensile strength than oil- and water-quenched samples, as has been observed by other investigators (28). The poor ductilities at faster cooling rates are thought to be due to a harder, less ductile ferritic matrix, a result of excess interstitials that are retained in solid solution during rapid cooling. This was confirmed by ferrite microhardness measurements made on Steel 1 after annealing at 760 C (Table II). The band for the slower-cooled samples is similar to the trade-off in properties observed in other heat-treated and as-hot-rolled dual-phase alloys (9,10,13,19).

Microstructural Variations

The microstructures produced by the various heat treatments were analyzed for ferrite grain size and the type and volume fraction of the second-phase constituents.

Since the ferrite grain size of annealed samples was found to remain fairly constant (4-6 µm range), this parameter was dropped from consideration. Retained austenite measurements were based on austenite (200) and (220) peaks compared to the intensity of the ferrite (200). These results varied significantly with the austenite peak considered, and no clear trends emerged with respect to cooling rate or annealing temperature. The amount of retained austenite generally increased with increasing carbon and some retained austenite was observed in all samples. However, because of the difficulty of producing precise results, this parameter was eliminated from further consideration in terms of possible effects on ductility.

In general, the amount of total second-phase increased as the cooling rate increased and changed from martensite-pearlite mixtures in most steels on air cooling to martensite at higher cooling rates. An example of the variation in microstructure with cooling rate is shown in Figure 6 for the 0.11% C steel annealed at 815 C. The mechanical properties for these structures were presented earlier in Figure 1. On air cooling (5°C/sec), the microstructure consisted of about 13% martensite and 5% pearlite (see arrows), but increasing the cooling rate to 12°C/sec (jet cooling), eliminated the pearlite and led to a microstructure containing about 22% martensite. The amount of martensite increased to about 30% on oil quenching and to about 38% on water quenching. Similar trends were observed for most of the steels and annealing temperatures considered. However, it is important to note that the lowest carbon steel contained no pearlite for any of the heat treatments.

The microstructural trends with respect to carbon content are illustrated in Figures 7 and 8. In Figure 7 the variation in total second phase (martensite + pearlite) and in martensite are shown as a function of carbon content and cooling rate for an annealing temperature of 815 C. Figure 8 compares the microstructure of three of the steels water-quenched and air-cooled from this temperature. This data show that:

● The amount of total second phase increased with both increasing carbon content and cooling rate. These changes in percentage of total second phase with cooling rate were greater at the higher carbon levels and were primarily responsible for the variations in tensile strength seen in Figure 2 and 4.

● The second phase produced by water and oil quenching was completely

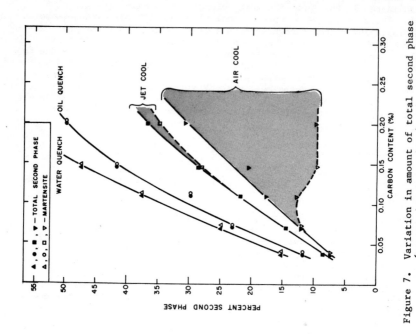

Figure 7. Variation in amount of total second phase (martensite + pearlite) and martensite with cooling rate and carbon content for an annealing temperature of 815 C.

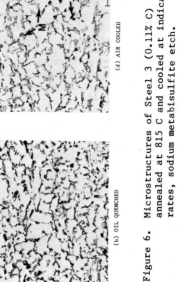

Figure 6. Microstructures of Steel 3 (0.11% C) annealed at 815 C and cooled at indicated rates, sodium metabisulfite etch.

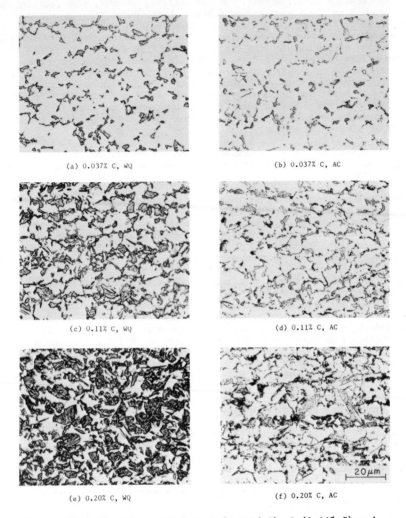

(a) 0.037% C, WQ

(b) 0.037% C, AC

(c) 0.11% C, WQ

(d) 0.11% C, AC

(e) 0.20% C, WQ

(f) 0.20% C, AC

20 μm

Figure 8.   Microstructures of Steels 1 (0.037% C), 3 (0.11% C) and 5 (0.20% C) on water quenching and air cooling from an annealing temperature of 815 C, sodium metabisulfite etch.

martensitic in all steels. On jet cooling, however, some pearlite was present in the higher carbon steels. Furthermore, in air-cooled samples, although no pearlite was observed at 0.037 and 0.07% C, increasing amounts of pearlite were produced with increasing carbon at higher carbon levels (the shaded area in Figure 7 represents the pearlite content). In fact, in the air-cooled samples the amount of martensite in steels with more than 0.07% C lies in the 10-13% range, the increasing volume fraction of total second phase with increasing carbon being due to an ever-increasing pearlite volume fraction. This result implies that in these steels the <u>average</u> hardenability of the austenite produced at an intercritical annealing temperature decreases with increasing carbon.

The effect of annealing temperature on the amount and nature of the second phase is shown for water quenching and air cooling in Figure 9. It is observed that:

- In water-quenched samples there were significant variations in the amount of martensite with carbon content at all annealing temperatures. Increasing annealing temperature resulted in increasing amounts of martensite at all carbon contents, the effect being greater at higher carbon levels. These trends in volume fraction martensite are similar to those observed for tensile strength (Figure 4).

- In the samples air-cooled after annealing there was little variation in the amount of total second phase with annealing temperature, although the amount of total second phase did increase with increasing carbon content at all temperatures. Similar variations in the amount of martensite and pearlite in this second phase were observed at all annealing temperatures; as noted earlier for 815 C, above about 0.07% C the amount of martensite present in the air-cooled samples leveled off or even decreased slightly with increasing carbon. The accompanying increases in second phase with increasing carbon were solely the result of increases in the pearlite volume fraction, and this may account for the much smaller effect of carbon on tensile strength observed in Figure 4 for air-cooled samples.

As noted earlier, the increasing fraction of pearlite in the second phase of air-cooled samples with increasing carbon content implies that the average hardenability of the austenite pools decreases with increasing carbon content of the steel. A possible explanation is that the average austenite composition, specifically the C, Mn or Si content, varied with carbon content of the steel. The approximate carbon content of the martensite, and hence the average carbon content of the austenite at intercritical annealing temperature, was calculated using a simple carbon mass balance (12). Inherent in these calculations are the assumptions that the carbon content of the ferrite is negligible and that the amount of martensite formed on water quenching is similar to the amount of austenite present at the annealing temperature. These assumptions are subject to dispute since the carbon content of the ferrite could be on the order of 0.01%, and as Matlock et al have pointed out (17), some growth of the existing ferrite* is always observed even during accelerated cooling from the intercritical annealing temperature.

---

*This new ferrite has been labelled "epitaxial ferrite" by Matlock et al (17) and was observed, but not quantified, in this investigation.

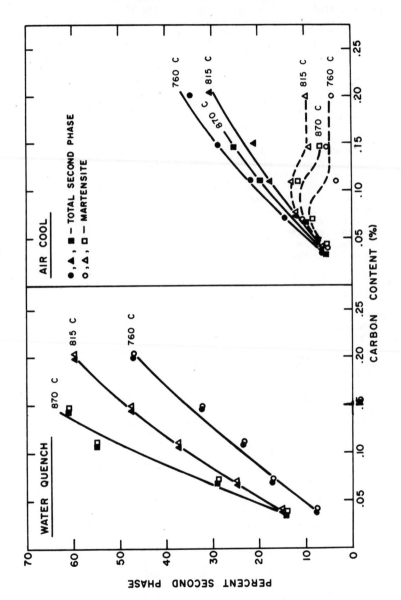

Figure 9. Variation in amount of total second phase (martensite + pearlite) and martensite with carbon content and annealing temperature for water-quenched and air-cooled samples.

125

As indicated by our mass-balance calculations, at a given annealing temperature the carbon content of the martensite formed on water quenching is not sensitive to the carbon content of the steel. This conclusion conflicts with the result of similar calculations by others (12) for C-Mn steels which showed that the martensite carbon content increases significantly with increasing steel carbon content. However, our calculations did indicate that, as expected, with increasing annealing temperatures the martensite, and hence the austenite, carbon content decreased, the average austenite carbon content being 0.45% at 760 C, 0.30% at 815 C, and 0.23% at 870 C.

Since the average austenite carbon content does not seem to vary significantly with the carbon content of the steel, the explanation for the decreasing average austenite hardenability is probably a result of variations in the average Mn and Si content of the austenite pools with increasing steel carbon content. It is suggested that while the steels all contain similar amounts of Mn and Si (on a weight percent basis), austenite pools form first at regions of locally increased alloy content (and thus in regions of higher hardenability). As increasing amounts of austenite form (e.g., with increasing carbon content at a given annealing temperature), regions of increasingly lower alloy content transform, and the average austenite hardenability decreases. Therefore, on air cooling, while the amount of martensite formed (from the higher alloy austenite pools) remains relatively constant (see Figure 9), the lower alloy pools transform to pearlite. As a result, the pearlitic fraction of the second phase increases with increasing steel carbon content.

Structure-Property Relations: Regression Analyses

Among the metallurgical factors that may influence the mechanical properties of dual-phase steels are ferrite grain size, ferrite strength, solid-solution alloy additions, and dispersed-phase strengthening. With the ferrite grain size and solid-solution strengthening increments being maintained at constant levels in the present study, the focus was on the amount and nature of the martensitic second phase. There is some controversy over how to adapt the theory of composite strengthening to predict the strength of dual-phase steels (29,30). Also, the question of whether the carbon content of the martensite significantly affects the strength of the ferrite-martensite composite is in dispute (7,12). Some of the previous studies limited themselves to a single alloy system or to one level of carbon, and hence imply associated variations in the martensite carbon content as the amount of martensite changes. Such limitations make a clear separation of the various factors difficult, and have contributed to the disagreement reflected in the literature.

In this study it was felt that variations in mechanical properties would be due to changes in the amount of martensite or pearlite, or in the carbon content of the martensite (as influenced by annealing temperature and cooling rate). Due to the uncertainty in the martensite carbon content calculations, this effect was not specifically considered. Instead the data were grouped by annealing temperature and cooling rate, since these heat treatment parameters should produce variations in martensite carbon content as we previously noted. To quantify the specific microstructural effects in these data sets, regression analysis was applied using the mechanical properties as the dependent variables, and martensite and pearlite volume fractions as the independent variables. Separate analyses were conducted for all the available data and for those samples which had a pearlite-free second phase.

126

The results of these regressions are shown in Table III for 3% flow stress, tensile strength, uniform elongation and total elongation. The groupings showed that there was no significant difference among all samples grouped by annealing temperature. Among ferrite-martensite samples grouped by cooling rate there were also no significant differences. However, in the samples containing pearlite (predominantly air- and jet-cooled specimens), the 3% flow stress, uniform elongation and total elongation all showed a significant cooling rate effect. The fact that grouping by annealing temperature showed no significant differences indicates that variations in the

Table III. Regression Constants and Coefficients for Ferrite/Martensite and Ferrite/Martensite/Pearlite Data Sets

Property		C	f_1	f_2	Variance Explained (%)
3% Flow Stress (MPa)	M/P	273.05	13.23	5.24	87
	M	286.83	13.44	-	90
Tensile Strength (MPa)	M/P	443.23	12.13	4.69	91
	M	452.54	12.13	-	92
Uniform Elongation (%)	M/P	20.75	-0.22	-0.10	71
	M	20.21	-0.23	-	83
Total Elongation (%)	M/P	31.15	-0.34	-0.04	76
	M	30.17	-0.35	-	82

(1) All equations are of the form,
    Property = C + f_1 (martensite) + f_2 (pearlite)
(2) Equations M/P are derived from all data, while equations M are for ferrite/martensite samples only.
(3) Range of independent variables considered: Martensite = 3.2 - 61%
                                                 Pearlite   = 0 - 31%

martensite carbon content in the range of about 0.23 to 0.45% have little effect on the mechanical properties. This finding is in agreement with Davies' suggestion (7). According to the equations listed in Table III:

● Martensite is about three times as effective a strengthener as pearlite (e.g., compare coefficients for both 3% flow stress and tensile strength). The effect of pearlite on tensile strength is similar to that reported in other studies of ferrite-pearlite steels - 2.5 to 4.4 MPa/1% pearlite (31-35).

● Martensite has a more deleterious effect on both uniform and total elongation than does pearlite on a volume percent basis. However, in terms of their relative strengthening increments, martensite and pearlite have similar effects on uniform elongation (-0.019%/MPa vs. -0.017%/MPa), but martensite has a much more deleterious effect than

127

pearlite on total elongation (-0.026%/MPa vs. -0.008%/MPa). The correlation coefficients for these equations are lower than those for flow stress and tensile strength.

- The coefficients for martensite in all the equations are not significantly altered by introducing pearlite-containing samples into the analysis. Thus, the effects of martensite and pearlite in these alloys are additive and little interaction occurs.

Several other investigators have developed linear regression equations specifically relating the tensile strength of dual-phase steels to the volume fraction of martensite (e.g., 7-10). The reported coefficients for martensite strengthening range from 8.5 to 13.0 MPa/1% martensite. No obvious trends are noted in these studies with regard to a limited range of martensite volume fraction influencing the magnitude of this coefficient. The fact that the constants differed from equation to equation can be attributed to variations in ferrite grain size, solid solution strengthening and ferrite interstitial content. Equations proposed by other investigators (12,36) for uniform elongation agree with our equation with the effect of martensite on uniform elongation being quite small.

Continuous Yielding Behavior in Dual-Phase Steels

While it was possible to relate the mechanical properties of the steels to the microstructural constituents in a linear fashion, the factors influencing the yield behavior of these steels proved to be more complex.

Influence of Cooling Rate. When annealed at 760 C and cooled to ambient at various rates, the 0.037% C steel provided a series of stress-strain curves in which the factors influencing the yield behavior of these steels could be isolated. For this steel and annealing temperature, the final microstructure was relatively insensitive to cooling rate. The microstructures produced at all cooling rates consisted of about 6-8% martensite in a fine-grained, polygonal ferrite matrix (Figure 10). Furthermore, transmission electron microscopy revealed no significant differences in ferrite substructure due to changes in cooling rate (Figure 11). All samples exhibited a high ferrite dislocation density, the distribution of dislocations being relatively homogeneous through the ferrite grains. Nevertheless, even with this structural similarity, distinct changes in mechanical properties and yield behavior were observed with variations in cooling rate (Figure 12). Although these tensile curves are similar in form to those for the 0.11% C steel shown earlier (Figure 1), the increase in flow strength with increasing cooling rate is smaller for the 0.037% C steel. The larger strength variations in the 0.11% C steel reflect the more significant microstructural changes with cooling rate in that alloy (Figure 6). With reference to Figures 10 to 12, the following observations can be made:

- There is a significant transition in yield behavior between the air-cooled (discontinuous flow) and jet-cooled (continuous yielding) samples. This difference in flow behavior is solely due to the presence of a yield point in the air-cooled sample. In fact, back extrapolation of the air-cooled stress-strain curve from beyond the yield point produces an exact match with the initial portion of the jet-cooled flow curve.

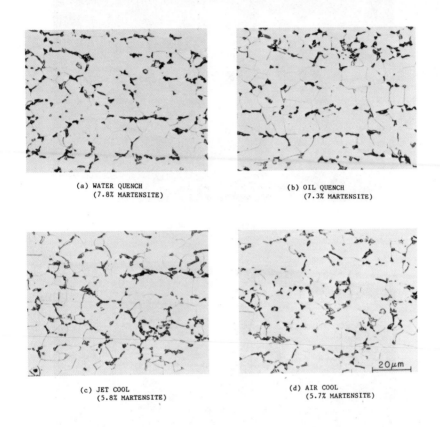

(a) WATER QUENCH
(7.8% MARTENSITE)

(b) OIL QUENCH
(7.3% MARTENSITE)

(c) JET COOL
(5.8% MARTENSITE)

(d) AIR COOL
(5.7% MARTENSITE)

20μm

Figure 10. Microstructures of Steel 1 (0.037% C) annealed at 760 C
and cooled at indicated rates, sodium metabisulfite etch.

(a) WATER–QUENCHED

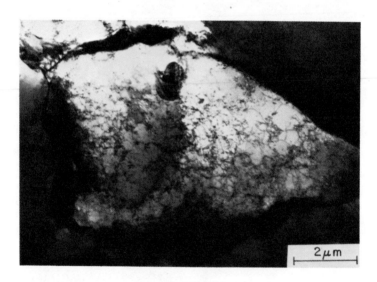

(b) AIR–COOLED

Figure 11.   Transmission electron micrographs of Steel 1 (0.037% C) annealed at 760 C and (a) water-quenched or (b) air-cooled from annealing temperature.

- This transition in yield behavior (discontinuous to continuous) also involves a reduced yield strength and an increased total elongation.

- The tensile strength increase of about 60 MPa with increasing cooling rate is due to a slight increase in the martensite content that accounts for 20-30 MPa and to an increasing ferrite strength (Table II).

Interrupted Cooling Experiments. To clarify the variation in yield behavior with cooling rate, a further series of experiments was performed. Samples of Steel 1 (0.037% C) were annealed at 760 C for 1 minute and then air-cooled. However, in these experiments the air cooling was interrupted at various temperatures and the continuous cooling was terminated by water quenching. The variation in yield strength, tensile strength and yield elongation with quench temperature shown in Figure 13 falls into three distinct regions:

Region 1. Between 760 C and 455 C, there is no change in the continuous yielding behavior of this steel. However, the yield and tensile strengths decrease continuously, because (a) on air cooling down to 455 C, the amount of polygonal ferrite increases slightly (metallographic examination reveals that the amount of martensite formed on water quenching from 455 C is equivalent to that formed on air cooling) and (b) softening of the ferrite occurs as the concentration of interstitials in solution decreases. Table IV shows that the ferrite microhardness of the samples quenched from 480 C is equivalent to that obtained in a sample air-cooled to room temperature.

Region 2. In the temperature range of 455 to 315 C, there is a gradual transition in the yielding behavior. This transition can be observed more clearly with the aid of the initial portion of the stress-strain curves for various quench temperatures (Figure 14) and mirrors the effect of cooling rate (Figure 12). At 455 C there is only a trace of a yield point in the flow curve (similar to that observed in this steel on jet-cooling from the annealing temperature), but by 315 C, a pronounced yield point elongation is observed. Over this temperature regime there is also a gradual increase in the yield strength (due to the transition from a 0.2% offset to a lower yield strength criterion), whereas the tensile strength decreases only slightly.

Region 3. From 315 C to room temperature there are no further changes in the microstructure or mechanical properties. For this cooling rate the processes that contribute to the mechanical property variations apparently cease below 315 C.

A similar transition in yielding behavior occurs on annealing the 0.037% C steel at 870 C (Figure 15). In this case, the transition to discontinuous yielding behavior begins at 540 C rather than 455 C. Regions 1 to 3 are once again present, the larger decreases in tensile strength being due to a higher amount of martensite formed on water quenching from 870 C compared to 760 C - respectively, 15% and 8%.

The study was extended to Steels 3 and 4 (Table I) which exhibited a ferrite-martensite-pearlite structure in the air-cooled condition. Their air-cooled structures and properties are shown in Figure 16. Steel 4 had approximately the same amount of martensite (about 5%) as steel 1 (in the air-cooled condition), but also had about 24% pearlite. Steel 3 had only 3% martensite in the air-cooled condition in addition to about 18% pearlite. Independent of these changes in final microstructure, a similar transition

131

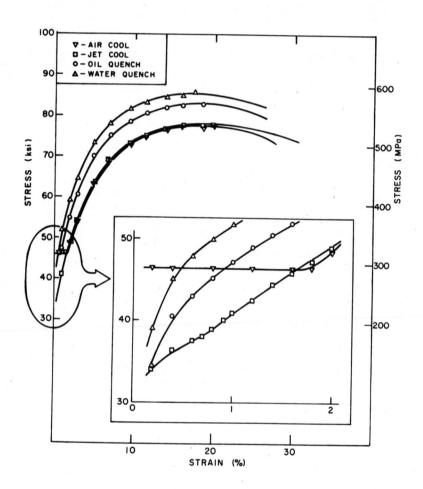

Figure 12.   Stress-strain curves for Steel 1 (0.037% C) annealed at 760 C
and cooled at indicated rates.

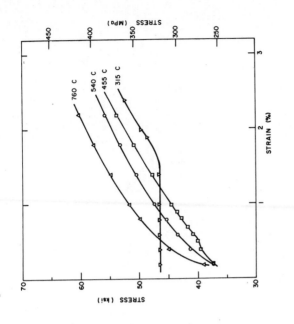

Figure 14. Comparison of initial region of stress-strain curves for Steel 1 (0.037% C) annealed at 760 C and air-cooled to the indicated quench temperature.

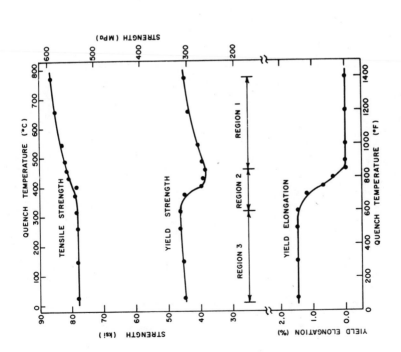

Figure 13. Variation in mechanical properties with quench temperature for Steel 1 (0.37% C) annealed at 760 C and air-cooled to the quench temperature.

133

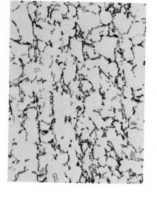

(a) STEEL 3 (0.11% C); 3% MARTENSITE AND 18% PEARLITE

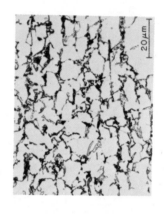

(b) STEEL 4 (0.145% C); 5% MARTENSITE AND 24% PEARLITE

Figure 16. Microstructures of (a) Steel 3 (0.11% C) and (b) Steel 4 (0.145% C) annealed at 760 C and air-cooled to room temperature, sodium metabisulfite etch.

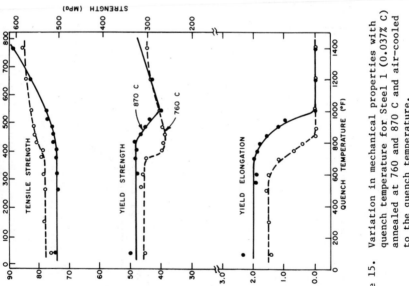

Figure 15. Variation in mechanical properties with quench temperature for Steel 1 (0.037% C) annealed at 760 and 870 C and air-cooled to the quench temperature.

134

in yield behavior was observed in these steels (Figure 17), with the transition from continuous to discontinuous yielding occurring at 455 to 480 C. As compared to Steel 1, the higher yield and tensile strengths of these steels are a result of the pearlite constituent. The greater decreases in yield and tensile strength between 760 C and 480 C for Steels 3 and 4 are a result of the much larger decrease in the amount of martensite present in the final microstructure over this temperature range.

Table IV. Ferrite Microhardness of 0.037% C Steel Air-Cooled From an Annealing Temperature of 760 C to Indicated Temperature and Water-Quenched

Quench Temperature	Ferrite Microhardness (MPa)
760 C	1672
650 C	1576
540 C	1523
480 C	1419
455 C	1411
Room Temperature	1411

The variations in microstructure during air cooling and interrupted quenching are illustrated in Figure 18 for Steel 4. On water quenching from 760 C, the microstructure is composed of 33% martensite. Air cooling to 650 C has little effect on the martensite content, but by 540 C the amount of martensite has decreased to 5% and about 13% pearlite is also observed. More pearlite is formed at 480 C, and the final air-cooled structure contains about 24% pearlite and about 5% martensite.

Dilatometry Experiments. Dilatometry was used in an attempt to determine the M_S temperature of the austenite pools formed during intercritical annealing of the C-Mn-Si steels. At the cooling rate employed (12°C/sec) the pearlite and martensite reactions overlapped in the higher carbon steels, and the start of the martensite transformation (M_S) could not be clearly defined. In the lower carbon steels, only a small amount of martensite formed during cooling, and therefore the overall specimen expansion associated with the austenite-to-martensite transformation was also small. Nevertheless, this slight expansion was detectable, and measurements on samples of Steels 1 and 2 cooled from various annealing temperatures indicated an M_S temperature of 480 C (±28 C for two standard deviations). This M_S temperature is in the temperature range in which there is a transition from continuous to discontinuous yielding during cooling from the annealing temperature (Figures 13,15 and 17).

In view of the small expansion associated with the martensite transformation in these steels, and thus the less-than-precise measurement of the M_S temperature, no clear effects of annealing temperature or carbon content on the M_S temperature could be defined. Based on the Steven and Haynes equation (37), the measured value of the M_S temperature corresponds to a

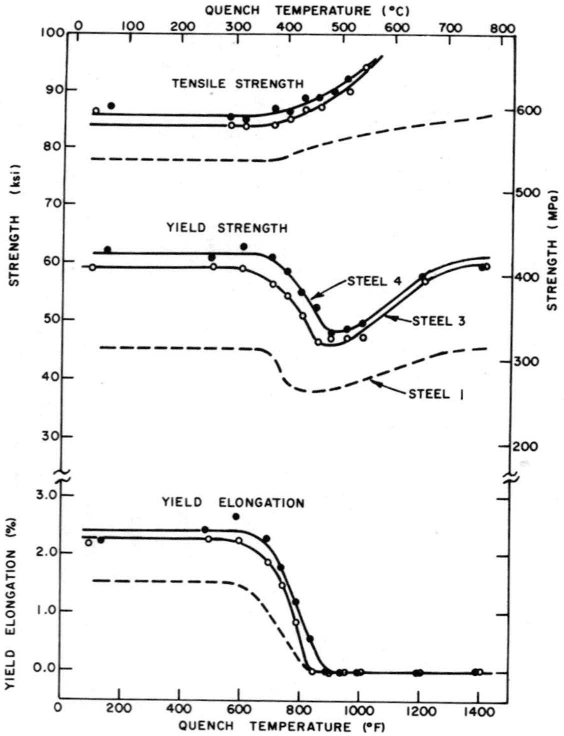

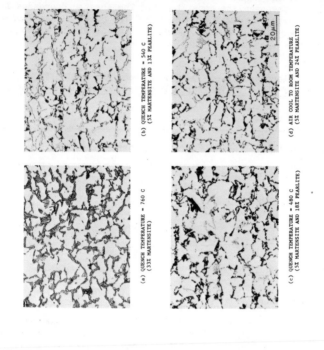

Figure 18. Microstructures of Steel 4 (0.145% C) annealed at 760 C and air-cooled to the indicated quench temperature, sodium metabisulfite etch.

(a) QUENCH TEMPERATURE = 760 C
(33% MARTENSITE)

(b) QUENCH TEMPERATURE = 540 C
(5% MARTENSITE AND 13% PEARLITE)

(c) QUENCH TEMPERATURE = 480 C
(5% MARTENSITE AND 18% PEARLITE)

(d) AIR COOL TO ROOM TEMPERATURE
(5% MARTENSITE AND 24% PEARLITE)

20μm

Figure 17. Variation in mechanical properties with quench temperature for Steels 1 (0.037% C), 3 (0.11% C) and 4 (0.145% C) annealed at 760 C and air-cooled to the quench temperature.

STRENGTH (MPa)

STRENGTH (ksi)

YIELD ELONGATION (%)

QUENCH TEMPERATURE (°C)

QUENCH TEMPERATURE (°F)

TENSILE STRENGTH

YIELD STRENGTH

YIELD ELONGATION

STEEL 4

STEEL 3

STEEL 1

136

carbon content of the austenite pools of 0.06%. This value is probably the M_s of the leanest alloy austenite pools formed during annealing, and because of the non-equilibrium nature of the intercritical annealing process many of the austenite pools are likely to have a significantly higher average carbon content and therefore a lower M_s temperature.

Rationale for Continuous Yielding in Dual-Phase Steels

On the basis of our experimental results and the existing theory of continuous yielding, two conditions must be fulfilled for a dual-phase steel to exhibit continuous yielding:

(1) There must be a sufficient density of mobile dislocations, i.e., the mobile dislocation density must exceed the critical density of approximately 10^6-10^8/cm^2 required for continuous yielding (23). Presumably, these dislocations are a result of the strains (due to the volume expansion) accompanying the austenite-to-martensite transformation. The relationship between ferrite dislocation density and the volume fraction of austenite transforming to martensite has not been investigated to date.

(2) The dislocations produced during the austenite-to-martensite transformation must remain mobile at room temperature. These dislocations can be pinned by interstitial elements if sufficient time at temperature is allowed for the necessary diffusion. Since these dislocations are generated below the M_s temperature, the thermal history below the M_s temperature is important. For a given steel and annealing temperature (and hence a given M_s temperature), there should be a critical cooling rate below which discontinuous yielding will re-appear.

The interrelationship between the M_s temperature and the cooling rate can be quantified in terms of a classical diffusion distance concept along the lines proposed by Shewmon (38). In this approach any thermal profile can be translated into a characteristic diffusion distance, $(Dt)^{\frac{1}{2}}$, by first converting the temperature-time profile to a diffusivity-time plot. The diffusion distance is then obtained by integrating the area (measured graphically using a planimeter) under the diffusivity-time curve for the particular time frame of interest. Therefore,

$$\overline{D}t = \int_{t_1}^{t_2} D(t).dt \tag{1}$$

where $t_1 \rightarrow t_2$ is the time interval of interest, and $D(t)$ is the diffusivity of the species of interest at time t.

In this study the diffusion of carbon in ferrite was considered to be controlling with regard to immobilization of dislocations, because above 260 C carbon diffuses faster than nitrogen. Therefore, at any temperature of interest the diffusivity of carbon in ferrite was determined (38) by

$$(D_c)_{ferrite} = 0.020 \exp\left[\frac{-10115.75}{T(^oK)}\right] \quad cm^2/sec \tag{2}$$

Then, for any given M_s temperature, the thermal profile below the M_s can be used to determine the critical $\overline{D}t$ above which sufficient carbon diffusion

137

will take place to produce a discontinuous yielding material. The M_s temperature for the austenite pools formed during intercritical annealing of Steel 1 was determined to be about 480 C, and the transition between continuous and discontinuous yielding behavior in this steel was observed at a cooling rate of about 12°C/sec. Therefore, we can estimate that for this steel, if the cooling profile below the M_s temperature (480 C) is such that a $\overline{D}t$ in excess of 20 x 10^{-8} cm^2 is produced, there will then be sufficient time for carbon diffusion in the ferrite to pin the mobile dislocations generated during the martensite transformation, the net result being discontinuous yielding. This $\overline{D}t$ value corresponds to a diffusion distance of about 4.5 m, similar to the ferrite grain size or the martensite interparticle spacing in this steel. In order to confirm this critical diffusion distance concept for the C-Mn-Si steels, a variety of discontinuous cooling profiles, including isothermal holds, were used below the M_s temperature after Steel 1 was annealed at 760 C. These data are gathered in the form of a yield point elongation-$\overline{D}t$ plot in Figure 19. Clearly, continuous yielding is the norm up to a $\overline{D}t$ of about 20-30 x 10^{-8} cm^2, but above this critical value we get discontinuous yielding, the extent of the yield elongation increasing continuously with increasing $\overline{D}t$. It is to be understood that while this concept should be applicable to any dual-phase alloy system, the critical $\overline{D}t$ in Figure 19 applies to the series of alloys in the present study and is likely to change with alloy system.

Continuous Yielding Behavior - Microstructural Effects

Pearlite exhibited no clear effect on the continuous yielding behavior of these steels. Indeed, the transitions in the higher carbon steels with up to 24% pearlite in the second phase were exactly similar to those in a pearlite-free steel (Figure 17). Consequently, the presence of pearlite does not by itself lead to discontinuous yielding in dual-phase steels. Rather, the presence of pearlite may be important only if it completely replaces the martensite and thereby eliminates the source of the mobile dislocations necessary for continuous yielding. In some other steels the presence of pearlite and the associated occurrence of discontinuous yielding in a ferrite-martensite-pearlite microstructure may be a symptom rather than the cause of discontinuous flow (10,18). In such cases its presence probably indicates an insufficiently rapid cooling rate below the M_s temperature and hence pinning of the dislocations produced by the martensitic transformation, as opposed to a direct effect of pearlite on the yield behavior.

Furthermore, in contrast to previous studies, the results of this investigation indicate that the transformation of as little as 3% martensite can produce continuous yielding in dual-phase steels, provided the cooling rate below the M_s temperature is sufficiently rapid to preclude pinning of the mobile dislocations introduced by this transformation. Of course, this effect may depend on steel composition and the heat treatment parameters. For example, in our experiments we used water quenching to minimize interstitial diffusion during cooling thereby preserving the mobility of enough of the dislocations introduced by the martensite transformation. When slower cooling rates below the M_s temperature are utilized, a significant fraction of the dislocations may be pinned by interstitials during cooling. In such cases, there would have to be a larger amount of martensite transformation to produce the density of mobile dislocations required for continuous yielding. Detailed studies of the several factors that control dislocation density in dual-phase steels will be required to quantify the interrelationship of amount of martensite transformation, M_s temperature, and the thermal profile below this temperature.

138

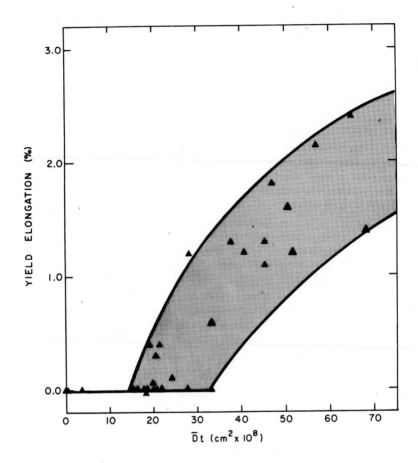

Figure 19.   Variation in yield elongation with $\bar{D}t$ for Steel 1 (0.037% C) annealed at 760 C and subjected to continuous and discontinuous cooling profiles below the M_S temperature.

Conclusions

The effects of intercritical annealing treatments of C-Mn-Si steels on the microstructure and mechanical properties in the resulting dual-phase steels were studied, and the key findings are:

- Increasing the cooling rate from the intercritical annealing temperature increases the volume fraction of martensite, resulting in increased tensile strengths and reduced ductilities.

- The tensile strength of water-quenched samples is very sensitive to annealing temperature and carbon content. For example, at 760 C the tensile strength increases about 310 MPa/0.1% C, and at 870 C the rate of increase is 480 MPa/0.1% C. Higher carbon steels are much more sensitive to annealing temperature variations; in a 0.037% C alloy, an increase in annealing temperature from 760 to 870 C increases the tensile strength by 69 MPa, whereas at 0.145% C this change in annealing temperature results in a strength increase of 345 MPa.

- Slower cooling rates reduce the sensitivity to annealing temperature and carbon content. For example, a change in annealing temperature from 760 to 870 C produces little change in mechanical properties for all steels when air-cooled. Also, similar mechanical properties result in the 0.10 to 0.15% C range for samples air-cooled from any of the annealing temperatures.

- The average hardenability of the austenite pools formed at intercritical annealing temperature decreases as the steel carbon content increases. Consequently, in air-cooled samples, the fraction of pearlite in the second phase increases as the carbon content increases.

- As indicated by the regression analysis, (1) increasing martensite increases the 3% flow stress and tensile strength and reduces the uniform and total elongation, (2) although the effect of pearlite on the mechanical properties is additive to that of the martensite, pearlite is only about one-third as effective as martensite in increasing the strength and also has a smaller negative effect on ductility.

The specific findings with respect to the yielding behavior are:

- The factors required for continuous yielding are an austenite-to-martensite transformation to produce a high dislocation density in the ferrite and a sufficiently fast cooling rate below the M_s temperature to preserve the mobility of enough of these dislocations to ensure continuous yielding.

- Transformation of as little as 3% martensite can produce continuous yielding behavior if a sufficiently fast cooling rate is employed below the M_s temperature.

- Pearlite does not directly affect the yielding behavior of ferrite-martensite aggregates

- A model coupling the M_s temperature and the thermal profile below this temperature can be used to predict the nature of yielding in C-Mn-Si steels.

140

Acknowledgments

The authors would like to thank A. O. Benscoter, F. J. Donchez, N. S. Naugle and C. Santos for their assistance with this investigation and B. S. Mikofsky for his editorial assistance. Thanks are also due to Professor Morris Cohen for a number of stimulating discussions.

References

1. S. Hayami and T. Furukawa, "A Family of High Strength Cold Rolled Steels," pp. 311-320 in Microalloying 75, Proceedings, Union Carbide Corporation, New York, N.Y., 1977.

2. M. S. Rashid, "GM980X - A Unique High Strength Sheet Steel with Superior Formability," SAE Reprint No. 760206, February 1976.

3. Formable HSLA and Dual-Phase Steels, A.T. Davenport, ed.; TMS-AIME, New York, N.Y., 1979.

4. Structure and Properties of Dual-Phase Steels, R. A. Kot and J. W. Morris, ed.; TMS-AIME, Warrendale, Pa., 1979.

5. G. Tither, A. P. Coldren and J. W. Morrow, "Continuous-Yielding Dual-Phase Strip Product," pp. 72-101, in Proceedings of 20th Mechanical Working and Steel Processing Conference, ISS-AIME, New York, N.Y., 1978.

6. S. S. Hansen and B. L. Bramfitt, "Hot-Strip Mill Processing of Dual-Phase Steels," pp. 1297-1307, in International Conference on Steel Rolling; Science and Technology of Flat Rolled Products, Iron and Steel Institute of Japan, Tokyo, Japan, September 29-October 4, 1980.

7. R. G. Davies, "Influence of Martensite Composition and Content on Properties of Dual-Phase Steels," Met. Trans., 9A (1978) pp. 671-679.

8. W. R. Cribb and J. M. Rigsbee, "Work-Hardening Behavior and its Relationship to the Microstructure and Mechanical Properties of Dual-Phase Steels," pp. 91-117 in Structure and Properties of Dual-Phase Steels, R. A. Kot and J. W. Morris, ed.; TMS-AIME, Warrendale, Pa., 1979.

9. J. H. Bucher, E.G. Hamburg and J. F. Butler, "Property Characterization of VAN-QN Dual-Phase Steels," ibid., pp. 346-359.

10. A. R. Marder and B. L. Bramfitt, "Processing of a Molybdenum-Bearing Dual-Phase Steel," ibid., pp. 242-259.

11. L. F. Ramos, D. K. Matlock and G. Krauss, "On the Deformation Behavior of Dual-Phase Steels," Met. Trans., 10A (1979) pp. 259-261.

12. G. R. Speich and R. L. Miller, "Mechanical Properties of Ferrite-Martensite Steels," pp. 145-182 in Structure and Properties of Dual-Phase Steels, R. A. Kot and J. W. Morris, ed.; TMS-AIME, Warrendale, Pa., 1979.

13. G.T. Eldis, "The Influence of Microstructure and Testing Procedure on the Measured Mechanical Properties of Heat Treated Dual-Phase Steels," ibid., pp. 202-220.

14. A. R. Marder, "Factors Affecting the Ductility of 'Dual-Phase' Alloys," pp. 87-98 in Formable HSLA and Dual-Phase Steels, A. T. Davenport, ed.; TMS-AIME, New York, N.Y., 1979.

15. J. M. Rigsbee and P. J. Vander Arend, "Laboratory Studies of Microstructures and Structure-Property Relationships in 'Dual-Phase' HSLA Steels," ibid., pp. 56-86.

16. M. S. Rashid, "Relationship Between Steel Microstructure and Formability," ibid., pp. 1-24.

17. D. K. Matlock, G. Krauss, L. F. Ramos and G. S. Huppi, "A Correlation of Processing Variables with Deformation Behavior of Dual-Phase Steels," pp. 62-90 in Structure and Properties of Dual-Phase Steels, R. A. Kot and J. W Morris, ed.; TMS-AIME, Warrendale, Pa., 1979.

18. T. Tanaka, M. Nishida, K. Hashiguchi and T. Kato, "Formation and Properties of Ferrite Plus Martensite Dual-Phase Structures," ibid., pp. 221-241.

19. T. Furukawa, H. Morikawa, H. Takechi and K. Koyama, "Process Factors for Highly Ductile Dual-Phase Sheet Steels," ibid., pp. 281-303.

20. J. M. Rigsbee, J. K. Abraham, A. T. Davenport, J. E. Franklin and J. W. Pickens, "Structure-Processing and Structure-Property Relationships in Commercially Processed Dual-Phase Steels," ibid., pp. 304-329.

21. A. H. Cottrell, Report of the Bristol Conference on Strength of Solids, p. 30; Physical Society, London, England, 1948.

22. J. J. Gilman and W. G. Johnston, "Behavior of Individual Dislocations in Strain-Hardened LiF Crystals," J. App. Phys., 31 (1960) pp. 687-696.

23. G. T. Hahn, "A Model for Yielding with Special Reference to the Yield-Point Phenomena of Iron and Related BCC Metals," Acta Met., 10 (1962) pp. 727-738.

24. D. S. Dabkowski and G. R. Speich, "Transformation Products and the Stress-Strain Behavior of Control-Rolled Mn-Mo-Cb Line-Pipe Steels," pp. 284-312 in Proceedings of 19th Mechanical Working and Steel Processing Conference, ISS-AIME, New York, N.Y., 1977.

25. D. Aichbhaumik and R. R. Goodhart, "Effect of Annealing Cycles on the Properties and Microstructures of Several Dual Phase Steels," SAE Reprint No. 790010, February 1979.

26. R. L. Miller, "A Rapid X-Ray Method for the Determination of Retained Austenite," ASM. Trans. Quart., 57 (1964) pp. 892-899.

27. A. R. Marder, R. R. Pradhan, Bethlehem Steel Corporation, unpublished research.

28. S. Hayami, T. Furukawa, H. Gondoh and H. Takechi, "Recent Developments in Formable Hot-and Cold-Rolled HSLA Including Dual-Phase Sheet Steels," pp. 167-180 in Formable HSLA and Dual-Phase Steels, A. T. Davenport, ed.; TMS-AIME, New York, N.Y., 1979.

29. I. Tamura, Y. Tomota and H. Ozawa, "Strength and Ductility of Fe-Ni-C Alloys Composed of Austenite and Martensite with Various Strengths," vol. 1, pp. 611-615, Proceedings 3rd International Conference on the Strength of Metals and Alloys, Cambridge, England, 1973.

30. R. G. Davies, "The Deformation Behavior of a Vanadium-Strengthened Dual-Phase Steel," Met. Trans., 9A (1978) pp. 41-52.

31. J. D. Baird and R. R. Preston, "Relationships Between Processing, Structure and Properties in Low Carbon Steels," pp. 1-42 in Processing and Properties of Low Carbon Steel, J. M. Gray, ed.; TMS-AIME, New York, N.Y., 1973.

32. R. W. K. Honeycombe and F. B. Pickering, "Ferrite and Bainite in Alloy Steels," Met. Trans., 3 (1972) pp. 1099-1112.

33. J. J. Irani, D. Burton, J. D. Jones and A. B. Rothwell, "Beneficial Effects of Controlled Rolling in the Processing of Structural Steels," pp. 110-122 in Strong Tough Structural Steels, Iron and Steel Institute Special Report No. 104, London, England, 1967.

34. F. B. Pickering and T. Gladman, "An Investigation into Some Factors Which Control the Strength of Carbon Steels," pp. 10-20 in Metallurgical Developments in Carbon Steels, Iron and Steel Institute Special Report No. 81, London, England, 1963.

35. J. D. Grozier and J. H. Bucher, "Correlation of Fatigue Limit with Microstructure and Composition of Ferrite-Pearlite Steels," J. Materials, 2 (1967) pp. 393-407.

36. A. R. Marder, "Deformation Characteristics of Dual-Phase Steels," submitted to Met. Trans. (1980).

37. W. Steven and A. G. Haynes, "The Temperature of Formation of Martensite and Bainite in Low-Alloy Steels," J. Iron and Steel Inst., 183 (1956) pp. 349-359.

38. P. G. Shewmon, Diffusion in Solids, pp. 30-31; McGraw Hill, New York, N.Y., 1963.

THE STRUCTURE-PROPERTY RELATIONSHIPS IN

CHROMIUM-BEARING DUAL-PHASE STEELS

A. R. Marder
Research Department
Bethlehem Steel Corporation
Bethlehem, Pennsylvania

A systematic study of the structure-property relationships in Cr-bearing dual-phase steels was made. The effects of annealing temperature and cooling rate on a range of C, Mn, Si, Cr alloys was evaluated. The results show that Cr dual-phase steels behave similarly to V and Mo dual-phase steels. Maximum hardenability for the transformation of austenite islands to martensite at the slowest cooling rate occurs at a particular Cr content in combination with specific C and Mn levels. Martensite volume fraction was found to be the major structural factor that controls the strength and ductility of these steels. The minimum in the 0.2% yield strength was found for certain alloys at specific cooling rates and this minimum is related to the lack of sufficient mobile dislocations, a condition that may be brought about by several factors. TEM studies of the effect of cooling rate on the structure show that mobile dislocations affect the continuous yielding of these steels.

Introduction

Intercritically annealed hot-rolled dual-phase steels can be characterized by the following features:

(a) High tensile strength together with a low yield strength or a Y.S./T.S. ≤0.5

(b) Absence of a yield point elongation

(c) High ductility and high work hardening rate.

Several previous investigations have been concerned with the additions of vanadium (1), molybdenum (2) and chromium (3-7) in enhancing the properties of dual-phase steels. Imamura and Furukawa (3) found that Cr improves the hardenability of dual-phase steels, and Ohashi et al (4) showed that a 0.5 Cr addition lowers the critical cooling rate at which dual-phase properties can be obtained. Goodhart et al (5,6) showed that the strength-ductility behavior of Mn-Si-Cr dual-phase steels compares favorably with that for Mn-Si-V dual phase steels. Tanaka et al (7,8) conducted a study on the effects of Cr additions as well as Mn and Mo additions on a 0.05 C steel and developed equations for the effect of alloying on the critical cooling rate necessary for the formation of a dual-phase structure.

The purpose of the present research was to systematically study the effects of annealing temperature, cooling rate and Cr additions on the dual-phase properties of a 0.1 C-1.5 Mn-0.5 Si alloy.

Table 1: Chemical Composition of Alloys Studied

	C	Mn	S	Si	Cr	Al
CR1	0.096	1.50	.0.009	0.60	0.55	0.091
CR2	0.11	1.40	0.014	0.53	0.22	0.026
CR3	0.11	1.45	0.012	0.56	0.76	0.050
CR4	0.14	1.46	0.006	0.56	0.52	0.046
CR5	0.086	1.46	0.007	0.55	0.52	0.019
CR6	0.13	1.46	0.009	0.55	0.26	0.066
CR7	0.083	1.46	0.006	0.55	0.25	0.031
CR8	0.082	1.46	0.011	0.98	0.24	0.017
CR9	0.13	1.19	0.011	0.54	0.52	0.067
CR10	0.11	1.39	0.012	0.49	0.47	0.068

Cu, Mo ≤ 0.02; P, Ni ≤ 0.01; V, Ti, Cb ≤ 0.005

Experimental Procedure

Ten different 300-lb ingots with variations in C, Mn, Si and Cr were evaluated, Table 1. The ingots were hot-rolled to a nominal 4 mm (0.1 in.) and were heat-treated to evaluate the effects of intercritical annealing temperature and cooling rate. Tensile specimens with a 50 mm (2 in.) gage length were heat treated in a tube furnace with a 300 mm (12 in.) constant-temperature zone. A chromel-alumel thermocouple was flash-welded to the center of the grip region of each specimen to record the complete thermal profile during heat treatment. After the specimens reached temperature they were held for 1 minute and cooled at rates ranging from 2 to 500 C/sec (3.6 to 900 F/sec). The later cooling rate approximated water quenching. The cooling rate between the annealing temperature and 260 C (500 F) was determined. All tensile tests were conducted at a strain rate of 0.05/min to 3% strain and then at a strain rate of 0.5/min to failure. Yield strength measurements were taken at 0.2% offset plastic strain. Standard light and electron metallography were conducted along with quantitative microscopy using a Leitz TAS quantitative microscope.

Results And Discussion

The results will be discussed in terms of the mechanical properties of the Cr alloys, and a statistical analysis of the effect of martensite on strength is presented. Finally, the influence of alloy level on the strength and microstructure is considered.

Mechanical Properties

The results for each alloy were plotted against cooling rate. For the three annealing temperatures studied, typical properties are seen for alloy CR2 (0.11 C, 1.40 Mn, 0.53 Si, 0.22 Cr) in Figure 1 and alloy CR10 (0.11 C, 1.39 Mn, 0.49 Si, 0.47 Cr) in Figure 2. As can be seen from this data, increasing the cooling rate generally increased the tensile strength and yield strength while eliminating the yield point elongation. At very low cooling rates in the lower chromium alloy CR2, a yield point elongation was found and a concomitant increase in the 0.2% yield strength occurred. Since the higher-Cr alloy CR10 did not have a yield point elongation at any of the cooling rates studied, the % YPE was not plotted in Figure 2.

The Cr dual-phase alloy data is quite similar to the vanadium and molybdenum results previously reported (2,9). Because of the exceptional work hardening behavior of dual-phase alloys at low strains (10) a low Y.S./T.S. value is expected. The reproducibility of the strength-ductility relationship for these steel is shown in Figure 3. The Cr results are plotted from the data for the 760 C (1400 F) and 816 C (1500 F) annealing temperatures for alloys CR1, CR4, and CR10 (Table I). These alloys were selected because they represent the best dual-phase properties over the entire cooling rate range. Also included in this figure are the strength-ductility results for V dual-phase steels (1) and Mo dual-phase steels (2). These results show overlap for all three alloy systems, except that the Cr alloys have a greater band width, indicating that for a given tensile strength a larger range in total elongation may be expected. At a given tensile strength, the band width is approximately 9% total elongation for the chromium alloys and 7% total elongation for the V and Mo alloys. It should be noted that these bands contain all the data and the data has not been subjected to statistical analysis.

Correlation of Strength with Volume Percent Second Phase

The effect of volume percent martensite (%M) on the ultimate tensile strength of the Cr alloys is shown in Figure 4. Except for the two low-carbon alloys (CR5 and CR7) all the Cr data appear to fit the same relationship previously reported for C-Mn-Si steel (11):

$$UTS \ (MPa) = 453 + 12.13 \ (\% \ M)$$
$$UTS \ (KSI) = 65.7 + 1.76 \ (\% \ M)$$

147

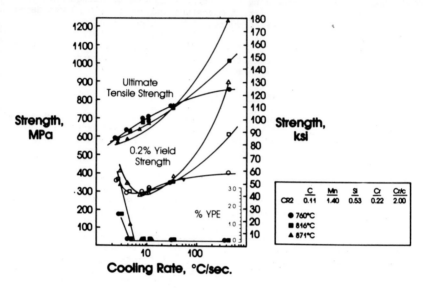

Figure 1.   The effect of cooling rate on the strength properties of alloy CR2.

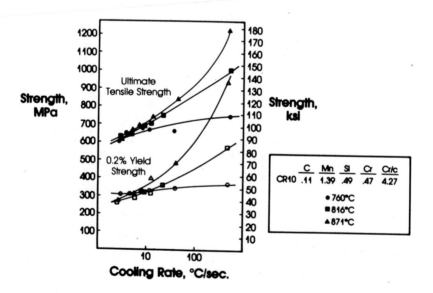

Figure 2.   The effect of cooling rate on the strength properties of alloy CR10.

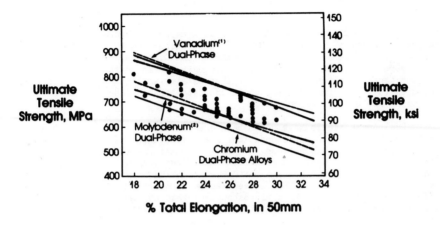

Figure 3.   The strength-ductility relationship in several dual-phase steels.

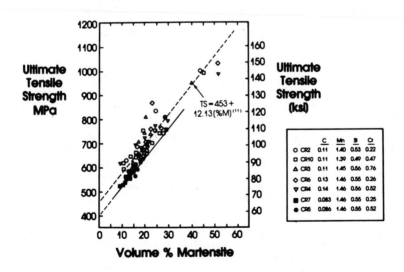

Figure 4.   The effect of volume percent martensite on the ultimate tensile strength of several Cr dual-phase alloys.

149

These results indicate that for 1.5 Mn, 0.5 Si-base composition small variations in C (0.11 - 0.14) and Cr (0.22 - 0.76) had no apparent affect on the UTS of these alloys. The strength is determined only by the volume percent martensite. A similar affect has been reported for C-Mn-Si-V and C-Mn-Si-Mo alloys (9) where:

$$UTS \ (MPa) = 480 + 9.10 \ (\% \ M)$$
$$UTS \ (KSI) = 69.6 + 1.32 \ (\% \ M)$$

Since the slope for the Cr alloys is greater, it appears that Cr additions allow the martensite to have a greater effect on UTS than Mo or V. The effect of volume percent martensite on the 0.2% yield strength is shown in Figure 5. A minimum in the yield strength was found at about 10 to 25 volume percent martensite, a result that is similar to the Mo and V alloy results (9). For the alloys studied the data appears independent of both chromium and carbon content.

Dual-phase steels have often been characterized by a low Y.S./T.S. ratio with no yield point elongation, i.e., continuous yielding. Figure 6 plots the relationship between the Y.S./T.S. ratio and yield point elongation for the Cr alloys. As seen from this figure, when Cr dual-phase steels have a Y.S./T.S. ratio above about 0.5, discontinuous yielding or a yield point elongation occurs. For these steels the onset of a yield point elongation correlates with a microstructure that has more than approximately 6% pearlite, Figure 7, and less than approximately 12% martensite, Figure 8.

Alloy Effects

The effect of Cr (at a constant 0.11 C) on the volume percent second phase formed after cooling at 3 C/sec (5 F/sec) and water quenching (500 C/sec or 900 F/sec) is seen in the lower half of Figure 9 for an intercritical heat treatment of 760 C (1400 F) for 1 minute. After water quenching, all of the second phase was martensite and increasing Cr slightly decreased the total volume percent transformed. Cooling at 3 C/sec (lower curves) produced different results. The higher Cr levels increased the amount of martensite, with only a slight change in the total volume percent of second phase constituents in these alloys. At the lowest Cr values most of the second phase is pearlite. That is, the difference between the total volume percent second phase and volume percent martensite curves will be the amount of pearlite. These results indicate that Cr improves the hardenability of the austenite pools.

The effects of carbon on the volume percent second phase can be seen in the top half of Figure 9 at a 0.5% Cr level. For an intercritical annealing temperature of 760 C (1400 F) for 1 minute, the volume percent martensite after water quenching increased as the carbon content was increased. At a cooling rate of 3 C/sec, some pearlite, as indicated by the difference between total volume percent and martensite volume percent values formed at the higher carbon contents.

Although the data is limited, increased Mn also increased the hardenability of the austenite pools formed at the intercritical annealing temperature, as seen in Figure 10 for both the 760 C (1400 F) and 816 C (1500 F) annealing temperatures.

These results can be rationalized on the basis of Mn and Cr segregation as a result of banding. Previous researchers (12,13) found that martensite

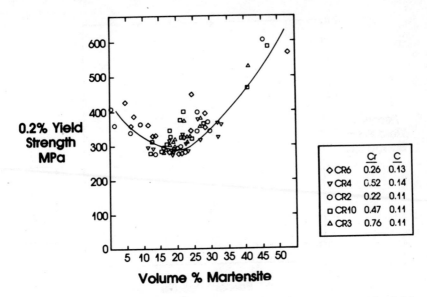

Figure 5.   The effect of volume percent martensite on the 0.2% yield
strength of several Cr dual-phase alloys.

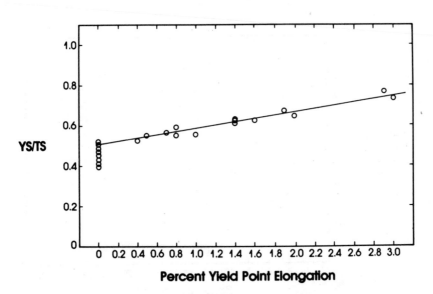

Figure 6.   The relationship between yield point elongation and the YS/TS
ratio.

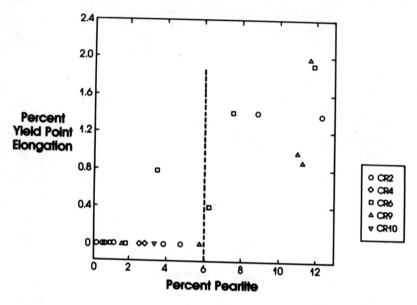

Figure 7.   The effect of pearlite on the yield point elongation.

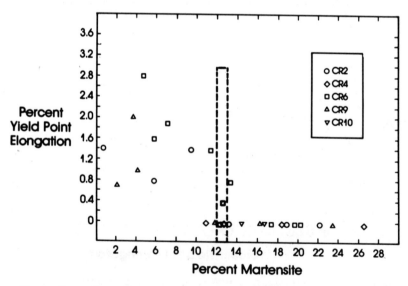

Figure 8.   The effect of martensite on the yield point elongation.

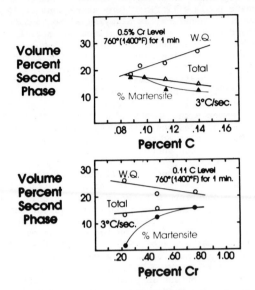

Figure 9.   The volume percent second phase produced after intercritical
             annealing at 760° C (1400°F) and water quenching or cooling at
             3°C/sec. (a) The effect of C in 0.5 Cr alloys. (b) The effect
             of Cr in 0.11 C alloys.

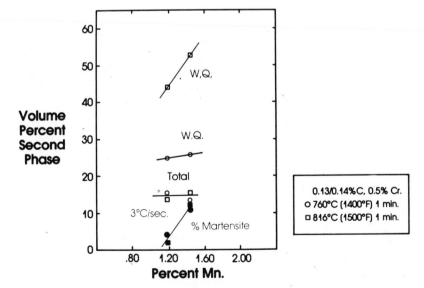

Figure 10.   The effect of Mn on the volume percent second phase produced
             after intercritical annealing at 760° C (1400° F) and 816° C
             (1500° F) and water quenching or cooling at 3°C/sec.

153

bands are generally located in the regions of higher manganese content. Chromium should also segregate in a similar manner. Hansen and Pradhan (11) proposed that austenite pools will first form at regions of locally increased alloy content. Therefore, as alloying elements such as Mn and Cr are increased, the hardenability of the localized pools will increase, producing an increased amount of martensite on slow cooling and virtually no pearlite. However, with increasing carbon content at a given annealing temperature, increasing amounts of austenite can transform from regions of lower alloy content, i.e., regions between the Mn and Cr segregation. Thus, the austenite hardenability of these pools will be less than that of the richer alloy pools. Accordingly, on slow cooling, the lower alloy pools transform to pearlite whereas the amount of martensite formed from the higher-alloy austenite pools remains relatively constant, e.g., Figures 9-10. As a result, the fraction of the second phase that is pearlitic increases with increasing carbon content.

This hypothesis allows for the rationalization of a critical Cr/C ratio. Figure 11 shows that the Cr/C ratio has a definite effect on eliminating the yield point elongation in these steels. In this figure % YPE_{max} refers to the maximum yield point elongation at the lowest cooling rate. Therefore, maximum hardenability, as defined by no yield point elongation at the lowest cooling rate, is dependent on the Cr/C ratio. Very limited data shows that low Mn (1.19% vs. 1.4/1.5%) may change the Cr/C ratio to higher values, whereas Si has no effect. Figure 11 shows that maximum hardenability is obtained in 1.5 Mn-0.5 Cr alloys when the Cr/C ratio exceeds 3.0. If the carbon content is high and the chromium content is low (Cr/C $\leq$ 3.0), there is excess carbon and austenite pools of low alloy content (Cr and Mn) will form on intercritical annealing, and on subsequent slow cooling these relatively alloy-depleted austenite pools can transform to pearlite.

Microstructural Characterization

Four alloys were selected for the microstructural study: a series of three Cr levels, 0.22, 0.47 and 0.76, having a constant 0.11 C content (CR2,CR10,and CR3) and an additional alloy (CR7) having a low C content (0.083 C) and a low Cr content (0.25 Cr). The strength properties of these alloys are plotted in Figure 12 along with the volume percents of martensite and pearlite. Cooling rates of approximately 3 C/sec and 9 C/sec were selected for the study, and the microstructure of these alloys can be seen in Figures 13 and 14, respectively.

At 3 C/sec the microstructure was essentially martensite in a matrix of ferrite, except for alloy CR2 (0.11 C, 0.22 Cr), in which the second-phase was pearlite. At the faster cooling rate, 9 C/sec, even the CR2 alloy had martensite as the predominant phase (Figure 14). Similar results can be seen in electron micrographs, Figures 15 and 16. At 3 C/sec only the CR2 alloy (0.11 C, 0.22 Cr) had a large amount of pearlite. The other alloys exhibited the typical dual-phase microstructure of martensite islands in a heavily dislocated ferrite. At the higher cooling rate (Figure 16) all the alloys had the typical dual-phase structure.

Although these results apparently show that pearlite causes discontinuous yielding in alloy CR2 (compare Figures 12, 13, and 15), this is not necessarily the case. It has been reported previously that in continuously cooled material a sufficient volume percent martensite produces a critical amount of mobile dislocations (9). The lack of mobile dislocations brought about by insufficient martensite formation causes a yield point elongation.

154

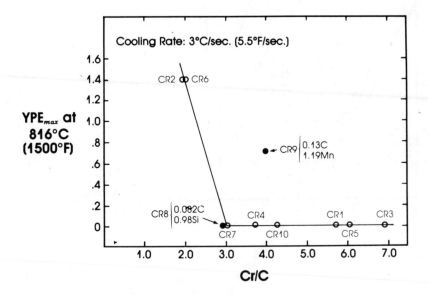

Figure 11. The effect of the Cr/C ratio on the maximum % yield point elongation found at 816° C (1500° F) for 1.5% Mn, 0.5% Si Steels.

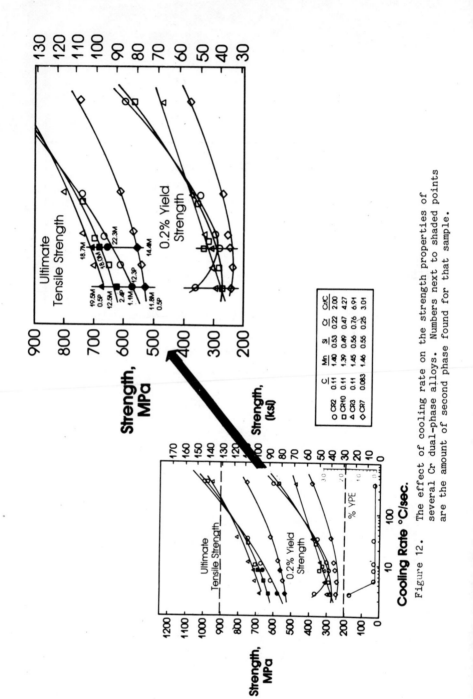

Figure 12. The effect of cooling rate on the strength properties of several Cr dual-phase alloys. Numbers next to shaded points are the amount of second phase found for that sample.

156

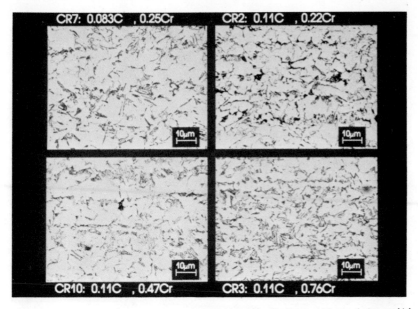

Figure 13.   Light microscopy of several Cr dual-phase alloys intercritically annealed at 816° C (1500° F) and cooled at 3°C/sec.

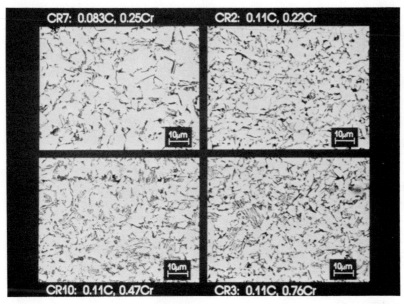

Figure 14.   Light microscopy of several Cr dual-phase alloys intercritically annealed at 816° C (1500° F) and cooled at 9°C/sec.

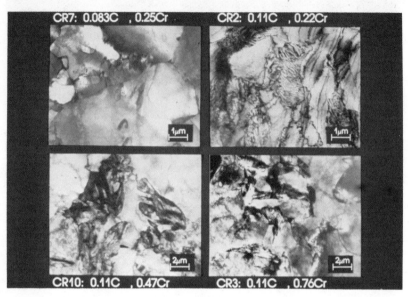

Figure 15.   Electron microscopy of several Cr dual-phase alloys intercritically annealed at 816° C (1500° F) and cooled at 3°C/sec.

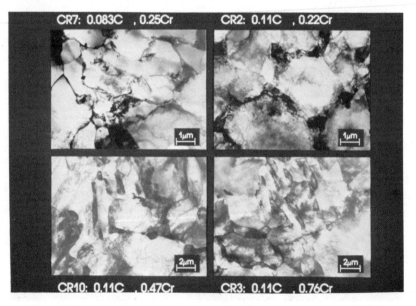

Figure 16.   Electron microscopy of several Cr dual-phase alloys intercritically annealed at 816° C (1500° F) and cooled at 9°C/sec.

In the materials of the present study, 12% martensite was found to be the critical amount necessary to prevent a yield point elongation (Figure 8). A comparison of the CR7 alloy (0.083 C, 0.25 Cr) with the CR2 alloy (0.11 C, 0.22 Cr) shows the effect of C on the hardenability of the prior austenite pools. With low-Cr contents in these two steels, increasing the carbon content to 0.11 lowered the average hardenability of the austenite and pearlite was formed on slow cooling (CR2 - Fig. 13 and 15). However, when the carbon content was low, 0.083 C, at an equivalent Cr level, very little pearlite was formed, since there was no excess of C to produce austenite pools away from regions of Mn and Cr segregation.

Conclusions

A study of a range of C, Mn, Si, Cr steels intercritically annealed at three temperatures and quenched at various cooling rates showed that:

- Cr dual-phase steels behave similarly to Mo and V dual-phase steels in that martensite is the predominant structural feature controlling the tensile strength of these alloys. In addition, a minimum in the yield strength occurs at approximately 10 to 25 volume percent mar tensite, values that are similar to these obtained with Mo and V alloy steels.

- Dual-phase properties, i.e., continuous yielding and Y.S/T.S ≤ 0.5, were obtained for these particular alloys when the structures contained more than 12% martensite and less than 6% pearlite. Although pearlite could be related to the onset of discontinuous yielding or a yield point elongation, it was found that for continuous yielding to take place there must be a critical amount of mobile dislocations resulting from the transformation of martensite.

- Increasing the Cr and Mn content in these steels increases the hardenablilty of the austenite pools during intercritical annealing. Increasing carbon content does not necessarily increase the average hardenability of the austenite pools, since some austenite may form in regions that are not enriched with Cr and Mn. Subsequent slow cooling of these regions can produce pearlite.

Acknowledgments

The author wishes to acknowledge S.S. Hansen and R.R. Pradhan for their discussions on dual-phase steels, C. Santos for his assistance in conducting the experiments and A. O. Benscoter for his metallographic work. Appreciation is also given to B. S. Mikofsky for editing the manuscript.

References

1. J. H. Bucher and E. G. Hamburg, "High-Strength Formable Sheet Steel," SAE Paper 770164, 1977.

2. A. R. Marder and B. L. Bramfitt, "Processing of a Molybdenum-Bearing Dual-Phase Steel," pp. 242-259 in Structure and Properties of Dual-Phase Steels, ed. by R. A. Kot and J. W. Morris, TMS/AIME, 1980.

3. J. Imamura and T. Furukawa, "Development of High-Strength, Dual-Phase Steel Sheets," Nippon Steel Tech. Report Overseas, No. 10, November 1977, p. 103.

4. N. Ohashi, I. Takahashi and K. Hashiguchi, "Processing Techniques and Formability of Age Hardenable, Low Yield, High Tensile Strength, Cold-Rolled Sheet Steels," Trans. Jap. ISI, Vo. 18, 1978, p. 321.

5. D. Aichbhaumik and R. R. Goodhart, "Effect of Annealing Cycles on the Properties and Microstructure of Several Dual-Phase Steels," SAE TP 790010, 1979.

6. D. A. Chatfield and R. R. Goodhart, "Method for Producing a Dual Phase Ferrite-Martensite Steel Strip," U.S. Patent No. 4,159,218, June 26, 1979.

7. T. Tanaka, M. Nishida, K. Hashiguahi, and T. Kato, "Formation and Properties of Ferrite Plus Martensite Dual-Phase Structures," pp. 221-241 in Structures and Properties of Dual-Phase Steels, ed. by R.A. Kot and J. W. Morris, TMS/AIME, 1980.

8. K. Hashiguchi, T. Kato, M. Nishida, and T. Tanaka, "Effects of Alloying Elements and Cooling Rate after Annealing on Mechanical Properties of Dual-Phase Sheet Steel," Kawaskai Steel Technical Report, No. 1, Sept. 1980, p. 70.

9. A. R. Marder, "The Effect of Heat Treatment on the Properties and Structure of Mo and V-Bearing Dual-Phase Steels, paper submitted to Met Trans.

10. D. K. Matlock and G. Krauss, "A Correlation of Processing Variables With Deformation Behavior of Dual-Phase Steels," pp. 62-90 in Structures and Properties of Dual Phase Steels, ed. by R. A. Kot and J. W. Morris, TMS/AIME, 1980.

11. S. S. Hansen and R. R. Pradhan, "Structure/Property Relations in C-Mn-Si Dual-Phase Steels," in this book.

12. G. R. Speich and R. L. Miller, "Mechanical Properties of Ferrite-Martensite Steels" pp. 145-182 in Structure and Properties of Dual-Phase Steels, ed. by R. A. Kot, and J. W. Morris, TMS/AIME, 1979.

13. T. Furukawa, H. Morikawa, H. Takechi and K. Koyama, "Process Factors for Highly Ductile Dual-Phase Sheet Steels," ibid, pp. 281-303.

PHASE TRANSFORMATION AND MICROSTRUCTURES OF INTERCRITICALLY ANNEALED

DUAL-PHASE STEELS*

Pierre Messien   Jean-Claude Herman   Tony Gréday

Centre de Recherches Métallurgiques

Liège - Belgium.

Intercritically annealed dual-phase steel sheets and strips with chemical compositions in the ranges : 0.05-0.10% C, 1-1.7% Mn, 0.2-0.7% Si have been examined in order to assess the effects of the annealing treatment parameters (temperature, holding time, cooling rate) on both the dual-phase microstructure and the mechanical properties.

An annealing treatment in the temperature range of 800-825°C (1472-1517°F) is likely to be the most suitable when the final cooling rate does not exceed 50°C/sec (90°F/sec). It promotes an optimal dispersion of martensite islands of about 1 island per 1-2 ferrite grains. Under these conditions and depending on the steel composition, 5-10 percent martensite, and some residual austenite are dispersed in a fine grain ferrite (part of which consists of secondary ferrite). Moreover this optimal microstructure, promoted by silicon, stabilizes the mechanical properties. UTS levels in the range 500-600 MPa (725-940 ksi) are obtained with optimum values of UTS x El_T for cooling rates of 10-70°C/sec (18-126°F/sec). The ductility is improved with carbon and phosphorus additions ; in opposition an excess of free nitrogen has a detrimental effect on the YS/TS ratio and reduces the ductility.

The ferrite in dual-phase steels is supersaturated in carbon. A model is proposed to estimate the carbon content of the two phases, ferrite and martensite ; it is then possible to determine critical cooling rates to be used to promote a given amount of martensite. Such calculations agree very well with the experimental results.

* This work was supported by IRSIA. Belgian Institute for Scientific Research in Industry and Agriculture.

Introduction.

The microstructure of dual-phase steels consists essentially of fine grained equiaxed ferrite in which are dispersed islands of a hard phase, most often martensite containing some retained austenite (1-6). The amount of martensite is an essential factor governing the mechanical properties of dual-phase steels (7-9) as well as their formability.

The dual-phase microstructure is easily produced by intercritical annealing (generally in a continuous annealing line) followed by a critical cooling rate which promotes the required amount of fine well dispersed martensite necessary to obtain the desired dual-phase mechanical properties. The carbon content of this martensite (10,11) is considered to be relatively low (0.3-0.6). The continuous yielding behavior of dual-phase steels is explained by the high dislocation density in the ferrite near the martensite ferrite interfaces (12).

When the cooling rate after annealing is limited to a maximum of 50°C/sec (90°F/sec), measured between 800 and 500°C (1472-932°F), use must be made of alloying elements such as manganese, chromium and molybdenum to reduce the critical cooling rate necessary for martensite formation. Other elements have a more complex action, e.g., silicon seems to increase the carbon content of the austenite formed during intercritical annealing (10-12) and to retard the nucleation of cementite in the ferrite phase. Other elements, such as carbon, nitrogen and phosphorus also effect the strength-ductility relationship of dual-phase steels (13).

It must be remembered that the dual-phase steel microstructure is not simple nor unique. As a consequence, a relatively wide range of mechanical properties can be covered (14) if the nature and the amount of the second phase as well as the physical state of the ferrite (density of dislocation and carbon content) can be controlled.

This research deals with the study of dual-phase sheets and strips obtained by intercritical continuous annealing. The objectives are :

1) to describe,as completely as possible, the optimum dual-phase micro-structure,

2) to study the effects of the annealing parameters (temperature, holding time and cooling rate) on the dual-phase microstructure and mechanical properties, and

3) to quantify the effects of alloying elements such as carbon, manganese, silicon, nitrogen and phosphorus on the properties of dual-phase steels.

Materials and Experimental Procedure.

A series of 18 low carbon (0.07-0.1% C) steels were prepared by air-induction melting of 36 kgs heats. The steels also contained (Table I) 1-1.7% Mn, 0.2-0.7% Si and were Al-killed. Moreover one heat was rephospho-rized (0.13% P) and three were renitrided (up to 0.025% N). The square sec-tion (115x115mm) ingots were forged to 20mm (.79 in) thick slabs. The pro-cessing schedule for hot rolling the slabs to strips 2.4, 3.2 and 4.0mm (.094, .125 and .157 in.) thick is indicated in Table II. A part of these strips were cold-rolled to 0.8, 1.05 and 1.4mm (.031, .041 and .055 in.) thick.

Table I. Chemical Composition Ranges of the Steels (Wt.%).

C	Mn	Si	P	N
0.05(2) 0.07-0.08(14) 0.10(2)	1.000 (2) 1.200 (3) 1.400 (12) 1.700 (1)	0.2 (2) 0.3 (9) 0.4-0.5 (3) 0.6-0.7 (4)	0.010 (17) 0.13 (1)	0.005 (15) 0.010 (1) 0.017 (1) 0.025 (1)

The number of heats is indicated between brackets.

All steels are Al-Killed (.05% Al) and the S content is in the
range of .005-.007%.

Table II. Rolling Conditions.

1. Ingots : 115x115mm (4.53 x 4.53 in.) - 36 kg (82 lbs)

2. Forged in blanks   : 20mm (.79 in.) thick.

3. Reheating temperature : 1250°C (2280°F) - 2 h.

4. Starting rolling temperature : 1100°C (2010°F) ;
   Finishing temperatures : 880 or 850°C (1615 or 1560°F);
   Passes : 4 ;
   Final Thicknesses : 2.4, 3.2 and 4mm (0.094, .126 and .157 in.).

5. Cold rolling : total reduction :65-70% ;
   Final Thicknesses : 0.8, 1.05 and 1.4mm (.031, .041 and .055 in.).

Table III. Intercritical Annealing Conditions.

Annealing temperatures : 700-850°C (1290-1560°F) by steps of 25°C (45°F) ;

Total heat treatment : 2 to 10 minutes ;

Heating rates : from 12.5°C/sec (22.5°F/sec) to 2.6°C/sec (4.5°F/sec)
               depending on the thickness of the sheet ;

Cooling rates between 800°C and 500°C (1470-930°F) :
               3.5°C/sec (6.3°F/sec) to 750°C/sec (1350°F/sec) depending
               on the thickness of the sheet   ; not measured after quen-
               ching of thinner gages (.8-1.05mm).

On both strips and sheets, intercritical annealing treatments were
performed in a laboratory furnace using an argon atmosphere. Various
temperatures, annealing times and cooling rates, which varied depending
on the sheet thickness, were used (Table III). The annealing cycle was
varied especially to provide information on the effects of manganese and
silicon.

Tensile samples 150mm (6 in.) long and 12.5mm (.5 in) wide were machined in the transverse direction to rolling. The mechanical properties were
measured by using a strain rate of 0.4mm/min (.016 in/min). Yield strength
(0.2% offset), ultimate tensile strength and total elongation [using a gage
length of 50mm (2 in.)] were evaluated in the conventional manner.

Extensive microstructural analysis was also performed. A special
etching technique adapted from the procedure proposed by LePera (15) was
used to obtain a sharp contrast between phases in an optical image.

163

An image analyser was used to quantitatively characterize the size and volume fraction of the martensitic phase. Moreover, microstructural examinations using electron microscopy (transmission and scanning), microhardness measurements and residual austenite evaluations (X-ray diffraction technique) were also used to identify the various phases and microconstituents.

Data and Results.

Effect of Intercritical Annealing Variables.

According to the annealing cycles performed (Table III), the maximum amount of martensite measured in the steels studied is of the order of 30 percent and is only obtained when using the maximum cooling rate, i.e., the quenching in cold water.

The volume fraction of martensite increases with both the maximum annealing temperature and the holding time in the intercritical range (Figure 1a and 1b). In the lowest part of that range : 725-750°C (1337-1382°F), and for a short annealing time (2 min) it has been observed by metallographic examinations that all the initial carbides may not be completely dissolved (Figure 1c). Consequently the amount of austenite is lower than the one corresponding to the equilibrium amount for the temperature and the carbon content of the steel; assuming now that with the high cooling rate used all the austenite transforms to martensite, the amount of martensite remains very low. The transformation from ferrite to austenite is initiated on those carbides (16-17) and, as it is observed, austenite grows mainly along the grains boundaries of the recrystallized ferrite. After a sharp increase in the amount of martensite in the temperature range of 750-775°C (1382-1427°F), a deceleration of this increase is observed when the annealing temperature is in the range of 775-800°C (1427-1470°F), mainly as the annealing time is short (2min). We interpret this effect as resulting from a lack of austenite nuclei as all the initial carbides, acting as preferential sites for the austenite nucleation, have been dissolved and from the slow diffusion rate of carbon through the $\gamma-\alpha$ interfaces restraining the amount of austenite to reach its equilibrium volume. When annealing is performed above this critical range of temperatures (775-800°C), more than 30 percent of austenite is formed which, owing to its lower carbon content, is partially transformed into bainite during the cooling state (Figure 1d).

When use is made of slower cooling rates in the range 15°C/sec to 48°C/sec (27°F/sec to 86°F/sec), the amount of martensite is highly dependent on the steel composition (Figure 2). It is observed that manganese promotes martensite formation. The microstructure for steels with different Mn levels are shown in Figure 3. Silicon also promotes martensite formation in the lower cooling rate range mentioned above. Martensite formation is also governed by the carbon content of the austenite regions, which depends on both the amount of austenite formed in the intercritical range and the temperature (diffusion of carbon). This explains the observed decrease of the martensite amount after slower cooling rates (Figure 2) and the appearance of bainite after quenching (Figure 1d) when the annealing temperature rises in the range of 775°-800°C.

The mean size of the martensite islands has been measured with an image analyser at a magnification of 1000 x. This mean size is highly dependent on the annealing temperature and reaches its maximum (4 square microns) when annealing at 775°C (1427°F). A stable mean size (1.5 square microns) is obtained when annealing in the temperature range 800-850°C (1472-1562°F).

164

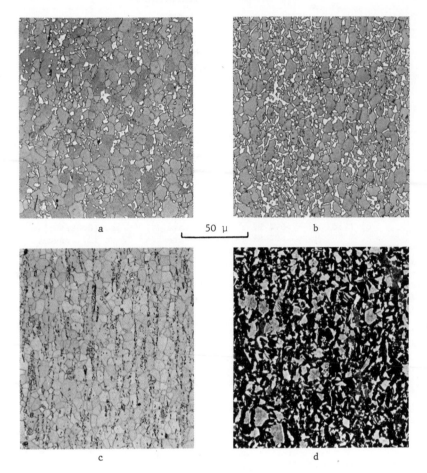

Fig. 1. Influence of the intercritical annealing temperature on the nature
and amount of the phases present after cold water quenching.
a. annealing temperature,750°C (1382°F) ; 10% martensite
b. annealing temperature,775°C (1427°F) ; 25% martensite
c. annealing temperature,725°C (1337°F) ; 0-5% martensite plus
                                          carbides
d. annealing temperature,850°C (1562°F) ; martensite plus bainite.

Steel composition : C=0.082% - Mn = 1.59% - Si=0.39%.

Etching : a,b;c : 2% picric acid, 0.5% sodium bisulfite in ethyl
                  alcohol, 10 sec.
          d   : 1.3% picric acid, 1.3% sodium bisulfite in ethyl
                  alcohol, 15 sec.

Magnification : 500 x.

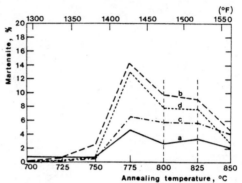

Fig. 2 - Influence of the intercritical annealing temperature on the amount
of martensite formed for various compositions and cooling rates.
Annealing time, 2 min. Sheet thickness, 0.8mm (0.03 in).
a. 1.03% Mn - .34% Si ; cooling rate : 48°C/sec (86°F/sec).
b. 1.59% Mn - .39% Si ; cooling rate : 48°C/sec (86°F/sec).
c. 1.50% Mn - .69% Si ; cooling rate : 48°C/sec (86°F/sec).
d. 1.50% Mn - .69% Si ; cooling rate : 15°C/sec (27°F/sec).

Moreover, the optimal dispersion of 1 martensite island per 1-2 ferrite
grains (Figure 4),which does not correspond to the maximum amount of martensite, is obtained when using annealing temperatures in the range of 800-
825°C (1472-1517°F). This range 800-825°C (1472-1517°F) becomes our optimal annealing temperature range. A favourable effect of silicon has also
been observed ; this element stabilizes the martensite dispersion against
variations of annealing temperature and time. As a conclusion, using an
annealing temperature in the range, 800-825°C (1472-1517°F),promotes the optimal dispersion of martensite islands of 1.5(microns)^2 in mean size,which
corresponds to a ratio 1 martensite island to 1 or 2 ferrite grains ; the
measured mean grain size of the ferrite being 4.5 microns (approx. 12.5ASTM).

Table IV. Microhardness of the Ferrite.

Heat treatment	Cooling rate	Ferrite microhardness H_v (2 gr)	
		as-annealed	80°C(176°F) 2h aged
775°C(1430°F) - 2 min.	Water quenching	143	87.2
775°C(1430°F) - 2 min.	48°C/sec(86°F/sec)	107	96.8
825°C(1520°F) - 2 min.	Water quenching	116	75.8
825°C(1520°F) - 2 min.	48°C/sec(86°F/sec)	118	90.6

When using a cooling rate in the range of 15-48°C/sec (27-86°F/sec),new
ferrite grains("secondary")ferrite are formed at the beginning of the austenite transformation. This "secondary" ferrite which appears when annealing
at a temperature above 750°C (Figure 5) refines the main ferritic microstructure, improving thus the steel ductility. Moreover, microhardness measurements show (Table IV) the supersaturation of carbon in the ferrite ;

166

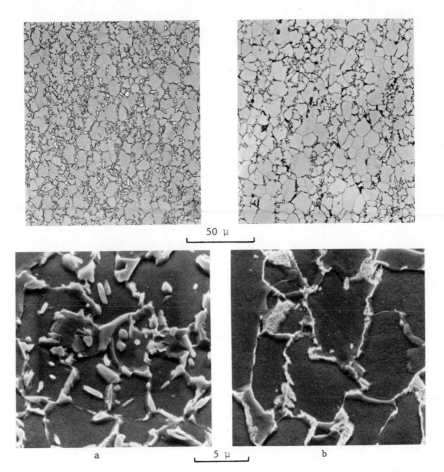

Fig. 3. Influence of manganese content on the nature of the second phase.
Intercritical annealing at 825°C (1517°F), 2 min.
Cooling rate : 48°C/sec (86°F/sec) ; etching the same as for Figure
1a. Magnifications : a and b, 500 x ; c and d, 4000 x.
a. Steel : 0.082 C, 1.59 Mn, 0.39 Si   b.Steel : 0.082C, 1.03Mn,0.34Si
   YS = 296 MPa, UTS = 590 MPa   YS = 304 MPa, UTS = 510 MPa
   EL_T= 29.3%, n = 0.248 EL_T= 30%, n = 0.211
   9.8% martensite.   2.7% martensite plus carbides.

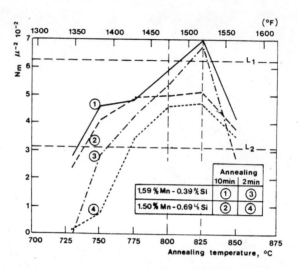

Fig. 4 - Changes in the martensite dispersion (number of martensite islands per square micron) vs. annealing temperature (cooling rate, 48°C/sec (86°F/sec)).
L_1 : level corresponding to 1 martensite island per 1 ferrite grain.
L_2 : the same but per 2 ferrite grains.

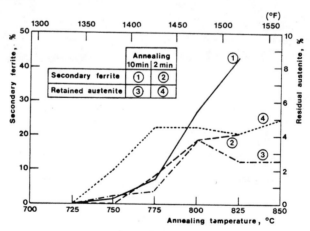

Fig. 5 - Effects of the annealing temperature on the amount of secondary ferrite and retained austenite (cooling rate, 48°C/sec (86°F/sec)). Steel composition, 0.082% C - 1.59% Mn - 0.39% Si.

this effect has been confirmed by examinations with the transmission electron microscope, revealing precipitation of carbide particles on dislocations (Figure 6a).

Retained austenite (Figure 6b) has been revealed by X-ray diffraction. Its amount varies between 1 to 5 percent depending on the cooling rate and on the annealing temperature. As it can be seen (Figure 5) retained austenite occurs simultaneously with secondary ferrite.

Level and Stability of the Mechanical Properties.

Hardness measurements have been used to estimate the tensile strength and to determine the level and the stability of tensile strength due to variations in annealing conditions. A large instability in hardness is noticed when the annealing temperature is below 800°C (1472°F) and the cooling rate is below 48°C/sec (86°F/sec). The hardness values are relatively constant with annealing conditions ranging in temperature from 800 to 825°C the holding time at temperature being between 1 and 9 minutes (Figure 7).

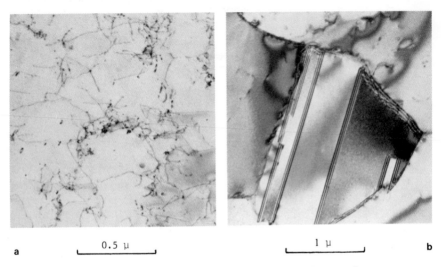

a   0.5 μ   1 μ   b

Fig. 6 - Microstructural aspects of dual-phase steels.
a. carbides precipitated in the ferrite
b. retained austenite.

As a general conclusion of our experiments, it is observed that :

1) 5 percent of martensite islands well dispersed in the ferrite promote a continuous yielding curve and a high "n" strain-hardening coefficient. These two properties are the two main keys to characterize dual-phase behavior (Figure 8).

2) using an annealing temperature in the range of 800-825°C (1472-1517°F) promotes a dual-phase microstructure where the appropriate amount (5 percent) of martensite is well dispersed in ferrite grains of both primary and secondary formation.

Examples of the mechanical properties for such a dual-phase microstructure
are given in Table V. Uniform elongations associated with the best dual-
phase microstructures range from 23% to 24%.

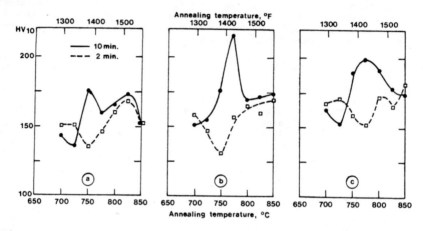

Fig. 7 - Influence of both the intercritical annealing parameters (T_{Max},
holding time, cooling rate) and the silicon content on the hardness (HV_{10}) of dual-phase steels.
a. 1.59% Mn, 0.39% Si ; cooling rate, 15°C (27°F)/sec
b. 1.59% Mn, 0.39% Si ; cooling rate, 48°C (86°F)/sec
c. 1.50% Mn, 0.69% Si ; cooling rate, 48°C (86°F)/sec.

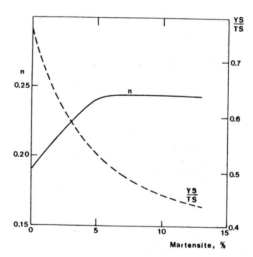

Fig. 8 - Influence of the amount of martensite on the work-hardening coefficient n and the $YS/_{TS}$ ratio. Annealing conditions, 825°C (1517°F)
for 2 min30sec ; cooling rate, 15°C/sec (27°F/sec) to 48°C/sec
(86°F/sec) ; sheet thickness, 0.8mm (0.031 in) and 1.4mm (0.055 in).

170

Table V – Mechanical Properties.

Range of mechanical properties(minimum strength, L, and maximum strength, H) obtained after intercritical annealing (T_{max} = 825°C (1517°F)) low carbon steels (0.07-0.08% C) with different manganese and silicon contents ; six thicknesses gages are considered with the corresponding ranges of cooling rates after annealing ; underlined values ,
——— $0.5 \leqslant YS/TS \leqslant 0.6$; ==== $YS/TS < 0.5$.

Sheet Thickness (mm) (in)	Cooling rate From to °C/sec °F/sec		1% Mn, 0.3 - 0.4% Si			1.4 - 1.6% Mn, 0.3 - 0.4%Si			1.4 - 1.6% Mn, 0.6-0.7% Si		
			YS - UTS MPa	EL_T %	n	YS - UTS MPa	EL_T %	n	YS - UTS MPa	EL_T %	n
0.8 (.031)	15 – 150	L	383-504	32.0	0.20	320-568	28.8	0.22	320-560	28.0	0.23
	27 – 270	H	321-560	28.6	0.23	288-617	28.8	0.25	305-656	28.2	0.25
1.05 (.041)	10 – 115	L	378-486	29.0	0.20	322-548	27.6	0.22			
	18 – 207	H	330-534	30.6	0.23	308-602	29.1	0.24			
1.4 (.055)	8.1 – 60	L	320-464	35.3	0.20	301-531	32.4	0.22	326-562	30.1	0.23
	14.6 – 108	H	328-509	29.1	0.21	265-594	32.8	0.24	286-639	29.6	0.25
2.4 (0.94)	6.3 –24.8	L	354-467	29.1	0.21	390-513	28.4	0.21			
	11.3 – 44.6	H	388-494	28.6	0.23	307-554	29.5	0.24			
3.2 (.126)	4.4 – 18.5	L	314-414	28.7	0.23	351-493	26.9	0.23	361-505	27.4	0.21
	7.9 – 33.3	H	298-456	31.9	0.23	280-548	28.8	0.24	267-541	29.9	0.24
4.0 (.157)	3.5 – 16.2	L	349-453	27.5	0.21	350-478	29.8	0.21	386-515	31.2	0.22
	6.3 – 29.2	H	343-465	30.0	0.22	290-532	32.7	0.25	301-555	30.6	0.26

171

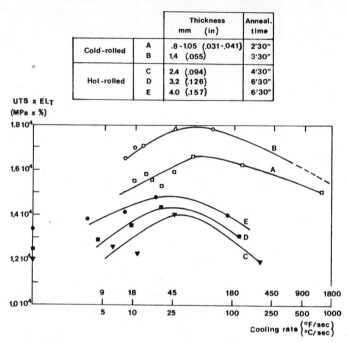

		Thickness		Anneal.
		mm	(in)	time
Cold-rolled	A	.8 -1.05	(.031-.041)	2'30"
	B	1,4	(.055)	3'30"
Hot-rolled	C	2,4	(.094)	4'30"
	D	3,2	(.126)	6'30"
	E	4.0	(.157)	6'30"

Fig. 9 - Influence of the cooling rate on the UTS x EL$_T$ product.
Intercritical annealing at 825°C (1517°F).

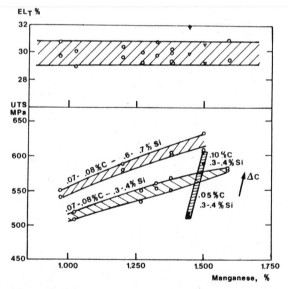

Fig. 10 - Influence of the steel composition (carbon, manganese and silicon) on mechanical properties. Annealing at 825°C (1517°F) for
2 min30sec and cooling at 48°C/sec (86°F/sec).

172

Effects of cooling rate and chemical composition on the mechanical properties.

A regression equation has been developed between the cooling rate and steel composition to obtain 5 percent martensite well dispersed in a fine ferrite (12-13 ASTM) microstructure

$$\log c_R^{5M} \; (°C/sec) = 4.93 - 1.70 \; [\% \; Mn] - 1.34 \; [\% \; Si] - 5.68 \; [\% \; C]$$

$$\sigma = .11 \qquad\qquad r = .972 \qquad\qquad (1)$$

Regression analysis has also been used to obtain an equation describing ultimate tensile strength in terms of chemistry, grain size and cooling rate for 0.8, 1.05 and 1.4mm (.031, .041 and .055 in.) thick sheets.

$$UTS \; (MPa) = 71.6 \; \log c_R + 955 \; [\% \; C] + 69.5 \; [\% \; Mn] + 131 \; [\% \; Si]$$
$$+ 9.4 \; d^{-1/2} + 70.8 \qquad\qquad (2)$$

$$\sigma = 8.4 \; MPa \qquad\qquad r = .980$$

These relationships are valid only for steels cold-rolled and continuously annealed as indicated and with chemical compositions in the following range : 0.05-0.1% C , 1-1.7% Mn and 0.2-0.7% Si.

It can then be concluded that :

1) Increasing the cooling rate raises both the UTS and the parameter UTS x EL_T (Figure 9) as long as the cooling rate is lower than 100°C/sec (180°F/sec). The use of a faster cooling rate not only increases the UTS but also produces an important loss of ductility. This behavior has been correlated with the replacement of equiaxed ferrite by bainite, the amount of which increases rapidly with carbon content.

2) Increases in the content of carbon, manganese and silicon lead to increases in U.T.S. without an important loss of ductility (Figure 10).

3) A better UTS x EL_T parameter is also obtained for the rephosphorized steel (0.13% P) (Figure 11) as a result of both an increased amount of secondary ferrite and a general refinement of the microstructure (Figure 12).

4) Increasing the nitrogen content increases the YS/UTS ratio and decreases UTS x EL_T product term (Figure 13). These detrimental effects will be eliminated if the nitrogen is combined with such strong nitrides forming elements as vanadium or titanium (13).

Discussion.

From the above results as well as from other results already published in the literature on the subject, it appears clearly that it is possible to characterize the microstructure of the dual-phase steels possessing the best combination of strength and ductility by using both scanning electron microscopy and quantitative metallography. So far, the choice of the chemical composition corresponding to the optimal microstructure cannot be physically justified because the chemical composition of the two main phases, martensite and ferrite, cannot be completely described (11). An important factor is the carbon content which not only affects the mechanical properties of each phase but also the deformation behavior of their interfaces.

Our results lead us to formulate some hypotheses about the carbon content of the martensite regions and its variation with the amount of martensite in the general microstructure of dual-phase steels. It can be concluded from the microstructural observations of this study that the carbon content of the martensite lies in the range of 0.4-0.8 percent. Moreover a supersaturation of carbon in the ferrite has been revealed by carbide precipitation on the

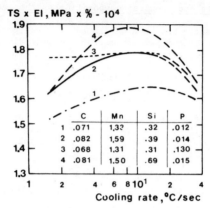

Fig. 11 - Influence of the steel composition (manganese,silicon, phosphorus)on the product TSxE1 in function of the cooling rate. Annealed 825°C - 2min30sec - Thickness :.8mm.

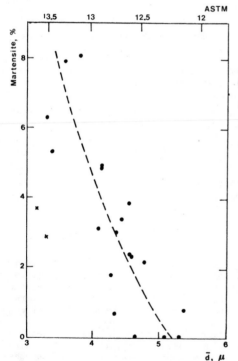

Fig.12- Relationship between the amount of martensite and the ferrite grain size Annealing conditions, 825°C(1517°F) for 2min30sec; cooling rate,18°C/sec(32°F/sec) and 25°C/sec(45°F/sec); high phosphorus (0.13%)steel results are indicated by the(x).

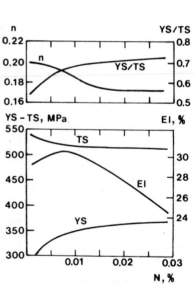

Fig.13 - Influence of the nitrogen content on the mechanical properties ; annealed : 825°C, 2min30sec; cooled:15°C/sec ; thickness:0.8mm.

174

dislocations after an aging treatment.

Assuming that the dual-phase microstructure (18,19) is a mixture of two phases of equal density and applying the law of mixtures, the distribution of carbon between these two phases is given by :

$$[C]_{tot} = [C]_m \cdot x_m + [C]_f \cdot (1-x_m) \qquad (3)$$

where x_m = amount of martensite

$[C]_m$, $[C]_f$, $[C]_{tot}$: carbon contents, respectively, in the martensite and ferrite and in the steel.

The same law of mixtures applied to the UTS gives :

$$(UTS)_{tot} = (UTS)_m \cdot x_m + (UTS)_f (1-x_m) \qquad (4)$$

In addition, the minimum cooling rate necessary to transform the austenite into martensite without any bainite formation is given (20) by the following formula :

$$\log t_C^B = 3.274 [C]_\gamma + 0.046 [Si] + 0.626 [Mn] - 1.818 \qquad (5)$$

where t_C^B is the time in seconds, between the annealing temperature and 500°C (932°F) ,

[C] is the carbon content of the austenite

[Mn] and [Si] are the mean chemical contents of the steel, assuming that during a short continuous annealing treatment no diffusion of those elements will occur (21).

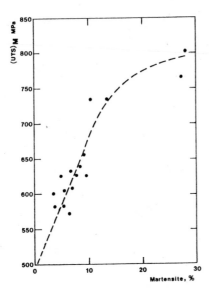 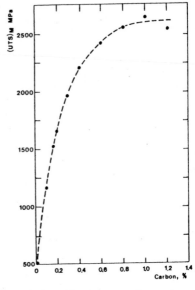

Fig. 14 a. - Increase of UTS with the amount of martensite. Intercritical annealing at 825°C(1517°F) for 2min30 steel composition, 0.082%C - 1.50% Mn-0.69% Si; cooling rate from 15°C/ sec(27°F/sec)to 150°C/sec(270°E/sec).

b. Relationship between the carbon content and $(UTS)_M$ (22).

175

Equations 3 and 4 can be used to calculate the carbon contents in the martensite and ferrite phases (Figure 14).

Various annealing treatments have been applied to a steel in order to obtain different dual-phase states characterized by different amounts of martensite. The UTS for the various states was measured and is shown in Figure 14 a. It can be concluded that the relationship between UTS and the amount of martensite is linear up to approximately 10 percent martensite. It is possible to extrapolate the linear portion of the curve to determine the $(UTS)_f$ value for the same steel without any martensite.

By inserting the $(UTS)_{tot}$ and the $(UTS)_f$ values in Equation 4 one gets the corresponding $(UTS)_m$ value. Figure 14b gives the relationship between $(UTS)_m$, calculated from hardness values, and the carbon content of the martensite (22). The $[C]_m$ value of martensite can then be inserted in Equation 4 in order to calculate $[C]_f$. The critical cooling rate corresponding to this martensite can then be derived from Equation 5.

This procedure for calculating $[C]_m$, $[C]_f$ and t_C^B has been applied with success to all the steels studied.

The main conclusions derived from this analysis are :

1) the carbon content of the martensite decreases as the amount of martensite increases ; this explains that a fully martensitic microstructure cannot be obtained when the amount of martensite exceeds 25-30% for the steels studied ;

2) the carbon content of the martensite increases with the manganese and the silicon contents of the steel (Figure 16) ;

3) ferrite is supersaturated in carbon, at least when the amount of martensite is low. The carbon content of the ferrite decreases as the silicon content of the steel increases (Figure 15 and 16).

Finally, the calculated theoretical critical cooling rate is consistent with the experimental results. In those steels where 3-4 percent martensite stabilizes at slow cooling rates (Figure 17), this calculation shows that

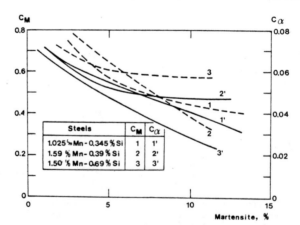

Fig. 15 - Changes in the carbon content of the martensite (C_M) and ferrite (C_α) with the amount of martensite.

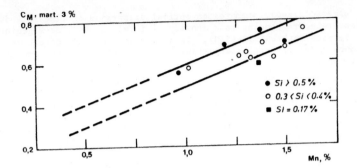

Fig. 16 - Influence of the manganese and silicon contents of the steel on the carbon content (C_M) of the martensite for an amount of 3 percent of martensite in the microstructure of dual-phase steels.

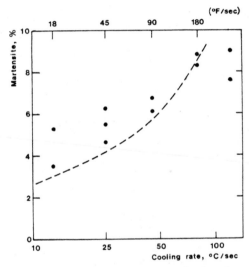

Fig. 17 - Comparison of the measured and calculated amounts of martensite as a function of cooling rate. Intercritical annealing at 825°C (1517°F) for 2min30sec ; steel composition, 0.082% C - 1.5% Mn - 0.69% Si.

such a stabilization results necessarily from an increase in the carbon content of the stabilized austenite just before the formation of martensite.

Conclusions.

The following conclusions can be drawn, on the basis of the experimental results for intercritically annealed dual-phase steel sheets and strips whose chemical compositions cover the ranges : 0.05-0.1% C, 1-1.7% Mn, 0.2-0.7% Si and Al-killed.

1) Annealing in the temperature range of 800-825°C (1472-1517°F) followed by a cooling whose rate between 800-500°C (1472-932°F) does not exceed

177

50°C/sec (90°F/sec), promotes an optimal dispersion of fine islands of martensite corresponding to a density of 1 island per 1 or 2 grains of fine ferrite (approx. 12 ASTM). This resultant microstructure and related mechanical properties are stable with variations in the annealing conditions.

2) The ideal dispersion of second phase consists of 5-10 percent martensite and a small amount (3-4%) of retained austenite. The mean size of the martensite islands is 1.5 square microns. Moreover, the ferrite phase includes a large amount (between 15 and 25 percent) of "secondary" ferrite (retransformed austenite formed on heating) which contributes to the refinement of the ferrite microstructure. Silicon promotes the above mentioned dispersion.

3) Applying the optimum annealing conditions to high manganese (1.3% and more), silicon (0.2% min) steels will produce a dual-phase behavior ($n \geqslant 0.23$; $\frac{Y.S.}{UTS} \leqslant 0.55$), the microstructure being characterized by a minimum amount of 5 percent of well dispersed martensite.

4) Increasing the manganese or silicon contents of the steel, increases the UTS level (up to 500-650 MPa) and lowers the YS/UTS ratios to values less than 0.5 without any important loss of ductility. Carbon and phosphorus behave in the same way, while free nitrogen is extremely detrimental to the ductility.

5) The optimum cooling rate (as measured between 800-500°C (1472-1562°F)) to produce the maximum value of the ultimate tensile strength times total elongation product term (UTS X EL_T) lies in the range of 10-70°C/sec (18-126°F/sec). This range is valid for all the thicknesses studied (0.8-4mm (0.031-0.157 in.)).

6) The ferrite phase of the dual-phase steels is supersaturated in carbon. A model, based on the law of mixtures, has been used to determine the carbon content of both the martensite and ferrite. It has also been possible to develop a regression equation to calculate the critical cooling rate necessary to produce a martensite volume fraction of 5%. These calculations are in good agreement with the results of our experiments.

References.

1. T. Furukawa, H. Morikawa, H. Takechi and K. Koyama, "Process Factors for Highly Ductile Dual-Phase Sheet Steels", paper presented at TMS-AIME Meeting, New Orleans, La., Feb. 1979.

2. J.Y. Koo and G. Thomas, "Thermal Cycling Treatments and Microstructures for improved Properties of Fe-0.12% - 0.5% Mn Steels," Materials Science and Engineering, 24 (1976) pp. 187-198.

3. J.Y. Koo and G. Thomas, "Design of Duplex Low Carbon Steels for Improved Strength : Weight Applications", in Formable HSLA and Dual-Phase Steels, A.T. Davenport, ed. ; AIME, New-York, N.Y. 1979.

4. R.G. Davies, "The Mechanical Properties of Zero-Carbon Ferrite-Plus-Martensite Structures", Metallurgical Transactions, 9A (1978) pp. 451-455.

5. N. Ohashi, I. Takahashi and K. Hashigushi, "Processing Techniques and Formability of Age-Hardenable, Low Yield, High Tensile Strength Cold Rolled Sheet Steels", Transactions of the Iron and Steel Institute of Japan, 18 (1977) pp. 321-329.

6. L.F. Ramos, D.K. Matlock and G. Krauss. "On the Deformation Behavior of Dual-Phase Steels", Metallurgical Transactions, 10A (1979) pp. 259-261.

7. S. Hayami and T. Fukukawa, "A Family of High Strength Cold Rolled Steels", pp. 311-321 in Microalloying 75, John Crane, ed. ; Union Carbide Corporation, New-York, N.Y., 1977.

8. D.F. Baxter, "G.M. Develops a Superformable HSLA Steel", Metal Progress, Aug. (1977) pp. 44-48.

9. J.H. Bucher and E.G. Hamburg, "High Strength Formable Sheet Steel", SAE paper, 1977, 770164.

10. G. Thomas and J.Y. Koo, "Developments in Strong, Ductile Duplex Ferritic-Martensitic Steels", paper presented at TMS-AIME Meeting, New Orleans, La., Feb. 1979.

11. J.Y. Koo, M. Raghavan and G. Thomas, "Compositional Analysis of Dual-Phase Steels by Transmission Electron Microscopy", Metallurgical Transactions, 11A, (1980) pp. 351-355.

12. P. Abramowitz and P.A. Moll, "Silicon-Carbon Interaction and Its Effect on the Notch Toughness of Mild Steel", Metallurgical Transactions, 1 (1970) pp. 1773 -1775.

13. H. Mathy, P. Messien and T.Gréday, "Aspects des Performances Potentielles des Aciers Multiphases pour Tôles de Diverses Epaisseurs", Revue de Métallurgie - CIT, 421 (1980) pp. 421-438.

14. K. Nakaoka, Y. Hosoya, M. Ohmura and A. Nishimoto, "Reassement of the Water-Quench Process as a Means of Producing Dual-Phase Formable Steel Sheets", paper presented at TMS-AIME Meeting, New Orleans, La., Feb. 1979.

15. F.S Lepera, "Improved Etching Techniques for Determination of Percent Martensite in HSLA Dual-Phase Steel", Metallography, 12, (1979) pp.263-268.

16. R.R. Judd and H.W. Paxton, "Kinetics of Austenite Formation from a Speroidized Ferrite-Carbide Aggregate", Transactions of the Metallurgical Society of AIME, 242 (1968) pp. 206-215.

17. G.R. Speich and A. Szirmae, "Formation of Austenite from Ferrite and Ferrite-carbide Aggregates", Transactions of the Metallurgical Society of AIME, 245 (1969) pp. 2063-2074.

18. J.Y. Koo, M.J. Young and G. Thomas, "On the Law of Mixtures in Dual-Phase Steels", Metallurgical Transactions, 11A (1980) pp. 852-854.

19. G.R. Speich and R.L. Miller, "Mechanical Properties of Ferrite - Martensite Steels", paper presented at TMS-AIME Meeting, New Orleans, La., Feb. 1979.

20. T. Kunitake and H. Ohtani, "Calculating the Continuous Cooling Transformation Characteristics of Steel from its Chemical Composition", The Sumitomo Search, n° 2 (1969) pp. 18-21.

21. J.B. Gilmour, G.R. Purdy and J.S. Kirkaldy, "Thermodynamics Controlling the Proeutectoïd Ferrite Transformations in Fe-C-Mn Alloys", TMS-AIME, 3 june (1972) pp. 1455-1464.

22. G. Krauss, "Martensitic Transformation, Structure and Properties in Hardenable Steels", pp. 229-248 in Hardenability Concepts with Applications to Steel, D.V. Doane and J.S. Kirkaldy, ed.; AIME, Warrendale, P.A., 1978.

PROCESSING, PROPERTIES AND MODELLING OF

EXPERIMENTAL BATCH-ANNEALED DUAL-PHASE STEELS

A.F. Crawley, M.T. Shehata, N. Pussegoda
C.M. Mitchell and W.R. Tyson
Physical Metallurgy Research Laboratories, CANMET,
Dept. of Energy, Mines and Resources, Ottawa, Canada

A laboratory study has determined the feasibility of producing dual-phase steels using C-Mn chemistries. A 3% Mn/0.06% C steel most closely approached the target specifications of 380 and 480 MPa for 0.2% and 3% offset yield strength respectively, 620 MPa ultimate tensile strength and 27% minimum total elongation. The effects of processing and compositions were correlated with microstructures. Optimum properties correspond to a microstructure of ferrite and 13-15% martensite.

The approach to modelling has been the direct measurement of the contributions of the individual phases in the 3% Mn/0.06% C steel. The results appear to confirm a re-examination of the significance of the stress states and the dislocation structure at the ferrite-martensite interface. It is concluded that the stresses do not act to lower the yield strength but only to generate mobile dislocations.

Introduction

This paper presents an overview of a laboratory investigation into the development of C-Mn dual phase steels by intercritical annealing in a batch process. Batch annealing is practical and economical because steel companies have existing facilities for this type of processing. Our objective has been to produce dual phase hot-rolled steels in thicknesses which would meet the following specifications.

Yield strength (0.2% offset)	380 MPa
Yield strength (3.0% offset)	480 MPa
Ultimate tensile strength	620 MPa
Total elongation in 25.4 mm	27%

Mould and Skeena have reported experiments with cold-rolled batch-annealed high manganese steels (1). Although typical dual-phase properties were obtained in tensile tests the combinations of strengths and ductility were inferior to those attainable by other processing routes.

The advantages of high-manganese steel are

(1) alloy costs are relatively low, and
(2) manganese lowers the critical transformation temperatures so that intercritical annealing temperatures are within the capability of existing batch annealing equipment.

The specifications were not all met using the selected chemistries as shown in this report. Over a limited composition and processing range, the strength properties were attained but elongation values were 1-2% low. In order to understand and improve these properties, experiments have been designed to test a micro-mechanical model.

This paper will describe the tensile properties and microstructures of a range of C-Mn steels, and the experimental approach and preliminary results on modelling.

Experimental

Table 1 shows the chemistries of the nine steels studied. These consist of three groups with manganese levels of about 2.35, 3.0, and 3.5 wt %, and three carbon levels of nominally 0.06, 0.08, and 0.1 wt % for each manganese level.

Production of sheet

Steels were produced as 227-kg aluminum-killed heats in an induction furnace. The heat was cast into three 68-kg, 175-mm diameter, ingots which were cut into two pieces for forging to suitable dimensions (64 mm thick and 254 mm long) for further processing on the pilot scale reversing mill (457-mm roll diameter). Before forging, the ingots were soaked for 2 h at 1200°C in an oil-fired furnace. Next, the forged ingots were soaked for 1½ h at 1200°C in an electric furnace before reduction by cross-rolling to plate about 25 mm thick. This material was then cut into rectangular sections 280 mm wide, 76-100 mm long and then surface milled to remove oxide scale. Finally the plates were soaked for 50 minutes at 1250°C and rolled to sheets 2.5 mm thick.

Table I. Analysis (wt %) of experimental C-Mn steels

Heat No.	C	Mn	S	P	Al	Si
11330	0.06	2.88	0.016	0.007	0.06	0.32
11327	0.075	3.15	0.016	0.006	0.02	0.30
11324	0.095	3.08	0.016	0.007	0.06	0.31
11357	0.06	3.53	0.020	0.007	0.05	0.25
11363	0.09	3.53	0.018	0.007	0.04	0.34
11369	0.10	3.48	0.017	0.007	0.03	0.30
11321	0.06	2.37	0.015	0.007	0.03	0.31
11343	0.075	2.30	0.015	0.008	0.03	0.26
11371	0.095	2.34	0.019	0.007	0.05	0.26

Rare earth silica (33% R.E.) - 137 g added to each 227 kg ingot.

During final rolling the temperature was monitored by a thermocouple inserted into the side of the plate. The reduction on the reversing mill was carried out in five passes with a nominal finishing temperature of 870°C. After the last pass the sheet was transferred to a salt bath at 600°C to simulate the treatment of commercial sheet between the last stand and the coiler. The sheets remained in the salt bath for 1 h, and were then removed, stacked on insulating brick and allowed to cool to room temperature.

Heat treatments

Five intercritical annealing temperatures were chosen, 677, 704, 732, 760 and 780°C for heat treating the machined tensile specimens. The selection of the lower temperature was based on the upper temperature limits of existing batch-annealing equipment and estimates of the Ac_1 temperatures of the steels. In order to determine the effect on the mechanical properties of different sections of the coil, the 3% Mn steels were annealed at each temperature for 5, 10 and 20 hours. The other compositions were annealed for 5 h only. No attempt was made to simulate the heating rate for industrial batch-annealing. After annealing, the specimens were cooled at 28°C/h which is a mid-range cooling rate for tight-coil annealing.

Evaluation of steels

Evaluation was based on tensile testing and microstructural examination. Tensile specimens, ASTM type E-8, were machined from blanks removed parallel to the rolling direction. Tests were conducted on duplicate specimens using an Instron Universal Tester at a cross-head speed of 1.25 mm/min up to 2% strain, and thereafter, at 12.5 mm/min until failure. Load and extension data were recorded continuously using a data acquisition system and microprocessor.

Microstructures were characterized principally by optical and scanning electron metallography. An image analysing computer (Quantimet 720 system) was used to measure the volume of second phase. The subconstituents of the second phase: martensite, pearlite, and bainite were identified by scanning electron microscopy and measured by standard point-counting procedures (ASTM E562).

Results and Discussion

Tensile Testing

The typical stress-strain curve for dual-phase steels at yield was obtained for each composition at all annealing temperatures above 704°C except for the 2.3% Mn steel at 788°C. Figures 1 and 2 show data obtained for the 3% Mn steels at three carbon levels. At annealing temperatures increasing above 704°C, both yield and ultimate tensile strengths increase and ductility decreases sharply. Ultimate tensile strength and elongation vary by about -3 MPa/°C and 0.13%/°C. At higher annealing temperatures, properties are less sensitive to temperature but they are then far removed from target specifications. The yield strength decrease between 677 and 704°C reflects the distinct yield point shown after annealing at 677°C.

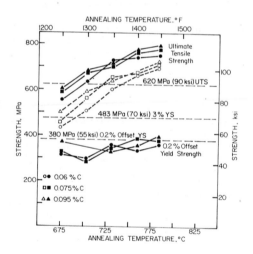

Fig. 1 - Strength properties of 3% Mn steels after intercritical annealing for 5 hours.

The 3% Mn steels most closely approached the target properties although over a limited range of chemistry and annealing temperatures. Specifically, a 3% Mn/0.06% C steel annealed at 704°C achieved the target strength properties but did not reach the elongation values. Table II lists the tensile properties for this steel. Figure 3 shows that time at temperature had no consistent effect on strength properties for the 3% Mn steels. Elongation data not shown, tended to decrease slightly as annealing time increased.

184

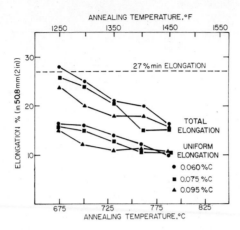

Fig. 2 - Ductility of 3% Mn steels after intercritical annealing for 5 hours.

Table II - Tensile Properties of Mn/0.06% C steel after intercritial anneals of 5 h

Annealing temperature °C	Yield strength 0.2% offset MPa	ksi	Ultimate tensile strength MPa	ksi	Elongation (%) in 25.4 mm Uniform	Total
678	317	46.0*	555	80.5	16.7	30.0
	317	46.0	557	80.8	16.2	26.0
704	299	43.4	636	93.3	18.9	26.8
	315	45.7	635	92.2	16.3	24.0
733	360	52.0	718	104.4	13.5	20.0
	361	52.3	750	108.8	15.0	22.0
760	335	48.6	746	108.2	12.4	22.3
787	361	52.2	755	109.5	9.2	17.8
	426	61.7	758	109.8	9.9	15.0

*Yield point

Results for the other steel groups, 2.35% and 3.5% Mn, are presented in Figures 4 to 7. They show the same general trend as for the 3% Mn steels but fail by wider margins to meet the target specifications.

Finally, Figures 8 and 9 sum up the effect of chemistry on steels annealed at 704°C. Figure 8 shows that both the ultimate tensile strength and elongation are linearly dependent on carbon content. Regression analysis of the data resulted in the following equations:

Ultimate tensile strength, MPa = -92.2 + 213.09 (% Mn) + 1257 (% C)   (1)
Total elongation (%)   = 53.86 - 8.03 (% Mn) - 86 (% C)   (2)

185

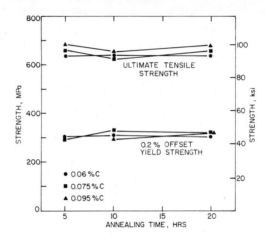

Fig. 3 - Effect of time of intercritical annealing at 704°C on the yield and tensile strength of 3% Mn steels.

The equations are plotted in Figure 9. Interpolation shows that the target properties are unattainable with the types of chemistries and processing considered in this investigation.

Microstructural Examination

Microstructural examination was reserved for the optimum composition, 3% Mn/0.06% C, judged by the tensile test results. Figure 10 shows the distribution of ferrite and second phase after annealing over the entire range of temperatures. The original banded structure of the as hot rolled steel remains evident after annealing at the lowest temperature, 678°C. At higher annealing temperatures the structure becomes more uniform as the volume of second phase increases. In Figure 10(e) much of the second phase is bainite and is in poor contrast with the ferrite under optical metallography. But even at magnifications of 1600X, the sub-constituent in the second phase cannot be distinguished by optical microscopy. Table III lists the volume fraction of second phase at each temperature and carbon level for 3% Mn steels, measured by the image-analysing system.

Figure 11 shows that SEM has revealed the sub-constituents of the second phase. At none of the annealing temperatures does the structure consist solely of the M/A constituent. After annealing in the temperature range of 705 to 733°C, which resulted in the best mechanical properties, the structure consisted mainly of M/A but with some pearlite. Annealing at 678°C was too low to cause complete dissolution of carbides. The structure shows a small fraction of M/A, an almost equal volume of pearlite and some free carbides (Fig. 11(a)). In contrast, in Figure 11(d) and 11(e) the predominant constituent of the second phase is bainite with small quantities of M/A. Volume fractions of the second phase and of the sub-constituents determined by point counting are plotted in Figure 12. The graph also shows that the volume of second phase determined by optical metallography is significantly, but not unexpectedly, underestimated.

186

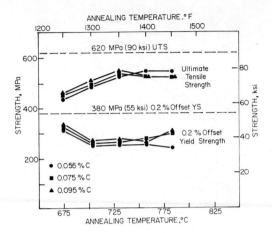

Fig. 4 - Strength properties of 2.35% Mn steels after intercritical
annealing for 5 hours.

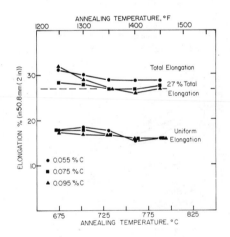

Fig. 5 - Ductility of 2.35% Mn steels after intercritical annealing for
5 hours.

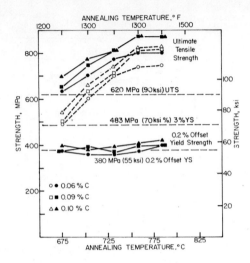

Fig. 6 - Strength properties of 3.5% Mn steel after intercritical annealing for 5 hours.

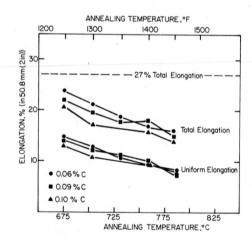

Fig. 7 - Ductility of 3.5% Mn steels after intercritical annealing for 5 hours.

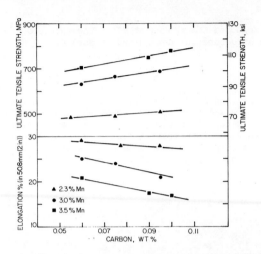

Fig. 8 - Effect of composition on the ultimate tensile strengths and
ductilities of C-Mn steels after intercritical annealing at
704°C for 5 hours.

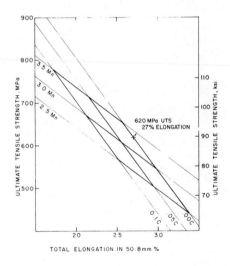

Fig. 9 - Interrelations among composition, ultimate tensile strength and
total elongation in intercritical annealed C-Mn steels based on
Equations (1) and (2).

189

This discrepancy is attributed to the poor contrast of bainite in the optical images.

For the 3% Mn/0.06% C annealed at 704°C the volume of retained austenite in the structure was determined to be 3-4%.

Modelling

The results of this study summarized in Figure 9 demonstrate that for the type of chemistry selected the target properties were unattainable. Optimum properties were obtainable with a restricted chemistry, 3% Mn/ 0.06% C, processed at an intercritical annealing temperature of 705-733°C. These properties correspond to a structure containing about 16% of the second phase which was largely M/A (Figs. 11 and 12).

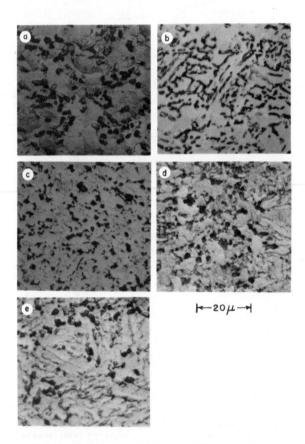

$\mapsto$ 20μ $\dashv$

Fig. 10 - Optical micrographs for 3% Mn - 0.06% C steel annealed at (a) 678°C (b) 704°C (c) 733°C (d) 760°C (e) 787°C; HF-acetic acid etch.

Table III. Volume % of second phase for 3% Mn steels by optical
metallography

Annealing Temperature °C	0.06% C	0.08% C	0.10%
678	12.3 ± 1.3	12.7 ± 1.3	10.6 ± 1.9
704	15.2 ± 1.9	15.2 ± 1.9	14.1 ± 1.1
733	16.2 ± 1.6	18.2 ± 1.1	13.2 ± 2.1
760	18.1 ± 1.6	19.4 ± 1.8	17.6 ± 1.8
787	14.9 ± 1.5	17.9 ± 1.4	11.9 ± 1.6

The precision at each measurement is given for the 95% confidence level.

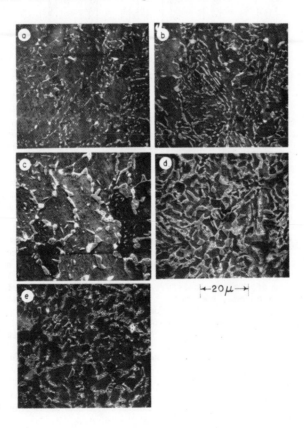

Fig. 11 - Scanning electron micrographs of 3% Mn - 0.06% C steel annealed
          at (a) 678°C (b) 705°C (c) 733°C (d) 760°C (e) 787°C; HF-acetic
          acid etch.

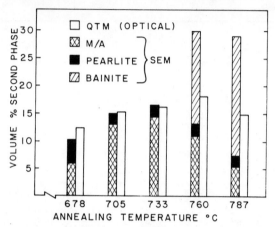

Fig. 12 - Volume percent of sub-constituent in the second phase in the
3% Mn - 0.06% C steels.

From both practical and fundamental considerations, batch annealing
merits further attention. First, batch annealing remains attractive
because of low capital investment. Second, the laboratory data failed by
only 1-2%, to meet specifications for total elongation. However, modelling of the factors controlling the mechanical properties of these steels
may lead to improved alloys and processes. A variety of models and possible mechanisms have been prepared which are applicable to dual phase
steels. The remainder of this presentation will describe our approach to
modelling and report some preliminary results.

We have focussed our efforts upon the most promising composition
(3% Mn/0.06% C) and have adopted an empirical approach to measure directly
the contributions of the individual phases. The method is to select an
Fe/Mn/Si/C alloy, determine from information available in the literature
the compositions of the terminal ferrite and austenite boundaries during
the intercritical anneal, and conduct tensile tests and microstructural
examinations. The relevant phase boundaries were calculated using the
method of Gilmour et al (2). Table IV lists the dual phase Fe/Mn/Si/C
alloy (DP) selected and the compositions of the terminal ferrite and
austenite phases of the DP alloy at 695°C. The materials were prepared
using techniques similar to those described in the earlier section on
experimental techniques.

Table IV - Compositions of alloys (wt %): DP (dual phase) and α, ν
(terminal phases ferrite, austenite)

Alloy	C	Mn	Si
DP	0.06	2.83	0.33
α	0.005	2.18	0.34
ν	0.35	5.94	0.23

The DP alloy was annealed for 5 hours at 695°C, quenched in brine and then immersed in liquid nitrogen for 30 minutes to minimize the volume of retained austenite. The resulting microstructure shown in Figure 13 contained about 16% by volume of martensite, and about 1% retained austenite, compared with 4% in the equivalent steel in the earlier program. The volume fraction of second phase is somewhat lower than predicted from the calculated phase diagram. Quantitative microanalysis by X-ray spectroscopy in the TEM and STEM showed that significant partitioning of manganese to the second phase took place after 5 hours of annealing at 695°C. The degree of manganese partition, somewhat lower than predicted from Table IV, increased slowly with annealing time; partition ratios of 1.7, 2.1, 2.3, and 2.5 were measured after 0.8, 5, 30, and 180 hours at 695°C. Figure 14 shows a STEM analysis of manganese levels across a martensite particle. We conclude that in using these high manganese steels for modelling, the choice of a five hour anneal is reasonable considering the kinetics of the partitioning process.

In an approach to modelling, we have re-examined the significance of the stress state and dislocation structure in the ferrite at the ferrite-martensite interface. Gerbase and others have proposed that internal stresses play an important role in the initial yielding behaviour of dual-phase steels (3). It has also been suggested that dislocations nucleated at martensite particles as a result of transformation stresses, significantly lower the yield stress (4). Such dislocation arrrays have been observed in the DP steel, Figure 15. We have attempted to account for their presence by calculating the distribution of internal stresses after quenching.

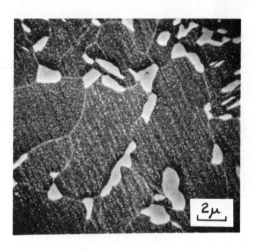

Fig. 13 - Microstructure of DP annealed at 695°C and quenched in brine.

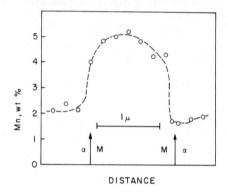

Fig. 14 - Variation of manganese concentration from STEM analysis a martensite particle. The positions of the martensite-ferrite interfaces are shown.

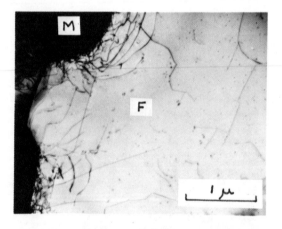

Fig. 15 - TEM micrograph showing a high density of dislocations in the matrix near martensite particles.

Internal stresses arise because of differential thermal contraction of the terminal phases and transformation strains from the austenite-martensite phase change. Using data in the literature on the effects of temperature and alloy additions, the lattice parameters of the alloys were estimated. From published CCT curves, M_S and M_F temperatures were estimated for the austenite-martensite transformation of the terminal austenitic phase. The results are illustrated in Figure 16 which shows a volume increase of about 3% during the transformation at 200°C. The net expansion will be reduced since a differential volume contraction between phases occurs during cooling to the M_S temperature due to the larger coefficient of thermal expansion of austenite (Fig. 16). Assuming that thermal stresses are relieved by plastic flow down to $T_{SR}\sim 600$°C, then the net volume change at room temperature would be about 2%.

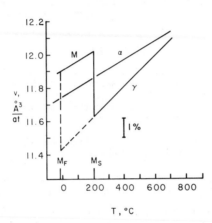

Fig. 16 - Atomic volumes of terminal alloys as a function of temperature, estimated from data in the literature. M_S and M_f are estimated from published CCT curves.

Internal stresses resulting from this net volume change may be readily calculated using the methods pioneered by Eshelby (5) and developed by Brown et al (6,7). Considering the martensite particles as misfitting spheres, the transformation strain generates stresses in the matrix which fall off as $1/r^3$ from a maximum shear stress of

$$\tau_{max} \simeq \mu \frac{(1 + \nu)}{3(1 - \nu)} \varepsilon \simeq 0.012\ \mu \qquad (3)$$

μ is the shear modulus, ν is Poisson's ratio ($\nu = 0.29$), and ε is the volume change (transformation strain) for the austenite-martensite reaction. This stress is certainly large enough to cause motion of existing dislocations to nucleate fresh dislocations. However, the only average stress in the matrix is an image stress (7) which is a hydrostatic tension

$$\sigma_m = \mu \frac{4(1 + \nu)}{5(1 - \nu)} \varepsilon\ f \simeq 0.002\ \mu \qquad (4)$$

195

where f is the volume fraction of martensite. This hydrostatic stress should have no effect on the magnitude of the stress required to cause yielding. We conclude, therefore, that the major effect of the transformation stress on the mechanical properties of the DP alloy is the generation of mobile dislocations in the vicinity of the martensite islands. Tensile results from the DP alloy and the ferrite phase also indicate no decrease in yield strength because of internal stresses at the ferrite-martensite interface, contrary to suggestions elsewhere (5).

Mechanical properties of the component phases were measured, taking care to account for the effect of ferrite grains size on the matrix properties. Results show yield strengths in the ratio 5:1 between the martensite and ferrite components. In future work, the mechanical properties of the DP material will be compared with predictions of the model of Tomota et al (8) for the tensile deformation of two-ductile-phase alloys which seems appropriate to the modelling approach presented in this paper.

Conclusions

1. Hot-rolled steels containing 3% Mn and about 0.06% C show dual-phase properties after intercritical annealing by a simulated batch process.

2. The total elongation is low by 1-2% compared to other similar steels produced by continuous annealing.

3. Microstructures corresponding to the optimum properties contained 13-15% martensite and about 2% of pearlite.

4. During intercritical annealing, partitioning of manganese from ferrite to austenite is approximately 80% complete after 5 hours at 695°C.

5. Calculation of the internal stresses present in dual-phase steels indicate that the major effect of the transformation stresses is to generate mobile dislocations adjacent to martensite particles. These stresses do not act to lower the yield stress, but facilitate yielding indirectly by causing generation of mobile dislocations.

References

1. P.R. Mould and C.C. Skeena, "Structure and Properties of Hot Rolled Ferrite-Martensite (Dual-phase) Steel Sheets"; pp 81-204 in Formable HSLA and Dual-Phase Steels, A.T. Davenport ed.; AIME, New York, N.Y., 1979.

2. J.B. Gilmour, G.R. Purdy and J.S. Kirkaldy, "Thermodynamics Controlling the Pro-eutectoid Tranaformation in Fe-C-Mn Alloys"; Metallurgical Transactions, 3:1455-1464; 1972.

3. J. Gerbase, J.D. Embury and R.M. Hobbs, "Thermochemical Behaviour of Some Dual Phase Steels - With Emphasis on the Initial Work Hardening Rate"; pp 118-144 in Structure and Properties of Dual-Phase Steels, R.A. Kot and T.W. Morris ed.; AIME, New York, N.Y., 1979.

4. J.M. Rigsbee and P.T. VanderArend, "Laboratory Studies of Micro-structures and Structure-Property Relationships in 'Dual-Phase' HSLA Steels"; pp 56-86 in Formable HSLA and Dual Phase Steels, A.T. Davenport, ed.; AIME, New York N.Y., 1979.

5. J.D. Eshelby, "The Determination of the Elastic Field of an Ellipsoidal Inclusion, and Related Problems"; Proceedings of the Royal Society (London), A241, 1957; 376-396.

6. L.M. Brown and W.M. Stobbs, "The Work-Hardening of Copper-Silica I. A Model Based on Internal Stresses, with no Plastic Relaxation"; Philosophical Magazine, 23:1185-1199; 1971.

7. L.M. Brown and D.R. Clarke, "Work Hardening due to Internal Stresses in Composite Materials"; Acta Metallurgica, 23:821-830; 1975.

8. Y. Tomota, K. Kuroki, T. Mori and J. Tamura, "Tensile Deformation of Two-Ductile-Phase Alloys: Flow Curves of $\alpha-\gamma$ Fe-Cr-Ni Alloys"; Material Science and Engineering, 24:85-94; 1976.

197

DEVELOPMENT OF AS-HOT-ROLLED DUAL-PHASE STEEL SHEET

Toshiyuki Kato, Koichi Hashiguchi, Isao Takahashi,
Toshio Irie and Nobuo Ohashi

Research Laboratories
Kawasaki Steel Corporation
Chiba, Japan

The effects of alloying elements and hot-rolling conditions were
investigated to develop as-hot-rolled dual-phase steels sheet. Mn-Si-Cr and
Mn-Si-Mo sttels were found to be the most suitable to produce the desirable
microstructure. The roles of Si, Mn and Cr on the formation of the dual-
phase microstructure were examined by the use of CCT diagrams. Controlling
the hot-rolling conditions to refine the austenite grain size, the cooling
patterns obtained on the hot-strip mill runout table and the coiling temperature have been found to be important factors. By using Mn-Si-Cr steels and
optimizing the above-mentioned rolling conditions, as-hot-rolled dual-phase
steel sheets having tensile strengths of 550-700 MPa can be produced
commercially.

Introduction

Dual-phase steel sheets have been shown to exhibit: (i) continuous yielding followed by a rapid work hardening, (ii) a low yield-to-tensile strength ratio, (iii) non-aging at ambient temperature, and (iv) good ductility (1-4). Hot- and cold-rolled dual-phase steel sheets were first produced by intercritical annealing in continuous-annealing lines (1). Recently, many efforts have been devoted to producing dual-phase steel sheets in the as-hot-rolled condition (6-10) by overcoming the following two difficulties. First, the transformation starts from a fully austenitic structure after rolling, which makes it difficult to get an optimum ratio of ferrite (α) and martensite (α') volume fractions after coiling. During the heat treatment process in a continuous annealing line, transformation occurs from a two-phase structure, $\alpha+\gamma$, which makes it much easier to control the α'/α ratio after cooling. Second, the hot-rolled steel sheet is rapidly cooled down to the coiling temperature on the runout table and then followed by slow cooling after coiling. It is difficult to control the cooling conditions uniformly throughout the whole length and width of the hot-rolled strip, compared to the continuous annealing operation where the cooling control is much simpler.

In the present study, the effects of chemical composition and processing variables on the formation of a ferrite plus martensite dual-phase microstructure were examined. The goal was to produce as-hot-rolled sheets with good ductility and an ultimate tensile strength of 550-700 MPa. The study consisted of two parts. In the first part, the effects of chemical composition and processing variables were investigated using laboratory ingots. In the second part, commercial hot-strip mill experiments based on the laboratory test results were conducted to determine the optimum processing conditions.

Effects of Alloying Elements and Processing Variables on the Formation of Dual-Phase Structures

In order to investigate the effects of alloying elements and processing variables, 100 kg ingots were vacuum-melted and hot-rolled to slabs of 12 mm thick. Chemical compositions of the slabs are shown in Table I. As shown in Figure 1, to simulate the actual hot-rolling process, slabs were heated to 1150°C, rolled in two passes to 2 mm thick sheet, cooled to the coiling temperature at a cooling rate of 4°C/sec, held for 40 min in a furnace at 500, 600 or 700°C, and furnace-cooled to ambient temperature at a cooling rate of 40°C/hr. Tensile testing was performed using a specimen with a gage section 12 mm wide by 25 mm long.

Table I. Chemical compositions of materials (wt%)

Steel	C	Mn	Si	Cr	Mo	Al
Mn-Si	0.059	1.71	1.42	–	–	0.030
Mn-Cr	0.047	1.71	–	1.00	–	0.036
Mn-Si-Cr	0.055	1.71	1.02	1.00	–	0.028
Mn-Mo	0.056	1.72	–	–	0.53	0.030
Mn-Si-Mo	0.053	1.72	1.01	–	0.50	0.028

In Figure 2, yield and tensile strengths, elongations and yield-to-

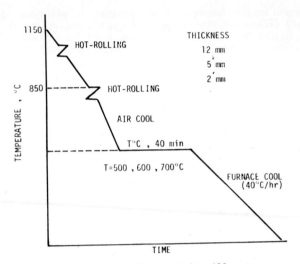

Fig. 1 - Simulation of hot-strip mill process.

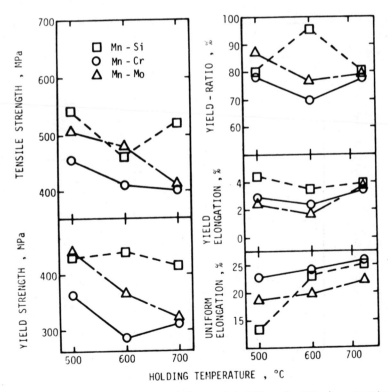

Fig. 2 - Effects of chemical composition and holding (coiling) temperature on mechanical properties in Mn-Si, Mn-Cr and Mn-Mo steels.

tensile strength ratios (thereafter abbreviated as yield-ratio) in Mn-Si, Mn-Mo and Mn-Cr steels are plotted against the holding (coiling) temperature after rolling. The steels held at 500 to 700°C have yield point elongations of more than 2% and yield-ratios of more than 70%. Figure 3 shows similar results for Mn-Si-Cr and Mn-Si-Mo steels. If the holding temperature is 500°C for the Mn-Si-Cr steel and below 600°C for the Mn-Si-Mo steel, yield-ratios less than 65% and no yield point elongation are obtained.

Figure 4 shows the optical micrographs of steels held at 500, 600 or 700°C. In the Mn-Cr and Mn-Mo steels, the microstructures consist of ferrite and pearlite or bainite instead of ferrite plus martensite phases. On the other hand, the microstructures in the Mn-Si-Cr and Mn-Si-Mo steels held at the temperature below the critical ones mentioned above, consist of ferrite grains dispersed with martensite islands, which is the so-called "dual-phase" structure.

These preliminary experiments indicate the importance of selecting the proper chemical composition and rolling parameters, especially, coiling temperature, to produce the desired α+α' dual-phase microstructure.

To investigate more precisely the effects of hot-rolling practice and cooling rate after coiling on the microstructure, the Mn-Si-Cr and Mn-Si-Mo steels were hot-rolled on a small scale mill according to the conditions

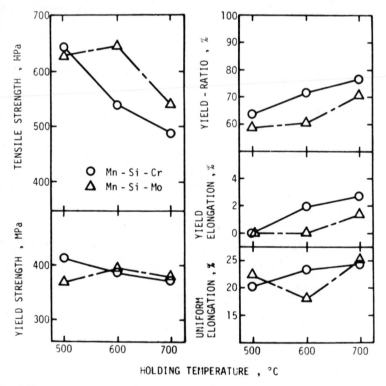

Fig. 3 - Effects of chemical composition and holding (coiling) temperature on mechanical properties in Mn-Si-Cr and Mn-Si-Mo steels.

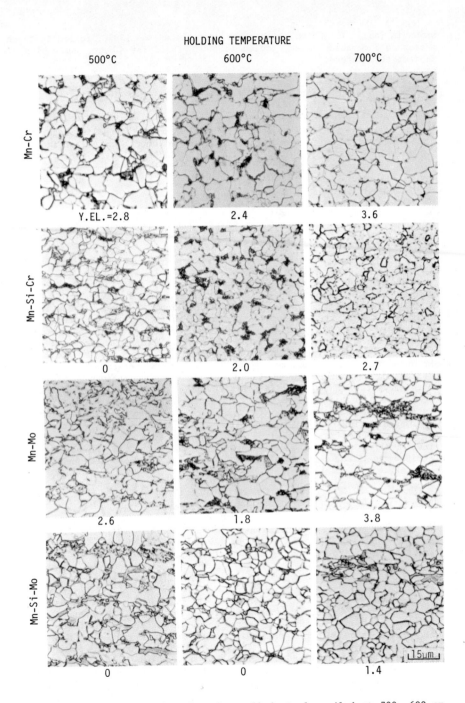

HOLDING TEMPERATURE

500°C   600°C   700°C

Mn-Cr

Y.EL.=2.8   2.4   3.6

Mn-Si-Cr

0   2.0   2.7

Mn-Mo

2.6   1.8   3.8

Mn-Si-Mo

0   0   1.4

Fig. 4 - Optical micrographs of as-hot-rolled steels coiled at 700, 600 or 500°C. Digits show yield point elongation.

shown in Table II. 25 mm thick slabs were heated to 1150°C, rolled in three
passes to 3 mm thick sheet, cooled to 500°C, held for 40 min and cooled to
ambient temperature.

Table II. Hot-rolling practice

Reheating temperature °C	Finish rolling temperature °C	Cooling rate after rolling °C/sec	Cooling rate after coiling °C/hr
1150	830	3.5	30
1150	800	3.5	30
1150	750	3.5	30
1150	810	6.5	30
1150	810	21.1	30
1150	810	3.5	100
1150	810	3.5	12.5

The tensile properties are given in Figure 5. Both the Mn-Si-Cr and
Mn-Si-Mo steels exhibit the dual-phase microstructure, giving yield-ratios
of less than 65% and no yield point elongation. However, the yield and
ultimate tensile strength depend on the hot-rolling schedule. When rolling
is finished at 750°C, both steels exhibit a higher strength and a larger
yield-ratio. This is due to finishing rolling below the Ar_3 temperature
which causes work hardening of the transformed ferrite grains. This shows
the necessity of finish rolling above the Ar_3 temperature to obtain the
desired microstructure and a low yield-ratio. The optimum cooling rate
which gives the highest ductility after rolling is around 6°C/sec. When the
cooling rate is higher than that, the diffusion of carbon from ferrite to
austenite is prevented and the austenite to ferrite transformation is
suppressed during cooling. The cooling rate after coiling which exhibits
the highest ductility is about 30°C/hr. It is supposed that rapid cooling
after coiling does not allow sufficient transformation of austinite to
ferrite and hence the formation of the desired $\alpha+\alpha'$ dual-phase microstructure.

Production Experiments

Experiments were conducted using a commercially melted Mn-Si-Cr steel
on an 80 inch wide hot-strip mill to investigate the optimum conditions for
producing as-hot-rolled dual-phase steel sheet.

Slab reheating temperature

Generally, there is a difference in austenite grain size between
laboratory rolled steels and commercially rolled steels. For hot-strip mill
rolling, grain refinement is caused by the effective deformation and
recrystallization of the austenite grains during multiple rolling passes
under large reduction ratio. Austenite grain refinement promotes the
dual-phase microstructure because it accelerates the austenite to ferrite
transformation by increasing the number of nucleation sites. However, if
the austenite grain size before hot-rolling is too large, bainite
transformation occures instead of the ferrite transformation. Therefore,
the as-soaked austenite grain size of Mn-Si-Cr steels having 0.05%C, 0.04%Al
and different amounts of Mn, Si and Cr was examined. Specimens were cut

from commercially melted, continuously cast and hot-rolled plates 30 mm thick. The specimens were heated in a temperature range between 1000 and 1300°C for 1 hr and quenched in water. Austenite grain boundaries were revealed by etching the specimens in a solution composed of picric acid and a surface active agent and the grain size was measured by a linear intercept method. Figure 6 shows the effect of the soak temperature on the austenite grain sizes of the steels. In most of the steels, coarsening of the

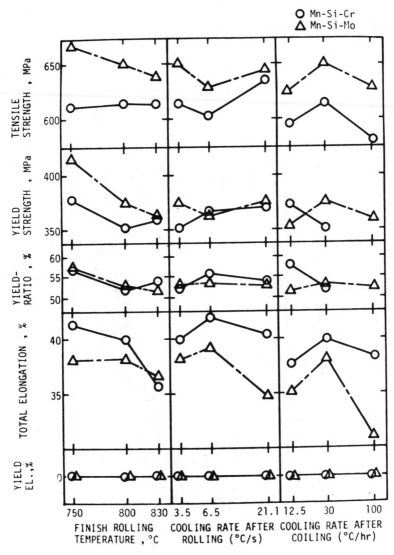

Fig. 5 - Influence of finish-rolling temperature and cooling rates prior to and after coiling on mechanical properties in Mn-Si-Cr and Mn-Si-Mo steels ; held for 40 min at 500°C.

austenite grains occurs above a temperature of 1200°C. This behavior agrees well with that observed in ordinary Al-killed steels and is due to the dissolution of AlN above 1200°C. In the 0.8%Mn-1.5%Si-1.4%Cr steel the coarsening temperature is higher than that for the other steels. This suggests that the effect of Si is to suppress the coarsening of the austenite grains. These experimental results show that the temperature range between 1150 and 1200°C is most suitable for reheating commercial steel slabs of relatively low Si content.

Cooling pattern and coiling temperature

The cooling rates after hot-rolling are also different between a laboratory rolling mill and a commercial hot-strip mill. The laboratory

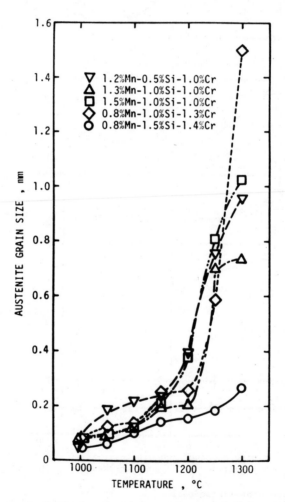

Fig. 6 – Effects of chemical composition and austenitizing temperature on austenite grain size.

test results showed that the optimum cooling rate after rolling, which gives the highest ductility in a 0.05%C-1.7%Mn-1.0%Si-1.0%Cr steel was around 6°C/sec. In hot-rolling on a commercial hot-strip mill, the strip is cooled rapidly on the runout table for less than 10 sec before coiling and the mean cooling rate is 20 to 30°C/sec. In general, the coiling temperature of the hot-strip is determined by controlled water cooling followed by air cooling. In addition, even if the coiling temperature is fixed, the mechanical properties and the microstructure of the strip largely depend on the cooling pattern on the runout table. To investigate the effect of the cooling pattern and the coiling temperature on the mechanical properties and the microstructure of these steels, experiments were conducted on a commercial hot-strip mill.

(I)   0.05%C-1.5%Mn-1.0%Si-1.0%Cr Steel. A heat having a chemical composition as shown in Table III was made in an 80 ton basic oxygen furnace and continuously cast to slabs 200 mm thick by 1000 mm wide. The slabs were hot-rolled to strips on a commercial hot-strip mill according to the hot-rolling practice shown in Table IV. The cooling patterns and coiling temperatures are shown schematically in Figure 7. Coiling temperature (CT) were varied in the range of 600 to 400°C by water cooling followed by air cooling. Cooling patterns were varied by changing the water cooling position on the runout table with a constant coiling temperature of 450°C. The cooling rate during water cooling was about 60°C/sec.

Table III.   Chemical composition (wt%)

C	Si	Mn	P	S	Al	Cr	Ce
0.049	1.01	1.54	0.021	0.004	0.042	1.03	0.013

Table IV.   Hot-rolling practice

Slab reheating temperature	1150°C
Thickness of sheet-bar	33 mm
Finishing delivery temperature (FDT)	800 – 830°C
Thickness of strip	2.9 mm

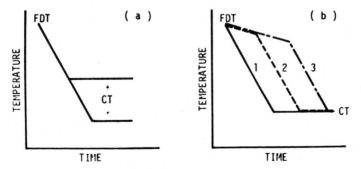

Fig. 7 – Schematic illustration of cooling patterns on a runout table : (a) coiling temperature is varied and (b) cooling pattern is varied.

The tensile properties for the center portion of the strips are plotted against coiling temperature in Figure 8. For coiling temperatures over 540°C, yield point elongations are observed. However, in strips coiled at temperature below 540°C, the yield point elongation is zero and the yield-ratio approaches 60%. The tensile strength increases with decreasing coiling temperature, but the yield strength exhibits a local minimum at 540°C where the yield point elongation has almost disappeared. Figure 9 shows the microstructure of the strips coiled at 500 and 400°C. The microstructure of the strip coiled at 500°C consists mainly of polygonal ferrite and martensite, whereas in the strip coiled at 400°C, a large amount of bainite is observed. Thus, when the coiling temperature is 500°C, the austenite transforms to ferrite during cooling on the runout table and the retained austenite transforms to martensite during further cooling. When the coiling temperature is 400°C, bainite transformation occurs after ferrite transformation.

Figure 10 shows the effect of cooling pattern after rolling on the tensile properties. In this figure, the coil numbers correspond to the cooling patterns shown in Figure 7. For all coils, no yield point elongation is measured and the yield-ratio is less than 65%. However, the yield and tensile strengths and the total elongation change with the cooling pattern. With an increase in delay time before water cooling, the yield and tensile strengths and the yield-ratio decrease and the elongation increases. The yield-ratio decreases to 55%, in the coil subjected to cooling pattern 3 which was water cooled from the lowest temperature on the runout table. Figure 11 shows optical micrographs of coils subjected to the three cooling patterns. The microstructures of coils 1 and 2 are composed of ferrite, bainite and martensite. The ferrite grain size in coil 1 is fine and the bainite in coil 2 is coarse. The ferrite grain size in coil 3 is slightly coarser than that in coil 1 but the volume fraction of ferrite is higher than that in coils 1 and 2. In coil 3, because the amount of ferrite

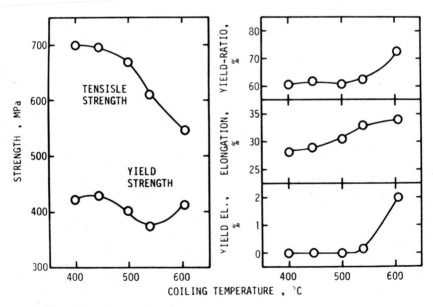

Fig. 8 - Effect of coiling temperature on tensile properties.

transformed before water cooling is larger than that in coils 1 and 2, the
concentration of carbon in the austenite is larger; thus, the austenite
transforms to martensite, not to coarse bainite.

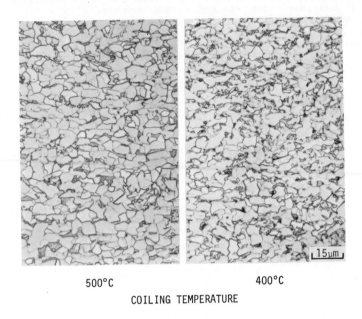

500°C   400°C

COILING TEMPERATURE

Fig. 9 - Optical micrographs of 0.05%C-1.5%Mn-1.0%Cr steel coiled at 500 or
400°C.

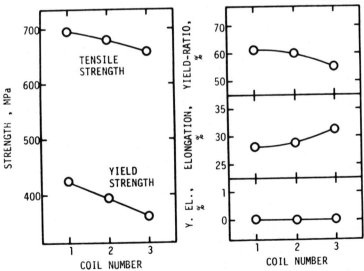

Fig. 10 - Effect of cooling pattern on tensile properties. Coil numbers
correspond to the cooling patterns shown in Figure 7b.

From the above-mentioned experiments, it was confirmed that when a 0.05%C-1.5%Mn-1.0%Si-1.0%Cr steel is used, the dual-phase microstructure and hence desirable tensile properties can be obtained when the coiling temperature is kept below 540°C and the cooling pattern is controlled. To produce a dual-phase steel having a microstructure with a high volume fraction of ferrite and exhibiting the tensile properties of low yield strength, low yield-ratio and high total elongation, water cooling on the runout table should be performed just before coiling and coiling temperature should be in the upper part of the temperatures range where the dual-phase microstructure can be obtained.

The same experiment as described above was performed on two steels with chemical compositions of 0.05%C-1.3%Mn-1.0%Si-1.0%Cr (steel A) and 0.05%C-1.2%Mn-0.5%Si-1.0%Cr (steel B) and the following results were obtained.

(1) In steel A, the highest coiling temperature which exhibits the dual-phase microstructure is about 500°C which is slightly lower than that observed in 0.05%C-1.5%Mn-1.0%Si-1.0%Cr steel examined previously.

(2) In steel B, the highest coiling temperature producing the dual-phase microstructure is below 300°C.

(3) In both steels, by delaying the time prior to water cooling on the runout table for the same coiling temperature, the yield strength and the yield-ratio decrease and the total elongation increases.

(4) In steel A, when water cooling is performed just after hot-rolling, the amount of coarse bainite in the microstructure is less than that observed in the 0.05%C-1.5%Mn-1.0%Si-1.0%Cr steel.

It is suggested from these results that in order to produce a dual-phase steel in the as-hot-rolled condition by controlling the coiling temperature above 400°C, the Si content must be 1% or higher and the Mn content must be decreased, in order to make the microstructure and tensile properties less sensitive to the change in cooling pattern on the runout table.

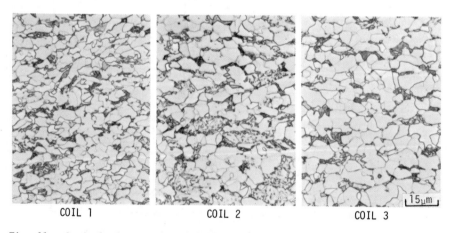

COIL 1   COIL 2   COIL 3

Fig. 11 - Optical micrographs of 0.05%C-1.5%Mn-1.0%Si-1.0%Cr steel subjected to three cooling patterns. The coil numbers correspond to the cooling patterns shown in Figure 7b.

(II)   0.05%C-0.8%Mn-1.5%Si-1.4%Cr Steel.   A heat with the chemical composition shown in Table V was made in an 80 ton basic oxygen furnace and continuously cast to slabs 200 mm thick by 1000 mm wide.   The slabs were hot-rolled to strips on a commercial hot-strip mill according to the conditions shown in Table VI.   Since the Ar_3 transformation temperature for this steel is higher than that for the steels described in the previous section, the aim finishing temperature was 860°C and the aim coiling temperature was in the range of 350 to 500°C.   The strips were water cooled either just after hot-rolling (A) or just before coiling (B).   Cooling patterns A and B are the same as those of 1 and 3 in Figure 7b.

Table V.   Chemical composition (wt%)

C	Si	Mn	P	S	Al	Cr	Ce
0.048	1.50	0.80	0.024	0.002	0.042	1.42	0.004

Table VI.   Hot-rolling practice

Slab reheating temperature	1210°C
Thickness of sheet-bar	33 mm
Finishing delivery temperature	850 - 870°C
Thickness of strip	2.9 mm

Significant changes in the tensile properties were not observed with changes in coiling temperature in the temperature range examined.   Figure 12 shows the changes in tensile properties along the length of the coils, which were cooled by patterns A and B.   In both coils, the tensile properties are uniform along the entire length.   Although not indicated in this figure, the properties were also uniform across the width of the coils.   The ultimate tensile strengths are in the range of 620 to 650 MPa, the yield-ratios are all below 60% and the total elongations are greater that 30%.   Figure 13 shows the microstructures at the middle location of the coils.   The micrographs for samples etched in nital and in picric acid containing $Na_2S_2O_5$ (11) are shown to clarify the distribution of the second phase.   The ferrite grain size is a little finer in the coil cooled on the runout table just after hot-rolling than in the coil cooled just before coiling.   In both coils, the volume fraction of polygonal ferrite is greater than 85% and coarse bainite is not observed.   In coil A, water cooled just after hot-rolling, the second phase regions are more in number but smaller in size than in coil B, water cooled just before coiling.   The volume fraction of second phase is almost the same between the two coils.   Figure 14 presents transmission electron micrographs of the two coils.   In both coils, most of the second phase is martensite and lies on ferrite grain boundaries and a larger dislocation density is observed at the ferrite-martensite interfaces. These microstructural features agree well with those observed in intercritically annealed dual-phase steels (4).

The same experiment was performed on a 0.05%C-0.8%Mn-1.0%Si-1.3%Cr steel, and confirmed that the microstructure and tensile properties are less sensitive to variations in cooling conditions than those in the 0.05%C-1.5%Mn-1.0%Si-1.0%Cr steel.   In the 0.8%Mn steel, however, if the coiling temperature is above 450°C, pearlite is observed and a yield point elongation is measured in the tensile test.

From the above mill trial results, the conditions to produce a dual-phase microstructure by controlling the coiling temperature above 400°C are summarized as follows:

(1)   For the main alloying elements, Si$\geq$1.0% and Mn+Cr$\geq$2.3% are needed.

(2)   In steels having higher Mn and Cr contents, water cooling on the runout table should be performed just before coiling to obtain desirable tensile properties such as low yield strength, low yield-ratios and high total elongation.

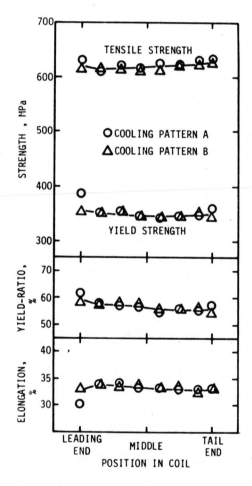

Fig. 12 - Change in tensile properties over the length of the coils. Cooling patterns A and B correspond to 1 and 3 in Figure 7b, respectively.

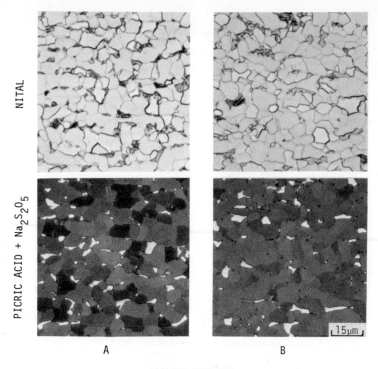

NITAL

PICRIC ACID + Na$_2$S$_2$O$_5$

⌞15μm⌟

A   B

COOLING PATTERN

Fig. 13 - Optical micrographs of 0.05%C-0.8%Mn-1.0%Si-1.4%Cr steel subjected to two cooling patterns.

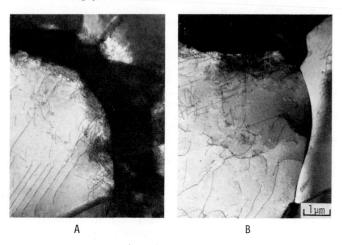

⌞1μm⌟

A   B

COOLING PATTERN

Fig. 14 - Transmission electron micrographs for the 0.05%C-0.8%Mn-1.5%Si-1.4%Cr steel subjected to two cooling patterns.

Discussion

To produce an as-hot-rolled dual-phase steel sheet by controlling the coiling temperature above 400°C, steels containing not only Mn-Cr or Mn-Mo but also Si must be used. If the Mn content in Mn-Si-Cr steel is increased, the mechanical properties become very sensitive to variations in the cooling pattern on the runout table. The present authors previously investigated the effects of Mn, Cr and Mo on the formation of dual-phase microstructures by intercritical annealing. It was demonstrated that the critical cooling rate, CR, to produce a dual-phase microstructure after cooling from the $\alpha+\gamma$ range (770°C) can be expressed in terms of a manganese equivalent, Mn_{eq} (4).

$$\log CR(°C/sec) = -1.73Mn_{eq}(\%) + 3.95 \qquad (1)$$

where $\qquad Mn_{eq}(\%) = Mn(\%) + 1.3Cr(\%) + 2.7Mo(\%) \qquad (2)$

The effects of Mn, Cr and Mo on the stability of residual austenite in ferrite for as-hot-rolled dual-phase steels can be also expressed by Equation 2. However, in producing dual-phase microstructures on the hot-strip mill, precise control of the $\gamma \rightarrow \alpha$ transformation is important because the transformation starts from the single austenite phase. Thus the effects of alloying elements and cooling rate on the $\gamma \rightarrow \alpha$ transformation were examined using dilatometry.

Figure 15 shows the CCT diagrams for the 0.05%C-1.7%Mn-1.0%Cr and 0.05%C-1.7%Mn-1.0%Si-1.0%Cr steels. The addition of 1%Si shifts the $\gamma \rightarrow \alpha$ transformation temperature to higher temperatures and increases the cooling rate necessary to produce the same amount of ferrite. In other words, if the cooling condition is the same, the amount of ferrite in the Si-added steel is larger than that in the Si-free steel. The addition of 1%Si creates a gap between the ferrite and bainite transformations in the CCT diagram which has been called the "coiling window" by Coldren and Tither (5). Si is known to increase the activity of carbon in ferrite and it inhibits carbide formation at the ferrite-martensite interface in a dual-phase steel produced by intercritical annealing (12). Si is used to obtain dual-phase microstructures in the as-hot-rolled condition by increasing the critical coiling temperature which accelerates the $\gamma \rightarrow \alpha$ transformation, resulting in an increase in the carbon content of the untransformed austenite.

Figure 16 shows the CCT diagrams for two steels with 0.05%C-1.0%Si-1.0%Cr and different Mn contents, i.e., 1.5% and 1.7%. With decreasing Mn content, the $\gamma \rightarrow \alpha$ transformation temperature increases and the pearlite transformation is accelerated. Therefore if the Mn content is too low, the pearlite transformation occurs during cooling on the runout table. For continuous cooling of the 1.5%Mn steel, there is an optimum cooling rate to obtain a ferrite-martensite microstructure instead of a ferrite-pearlite or ferrite-bainite microstructure. Therefore, if the steel strip is cooled at this optimum rate down to below a critical temperature, the microstructure is not sensitive to variations in the coiling temperature. The optimum cooling rate for the 1.7%Mn steel is so small that it cannot be obtained on a hot-strip mill. Figure 17 shows the CCT diagram after a deformation of 20% at 850°C for the 0.05%C-1.5%Mn-1.0%Si-1.0%Cr steel compared with the CCT diagram without deformation. The deformation makes the ferrite transformation easier and shifts the bainite transformation region to the left side. The cooling curves for the coils processed in the hot-strip mill experiments discussed earlier are also shown by the dotted lines in Figure 17. The cooling curve for the strip water cooled just before coiling runs through the ferrite transformation region longer than that for the strip

214

water cooled just after hot-rolling. In rolling on the hot-strip mill, the bainite region is expected to shift to shorter times because of the large deformation. Therefore, a dual-phase microstructure with a small amount of bainite is obtained in this steel when water cooled just before coiling.

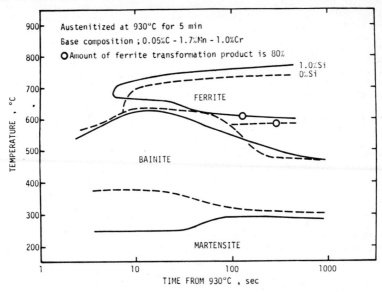

Fig. 15 - Effect of Si content on the continuous-cooling transformation diagram.

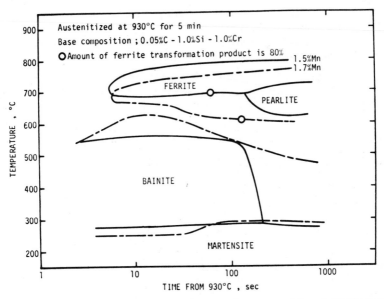

Fig. 16 - Effect of Mn content on the continuous-cooling transformation diagram.

215

Figure 18 shows the CCT diagrams, with of without a 50% deformation at 900°C, for the 0.05%C-0.8%Mn-1.5%Si-1.4%Cr steel which does not show a change in mechanical properties with variations in the cooling pattern on the hot-strip mill. Ferrite transformation in this steel occurs in a short time. In particular, in the deformed state, 80% ferrite forms for a cooling rate of 70°C/sec. In this steel, similar to the results shown in Figure 17, the bainite region also shifts to the left with deformation. Without deformation, there is no cooling rate which will produce a ferrite-martensite microstructure, but for the deformed state, there exists a wide range of the cooling rates. Therefore, for this steel, even if the strip is water cooled just after hot-rolling, a large amount of ferrite is obtained and the dual-phase microstructure can be obtained readily by controlling only the coiling temperature below a critical value.

The CCT diagrams reveal the effects of the alloying elements, the cooling rate and deformation on the transformation behavior obtained during cooling between hot-rolling and coiling. The transformation after coiling, however, can not be explained by the CCT diagrams, and the stability of austenite during the slow cooling stage after coiling should be analyzed. Figure 19 shows the effects of alloying elements on the TTT diagram of a 1021 steel (13). Mn, Cr and Mo markedly delay the decomposition of austenite in the temperature range of 500 to 600°C. It is suspected that Mn, Cr and Mo, and also the enrichment of carbon content in austenite make the austenite phase stable before the initiation of the martensite transformation. From the CCT diagrams the M_s temperature is estimated to be about 300°C. A line scan analysis was carried out using an electron probe microanalyzer over both the ferrite and martensite phases in the as-hot-rolled dual-phase steels. This analysis indicated no partition of Mn, Si or Cr content between the two phases. Assuming that substitutional solute concentration remains the same between the ferrite and martensite

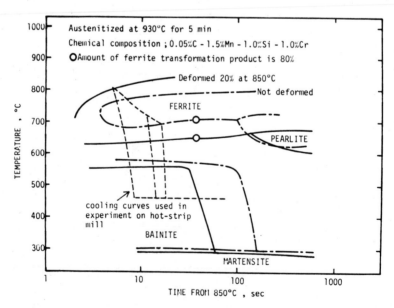

Fig. 17 – Effect of deformation on continuous-cooling transformation.

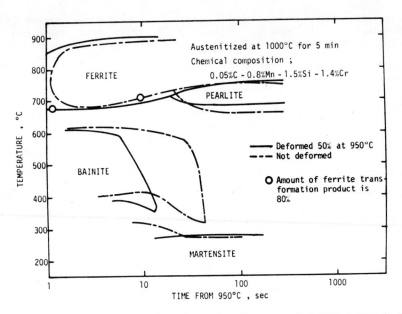

Fig. 18 - Continuous-cooling-transformation diagrams of 0.05%C-0.8%Mn-1.5%Si
-1.4%Cr steel.

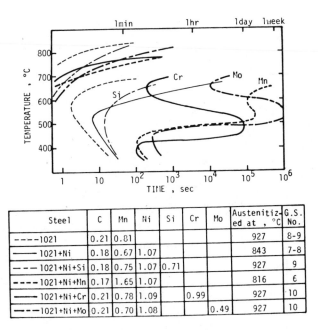

Steel	C	Mn	Ni	Si	Cr	Mo	Austenitized at , °C	G.S. No.
----1021	0.21	0.81					927	8-9
——1021+Ni	0.18	0.67	1.07				843	7-8
---1021+Ni+Si	0.18	0.75	1.07	0.71			927	9
-·-·-1021+Ni+Mn	0.17	1.65	1.07				816	6
——1021+Ni+Cr	0.21	0.78	1.09		0.99		927	10
-··-1021+Ni+Mo	0.21	0.70	1.08			0.49	927	10

Fig. 19 - Effect of Mn, Si, Cr or Mo on TTT-diagrams (13).

217

phases and using M_s temperature formulae (14), the carbon content in the martensite is calculated to be about 0.4%. This means that a marked concentration of carbon in untransformed austenite occurs during cooling.

By using Mn–Si–Cr steels and optimizing the above-mentioned rolling practices, as-hot-rolled dual-phase steel sheets having tensile strengths between 550 and 700 MPa have been produced in commercial hot-strip mills. Figure 20 shows the relationship between the total elongation and the ultimate tensile strength for the experimental coils produced. Among the alloying elements, Si has the largest effect on ductility. Between the 0.5%Si steels and the 1.0%Si steels, there exists a difference in elongation of 5% for the same tensile strength. For the 0.5%Si steels, the results of strips having a ferrite-pearlite structure are also shown in this figure. If the Si content is the same, the dual-phase steel has better ductility than the ferrite-pearlite steel. Various formability experiments were undertaken using the as-hot-rolled dual-phase steels and they were shown to have equal or better formability than the dual-phase steels produced by intercritical annealing and better formability than HSLA steels containing alloying elements, such as Nb, V and Ti.

Conclusions

To develop as-hot-rolled dual-phase steel sheet, the effects of alloying elements and hot-rolling conditions were investigated. The results are summarized as follows:

(1) Dual-phase steel sheet can be produced in the as-hot-rolled condition, when Mn–Si–Cr and Mn–Si–Mo steels are water cooled after rolling and coiled in a controlled temperature range.

(2) The upper critical coiling temperature decreases with decreasing content of Mn, Si, Cr or Mo.

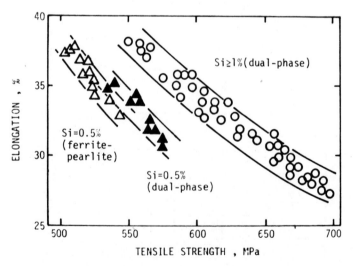

Fig. 20 – Relationship between tensile strength and elongation for the high strength steels processed on hot-strip mill.

(3) In a 0.05%C-1.5%Mn-1.0%Si-1.0%Cr steel, the upper critical coiling temperature for producing a dual-phase microstructure is 540°C. For hot-strip mill experiments using this steel, the yield strength and the yield-ratio increase and the total elongation decreases with decreasing coiling temperature. In this steel, if the coiling temperature is constant, the tensile properties vary with the change in cooling pattern after hot-rolling. By delaying water cooling on the runout table, the yield strength and the yield-ratio decrease and the total elongation increases. This is due to the increase in ferrite and the decrease in bainite volume fraction.

(4) In a 0.05%C-0.8%Mn-1.5%Si-1.4%Cr steel, the mechanical properties are insensitive to cooling pattern changes after hot-rolling and a dual-phase steel having low yield strength, low yield-ratios and high total elongations can be produced. This can be explained in terms of the rapid ferrite transformation after hot-rolling in this steel.

(5) In Mn-Si-Cr steels, Si accelerates the ferrite transformation and stabilizes the austenite, through enriching the carbon content. Mn and Cr stabilize the austenite, but when the Mn content is too large, the ferrite transformation is suppressed.

(6) Hot-rolling accelerates ferrite transformation and suppresses bainite transformation.

(7) Si improves the ductility of dual-phase steels.

References

1) M. S. Rashid, "GM980X - A Unique High Strength Sheet Steel With Superior Formability," SAE paper 760206 presented at Automotive Engineering Congress and Exposition, Detroit, Michigan, 1976.

2) M. Nishida, K. Hashiguchi, I. Takahashi, T. Kato and T. Tanaka, "Age-Hardenable, Low Yield and High Tensile Strength Sheet Steels for Automotives," pp.211-213 in Sheet Metal Forming and Formability, IDDRG 10th Biennial Congress, Univ. of Warwick, England, 1978.

3) N. Ohashi, I. Takahashi and K. Hashiguchi, "Processing Techniques and Formability of Age-Hardenable, Low Yield, High Tensile Strength Cold-Rolled Sheet Steels," Transactions of Iron and Steel Institute of Japan, 18 (1978) pp.321-329.

4) T. Tanaka, M. Nishida, K. Hashiguchi and T. Kato, "Formatin and Proper-ties of Ferrite Plus Martensite Dual-Phase Structures," pp.221-241 in Structure and Properties of Dual-Phase Steels, R. A. Kot and J. W. Morris, ed; AIME, New York, N.Y., 1979.

5) A. P. Coldren and G. Tither, "Development of a Mn-Si-Cr-Mo As-Rolled Dual-Phase Steel," Journal of Metals, 30(4) (1978) pp.6-9.

6) K. Hashiguchi, M. Nishida, T. Kato and T. Tanaka, "Development of As-Hot-Rolled Dual-Phase Sheet Steel," Tetsu-to-Hagane 64 (1978) pp.S257.

7) K. Kunishige, M. Takahashi, S. Sugisawa and Y. Masui, "As-rolled Dual-Phase Steel with Good Ductility and High Strength," Tetsu-to-Hagane 65 (1979) pp.1916-1925.

8) T. Furukawa, H. Morikawa, H. Takechi and K. Koyama, "Process Factors for Highly Ductile Dual-Phase Sheet Steels," pp.281-303 in Structure and Properties of Dual-Phase Steels, R. A. Kot and J. W. Morris, ed; AIME, New York, N.Y., 1979.

9) T. Greday, H. Mathy and P. Messien, "About Different Ways to Obtain Multiphase Steels," pp.261-280 in Structure and Properties of Dual-Phase Steels, R. A. Kot and J. W. Morris, ed; AIME, New York, N.Y., 1979.

10) J. M. Rigsbee, J. K. Abraham, A. T. Davenport, J. E. Franklin and J. W. Pickens, "Structure-Processing and Structure-Property Relationships in Commercially Processed Dual-Phase Steels," pp.304-329 in Structure and Properties of Dual-Phase Steels, R. A. Kot and J. W. Morris, ed; AIME, New York, N.Y., 1979.

11) F. S. LePera, "Improved Etching Technique to Emphasize Martensite and Bainite in High-Strength Dual-Phase Steel," Journal of Metals, 32(3) (1980) pp.38-39.

12) G. Thomas and J. Y. Koo, "Development in Strong, Ductile Duplex Ferritic-Martenstic Steels," pp.183-201 in Structure and Properties of Dual-Phase Steels, R. A. Kot and J. W. Morris, ed; AIME, New York, N.Y., 1979.

13) Atlas of Isothermal Transformation Diagrams, U. S. Steel, Pittsburgh, Pa., Third Edition, 1963.

14) G. Krauss, "Martensitic Transformation, Structure and Properties in Hardenable Steels," pp.229-245 in Hardenability Concepts with Applications to Steel, D. V. Doane and K. S. Kirkaldy, ed; AIME, Warrendale, Pa., 1978.

STRUCTURE FORMATION AND MECHANICAL PROPERTIES

OF INTERCRITICALLY ANNEALED OR AS-HOT-ROLLED DUAL-PHASE STEELS

T. Furukawa and M. Tanino

Fundamental Research Laboratories, Nippon Steel Corporation
Nakahara-ku, Kawasaki, Kanagawa 211, Japan

For a steel of rather lean chemical composition, martensite formation with a concomitant ductile ferrite matrix can be effectively attained by using a two-stage cooling procedure after intercritical annealing. The two-stage cooling procedure consists of a mild cooling in the higher temperature region and a rather fast or rapid cooling in the lower temperature region. The values for the cooling rates and the temperatures at which the change in cooling rate occurs may vary depending on the chemical composition. The austenite hardenability of a steel heated for a short time in the intercritical temperature range can be modified by using a high coiling temperature in the hot rolling schedule prior to the intercritical heat treatment. The martensite transformed from the austenite with such a modified hardenability shows a higher resistance to auto-tempering when formed using the two-stage cooling procedure.

To obtain an as-hot-rolled dual-phase steel, the finishing temperature should be selected so as to promote the gamma-to-alpha phase transformation during or immediately after finishing and to avoid heavy deformation of the transformed alpha. This requirement determines an optimum finishing temperature range for the process which has been called the DPR (Dual-Phase Rolling) process. A silicon addition is quite beneficial for the DPR process since it greatly enlarges the optimum finishing temperature range and enhances the hardenability of the partitioned austenite.

Introduction

It is well-known that a dual-phase steel has outstanding features such as a lower yield-to-tensile strength ratio and a higher work-hardening rate, compared with conventional high-strength steels. These features have been believed to stem from the combined effects of a "clean" ferrite matrix with respect to solute interstitials, the presence of martensite particles and possibly the presence of a retained austenite phase. The formation of these microstructures has been fairly well analyzed (1-7) in connection with steel composition and cooling rate after an intercritical heat treatment. On the basis of these studies and analyses, it has been widely recognized that the simultaneous attainment of both a uniformly dispersed martensite phase and a "clean" ferrite matrix becomes more difficult as the alloy content becomes leaner, due to the decreased austenite hardenability. Considering the importance of the cooling rate (or practice) after an intercritical heat treatment on microstructure formation, a "two-stage" cooling concept has been introduced to overcome the above difficulty.

In an as-hot-rolled dual-phase steel produced by a rolling practice which involves a low finishing temperature and a very low coiling temperature (DPR process) (6, 8), the formation of the microstructure is not precisely understood as yet. Experiments have shown that the alpha-gamma phase separation without heavy deformation of the alpha phase during finishing is very important. A silicon addition helps to widen the optimum finishing temperature range in which this requirement is fulfilled.

Part 1. Structure and Properties of Intercritically Annealed

Dual-Phase Steels

Continuous Cooling

As is already known, mechanical properties of an intercritically annealed dual-phase steels are affected by the cooling rate after annealing. Figure 1 shows as an example the relationship between mechanical properties and cooling rate for two steels. A tentative strength-ductility parameter, ultimate tensile strength x total elongation (TS x T-El, MPa %), sharply decreases at a very high cooling rate, i.e. water quenching. On the other hand, the yield-to-tensile strength ratio (YS/TS) cannot be maintained at a low level for a low rate of cooling, 10°C/second, unless the chemistry provides sufficient austenite hardenability (compare the 1.4% Mn steel with the 2.0% Mn steel).

Although a ferrite-martensite dual-phase microstructure can easily be formed by rapid quenching, a large amount of solute carbon is retained in the ferrite, resulting in a loss of ductility (1). Ductility is improved by using a mild cooling rate which, however, makes martensite formation more difficult as the steel chemistry becomes leaner thus causing a decrease in the ultimate tensile strength and an increase in the yield-to-tensile ratio. Consequently, there is a compromise, or trade-off, between chemistry and cooling rate. It is important to understand the metallurgical phenomena going on during the cooling process; a more rational, or improved "trade-off" to obtain better dual-phase steels may be possible through this understanding.

Two-Stage Cooling

The experimental thermal cycle, indicated in Figure 2, was applied to 0.12% C - 2% Mn and 0.08% C - 1.4% Mn steel sheets cold rolled to a 70%

222

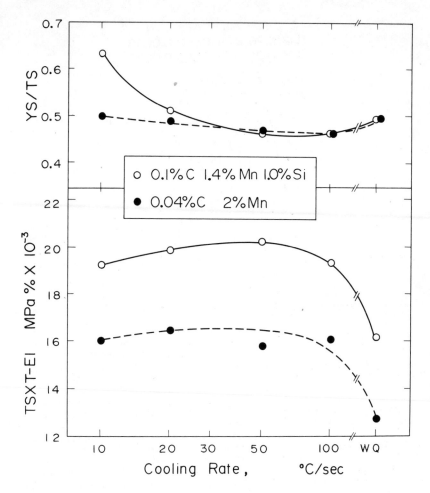

Fig. 1 - Effect of cooling rate on yield-to-tensile ratio and ductility
parameter, after heating at 780°C for 1 minute.

reduction in order to observe microstructural variations and property
changes. The thermal cycle involves a short time intercritical anneal
followed by air cooling at a rate of about 10°C/second to various selected
temperatures and then water quenching. The resulting microstructures and
tensile properties were examined.

Figure 3 shows the results for the 2% Mn steel, which have already been
reported in a previous paper (6), indicating that: 1) the gamma-to-alpha
transformation proceeds down to about 400°C on air cooling, as evidenced by
a decreasing second phase volume fraction, 2) the amount of retained austenite tends to increase on air cooling down to a temperature lower than about
400°C, 3) the structure consists mainly of ferrite, martensite (α') and
retained austenite (γ_R) even after air cooling down to ambient temperature,

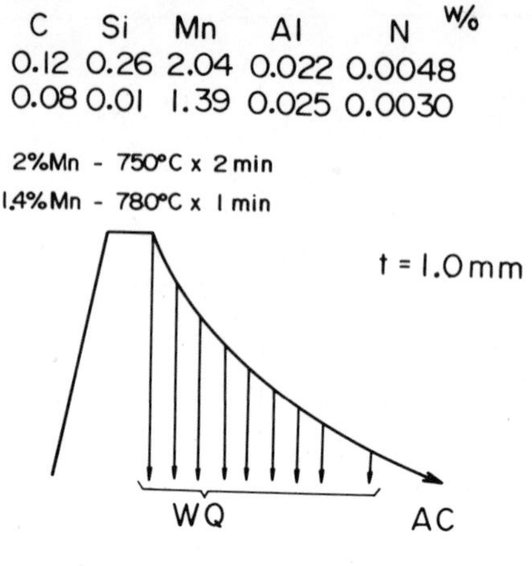

C   Si   Mn   Al   N   W/o
0.12 0.26 2.04 0.022 0.0048
0.08 0.01 1.39 0.025 0.0030

2%Mn - 750°C x 2 min
1.4%Mn - 780°C x 1 min

t = 1.0 mm

WQ   AC

Fig. 2 - Materials and thermal cycle employed to examine microstructures and mechanical properties.

and 4) the strength-ductility parameter can be optimized by the air cooling to a temperature of 400°C or less.

Figure 4 indicates the results for the 1.4% Mn steel. Judging from the second phase volume fraction or the ultimate tensile strength as a function of quenching temperature, the gamma-to-alpha transformation appears also to proceed on air cooling down to 500 to 400°C. The amount of retained austenite peaks at a quenching temperature of 600°C, indicating that some carbon enrichment into the untransformed austenite has taken place during air cooling down to 600°C. However, the untransformed austenite enriched with carbon is not as stable as the austenite in the 2% Mn steel, since the amount of retained austenite falls to zero percent on air cooling below 500°C. In agreement with this, the resultant microstructure tends to consist of ferrite-pearlite instead of ferrite-martensite when air cooling is carried out to a temperature of 500°C or lower. The resultant microstructural changes that occur with changes in quench temperature are shown in Figure 5.

It can be seen in Figure 4 that a minimum yield strength and a maximum strength-ductility parameter are obtained by using a quenching temperature of 600°C. This means that the mechanical properties of the 1.4% Mn steel can be optimized, as a dual-phase steel, through a sequence of a mild cooling stage followed by a rapid cooling stage. Each stage has its own metallurgical effect. The mild cooling stage in the higher temperature region seems to promote carbon enrichment into the untransformed austenite along with the gamma-to-alpha diffusional transformation process. The ferrite solute carbon content will be decreased also in this stage. The rapid cooling stage in the lower temperature region should help to avoid decomposition of the untransformed austenite to pearlite, so as to produce martensite and possibly some retained austenite.

The proper cooling rate for each stage could vary depending on the austenite hardenability which is controlled by the steel composition. In case of the 2% Mn steel described in Figure 3, for example, air cooling

224

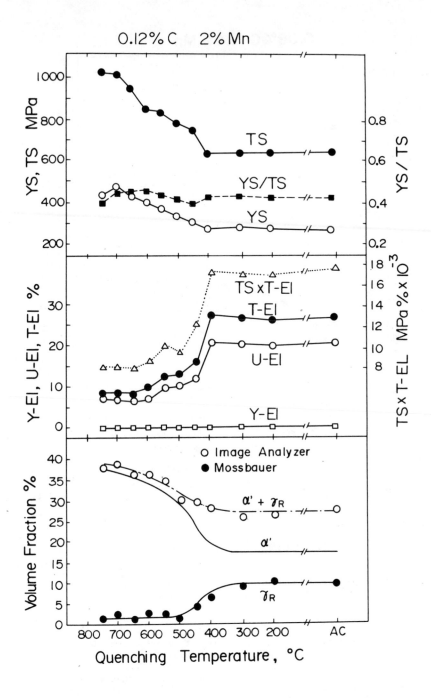

Fig. 3 – Mechanical properties and the second phase volume fraction as a function of quenching temperature.

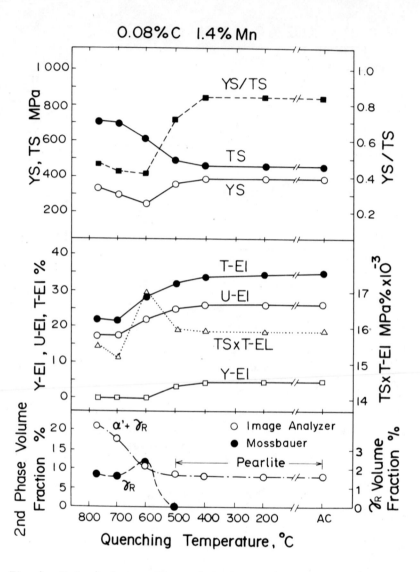

Fig. 4 – Mechanical properties and the second phase volume fraction as a function of quenching temperature.

in the lower temperature region is sufficient to attain the martensitic transformation. Comparing Figure 4 with Figure 3, especially with reference to the changes in yield-to-tensile ratio and the strength-ductility parameter with quenching temperature, it can be concluded that a higher cooling rate for the lower temperature region and a higher transition temperature for the start of the second stage are required as the steel composition becomes leaner.

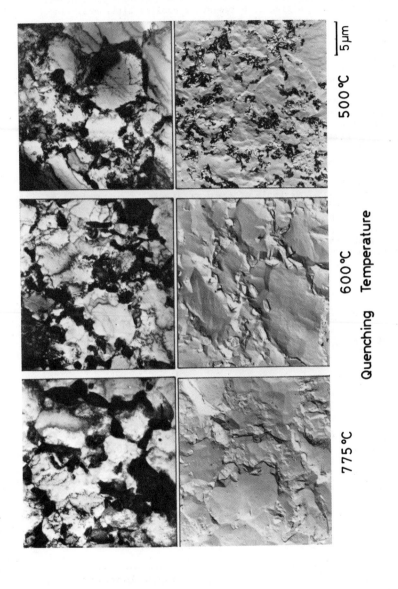

775°C   600°C   500°C   5μm

Quenching   Temperature

Fig. 5 – Transmission (upper) and extraction-replica (lower) electron micrographs of the steel in Figure 4, showing that quenching from 600°C or higher produces a ferrite–martensite microstructure whereas quenching from 500°C or lower results in a ferrite–pearlite microstructure.

As previously reported (6), austenite hardenability is also controlled
to some extent for a given steel composition, by using a high coiling
temperature in the hot rolling process prior to the intercritical annealing
heat treatment. A good method for producing a dual-phase steel with a
lean composition can be obtained by combining the austenite hardenability
modification during hot rolling with two-stage cooling after annealing.
This technique will be described in the next section.

Austenite Hardenability Modification plus Two-Stage Cooling

Figures 6-8,which have already been described in the previous paper
(6), demonstrate that the austenite hardenability of a steel having a given
chemical composition can be modified by high temperature coiling which gives
rise to an obvious enrichment of carbon and manganese in the second phase in
the coiled sheet (Figure 7), resulting in a marked decrease in the eventual
yield strength after an intercritical heat treatment (Figure 8), due to the
formation of a ferrite-martensite dual-phase microstructure.

Experiments were conducted, using the same steel composition shown in
Figure 6 to find out whether any additional effects are present when the
method for austenite hardenability modification is combined with the two-
stage cooling process. An example of the experimental results obtained when
using this technique is shown in Figure 9. It can be seen that the improve-
ments in mechanical properties when using a high coiling temperature (750°C)
compared with the normal coiling temperature (600°C) for a continuous cooling practice (data in leftmost column) are more pronounced if a two-stage
cooling practice (second leftmost column) is utilized. Also it should be

(wt%)

C	Si	Mn	P	S	Al	O	N
0.049	0.02	1.49	0.001	0.006	0.028	0.0030	0.0019

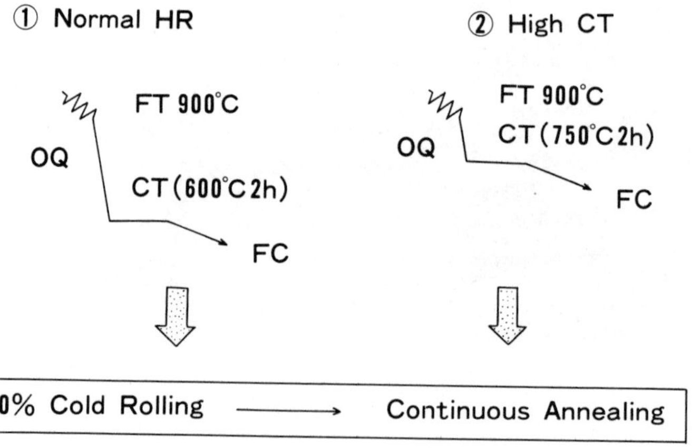

① Normal HR

② High CT

Fig. 6 - Material and experimental procedure for a low-tensile strength
dual-phase steel.

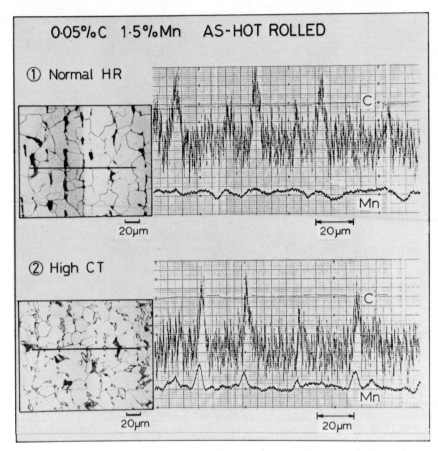

Fig. 7 - Enrichment with carbon and manganese in the second phase due to coiling at a high temperature.

noted that holding* at a low temperature, 300°C for 5 minutes, causes no serious increase in the yield-to-tensile ratio when the high coiling temperature steel undergoes the two-stage cooling practice (rightmost column).

It is hypothesized from the changes in the ultimate tensile strength and the yield-to-tensile ratio described in Figure 9, that both the high temperature coiling treatment and the two-stage cooling process help to enhance the martensite content of the total second phase. The total second phase volume fraction changes due to the coiling and/or the cooling pattern practices were not clearly discernible by an image analyzer probably owing to the lean chemical composition and to the mild cooling stage in the higher

---

* This kind of holding temperature profile is sometimes unavoidable when continuous annealing production equipment used for regular cold-rolled steels is employed to produce the dual-phase steels.

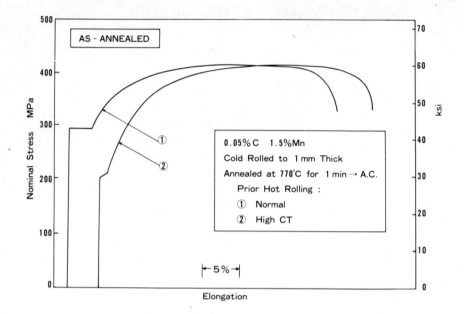

Fig. 8 - Stress-elongation curves showing the effect of hot-coiling
practice on the final tensile properties.

temperature region. If it is assumed that the holding temperature (300°C)
is well below the M_S temperature of the austenite*, the martensite transformed from the austenite with increased hardenability seems to have an
improved resistance to auto-tempering. This is based on the fact that the
increase in the yield-to-tensile ratio due to holding at 300°C is small
for the high coiling temperature plus two-stage cooling process.

---

* The M_S temperature is calculated (9) to be about 367°C assuming that
the partitioned austenite contains 0.3% C and 1.5% Mn (10). If
higher values are assumed, 0.4% C (due to the slower cooling stage in
the two-stage process) and 1.6% Mn (due to the high coiling temperature), the calculated M_S temperature is about 321°C.

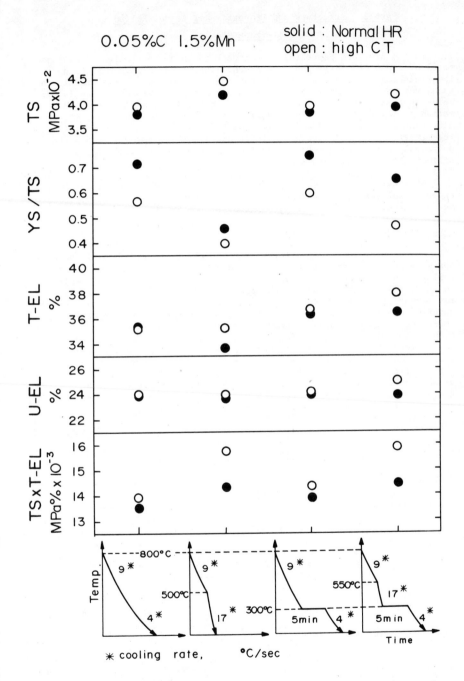

Fig. 9 – Mechanical property changes for various coiling practices and cooling patterns.

231

Part 2. Structure and Properties of As–Hot–Rolled
Dual–Phase Steels

In the previous paper (6) an approach to an as–hot–rolled, dual–phase steel of plain chemistry was introduced. The method consists of using a low finishing temperature (at or about the Ar_3 temperature measured in the usual way by dilatometry) and a very low coiling temperature which is well below the estimated M_S temperature of the partitioned austenite. This method is called the Dual–Phase Rolling (DPR) process. This is quite different from what Coldren et al. proposed (11), in which a normal finishing temperature in the austenite range and a coiling temperature in the polygonal ferrite plus retained austenite range are employed, for a silicon–chromium–molybdenum steel.

This part of the present paper describes the DPR process laboratory experimental results, with emphasis on the effect of finishing temperature on the structure–property relationships.

Materials and Experimental Procedure

The steels used in this investigation are listed in Table I. Each of the steels was vacuum–melted and cast in a 25 kilogram ingot, then hot rolled to a 25 mm thick slab and air cooled. The DPR process was applied to these slabs. The standard DPR process consisted of heating at 1100°C for 1 hour, followed by three successive rolling passes, the final pass of which was performed at various finishing temperatures (FT), and then oil quenched. Some specimens underwent interrupted oil quenching and coiling simulation which consisted of holding in a furnace at a simulated coiling temperature (CT) for 1 hour and furnace cooling. Also the effects of cooling rate after rolling on the mechanical properties were investigated, for practices involving selected finishing temperatures followed by coiling at ambient temperature. The various experimental DPR processes are illustrated in Figure 10. After the DPR process treatments, microstructures and mechanical properties were examined.

Table I. Steels used in the DPR experiments

series	nominal	C	Si	Mn	P	S	Al
	1 Mn	0.060	<0.01	0.98	0.003	0.004	0.031
C–Mn	1.2 Mn	0.062	<0.01	1.19	0.003	0.004	0.027
	1.4 Mn	0.061	<0.01	1.38	0.004	0.005	0.030
	1.7 Mn	0.063	<0.01	1.71	0.003	0.005	0.028
	0 Si	0.061	<0.01	1.38	0.004	0.005	0.030
C–Si–Mn	0.5 Si	0.062	0.49	1.39	0.002	0.004	0.029
	0.7 Si	0.069	0.71	1.42	0.003	0.004	0.029

Chemical compositions in wt. percent

232

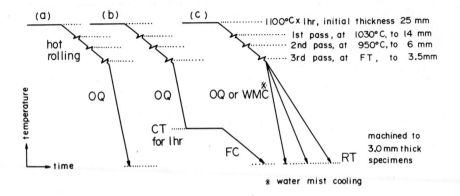

Fig. 10 - Schematic representation of DPR process experiments, (a) standard, (b) with coiling simulation, and (c) with various cooling.

Effects of Finishing Temperature

The effects of finish temperature (FT) on the mechanical properties are shown in Figures 11 and 12, for the carbon-manganese (C-Mn) and carbon-silicon-manganese (C-Si-Mn) steels, respectively. These results were obtained for the treatments described in Figure 10a. For the C-Mn steels, an optimum finishing temperature range for each manganese level is found which produces a minimum in the yield-to-tensile ratio. For example, a range of FT from 750°C to 810°C will be optimum for the 1.4% Mn steel. A manganese content of 1.4% or more is necessary for the 0.06% carbon content, to attain a yield-to-tensile ratio of 0.6 or less. The ultimate tensile strength tends to decrease with increasing FT. The total and uniform elongation appear to be either insensitive to FT or exhibit an optimum value at the FT that produces a minimum in the yield-to-tensile ratio as is the case for the 1.0% or 1.2% Mn steels. For the C-Si-Mn steels, as shown in Figure 12, the general trend in mechanical properties as a function of FT is roughly similar to that for the C-Mn steels. However, two notable differences are observed. The optimum FT range is much wider compared to the silicon free series and the ultimate tensile strength is insensitive to FT. Also, the broad peaks for the ductility parameters, in the optimum FT range, are associated with minimum values in the yield-to-tensile ratio.

It is interesting to note that the optimum FT seems to be related to the Ar$_3$ temperature measured by dilatometric methods for unstrained material. Figure 13 shows equi-YS/TS curves obtained for the C-Mn steels, for the range of manganese contents and FT studied. The FT for the minimum YS/TS follows Ar$_3$ (in the unstrained condition) which decreases as the manganese content is increased. A similar description for the C-Si-Mn steels, with varying silicon content, is presented in Figure 14. Again, the FT corresponding to the minimum YS/TS coincides with Ar$_3$ which increases with increasing silicon content. Note also that the range of FT for a low YS/TS increases as the silicon content is increased up to 0.7%.

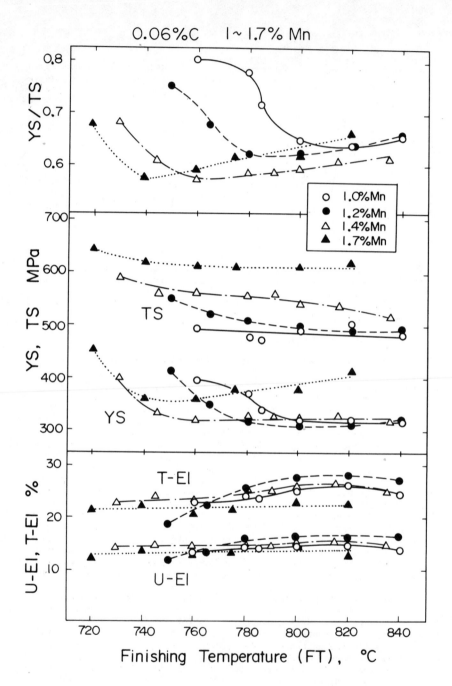

Fig. 11 – Mechanical properties of C-Mn DPR processed steels as a function of finishing temperature.

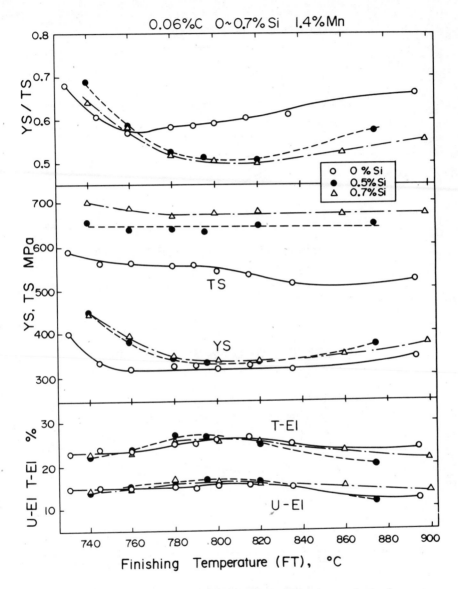

Fig. 12 – Mechanical properties of C-Si-Mn DPR processed steels as a function of finishing temperature.

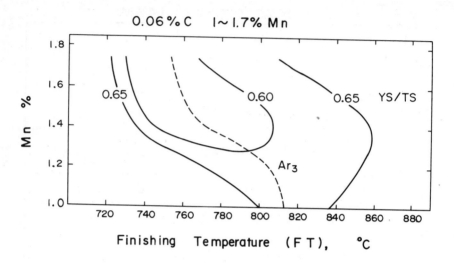

0.06 % C   1~1.7% Mn

Mn %

1.8
1.6 — 0.65   0.60   0.65   YS/TS
1.4
1.2   Ar₃
1.0
   720   740   760   780   800   820   840   860   880

Finishing   Temperature   (F T),   °C

Fig. 13 – Equi–YS/TS curves for various manganese contents and finishing temperatures.

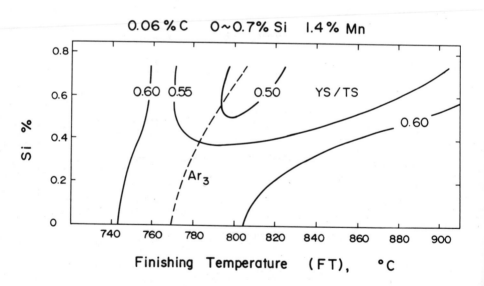

0.06 % C   0~0.7% Si   1.4 % Mn

Si %

0.8
0.6 —   0.60 0.55   0.50   YS / TS
0.4   0.60
0.2   /Ar₃
0
   740   760   780   800   820   840   860   880   900

Finishing   Temperature   (F T),   °C

Fig. 14 – Equi–YS/TS curves for various silicon contents and finishing temperatures.

236

Examples of microstructural changes with FT are given in Figure 15, for a C-Mn steel and a C-Si-Mn steel. If the FT becomes too high, the resultant microstructure tends to involve coarse-grained quenched-in phases, like acicular ferrite. This tendency is more notable for the C-Mn steel than in the C-Si-Mn Steel. On the other hand, too low a FT gives rise to a microstructure in both the C-Mn and C-Si-Mn steels, consisting mainly of deformed and rather coarse-grained ferrite, with a small volume fraction of martensite. A dual-phase microstructure consisting mainly of polygonal ferrite and well-dispersed martensite is developed by using the DPR process with a proper finishing temperature.

Some transmission electron micrographs of the second phases obtained for the various finishing temperatures are presented in Figure 16. In the C-Mn steel, the second phase varies considerably depending on the FT, from a bainitic structure (840°C) to a lath (760°C) or twinned (730°C) marten site structure. The C-Si-Mn steel contains lath martensite, regardless of FT. A fine grained, recovered substructure, as shown in Figure 17 is found in the ferrite matrix when the steels are finished at a low temper- ature. These structure changes presumably are reflected in the mechanical property changes with FT. The wide range of FT associated with a low YS/TS in the C-Si-Mn steel is consistent with the greater tendency to form martensite at all FT.

Diffraction and dark field electron microscopy revealed the presence of retained austenite films between the martensite laths, as shown in Figure 18, which is quite similar to that found by Koo and Thomas (12) in a quenched, low carbon steel. Unlike heat-treated dual-phase steels (6, 13), isolated retained austenite particles were rarely observed.

Effects of Simulated Coiling Temperature

It has already been pointed out in the previous paper (6) that the coiling temperature is of vital importance in DPR process, because both the aging of the ferrite matrix and the tempering of the martensite should be suppressed in order to obtain an as-hot-rolled dual-phase steel with a low YS/TS.

Figure 19 shows the mechanical properties of the C-Mn and C-Si-Mn steels as a function of simulated coiling temperature (CT). Experiments were carried out in accordance with Figure 10b. Finishing temperatures of 760°C for the C-Mn steel and 800°C for the C-Si-Mn steel were selected. It can be seen that the coiling temperature to obtain a low YS/TS of 0.6 or lower is about 200°C for the C-Mn steel, and about 250°C for the C-Si-Mn steel. The ultimate tensile strength (TS) gradually decreases with in- creasing CT and the yield strength (YS) shows a rather sharp increase as the CT increases above 200°C. An increase in the YS/TS for CT of more than 200°C is due to the combined effects of changes in YS and TS with CT. The yield-point elongation (Y-El) starts to appear at a CT of 200°C, and increases with increasing CT. The gradual decrease in TS with increasing CT may be explained in terms of the auto-tempering of the martensite leading to a loss of martensite strength. The increase in YS occurring at a CT of more than 200°C would be due to the appearance of carbide pre- cipitates in the ferrite matrix, and also due to some loss of martensitic transformation strain (inasmuch as the martensitic transformation strain is considered to help to lower the yield strength of ferrite matrix) as a result of the auto-tempering of martensite (6).

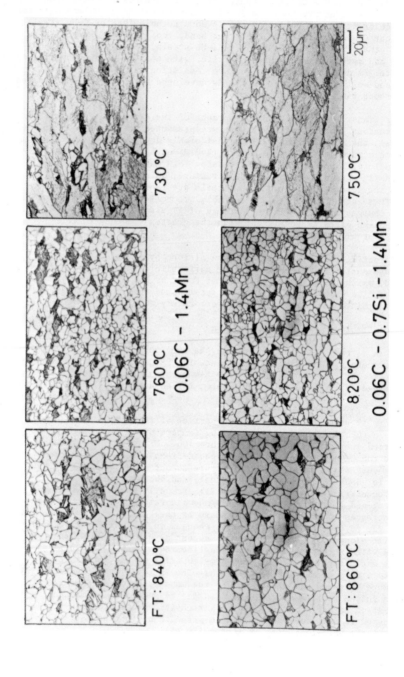

Fig. 15 – Optical microstructural variation with changes in finishing temperature.

730°C

750°C

760°C

820°C

FT: 840°C

FT: 860°C

0.06C – 1.4Mn

0.06C – 0.7Si – 1.4Mn

20μm

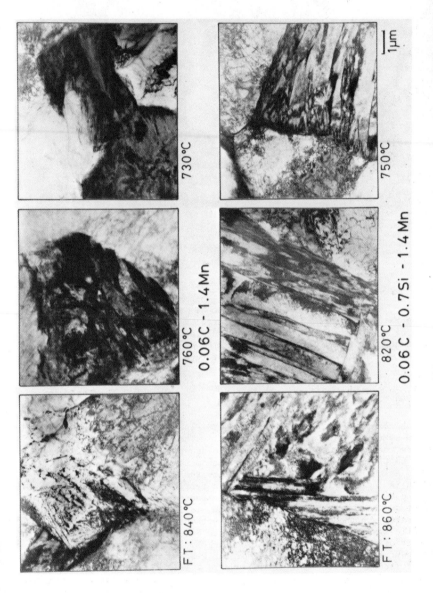

730°C

750°C

760°C
0.06C - 1.4Mn

820°C
0.06 C - 0.7Si - 1.4 Mn

FT: 840°C

FT: 860°C

1μm

Fig. 16 – Transmission electron micrographs of the second phases in the samples corresponding to Figure 15.

239

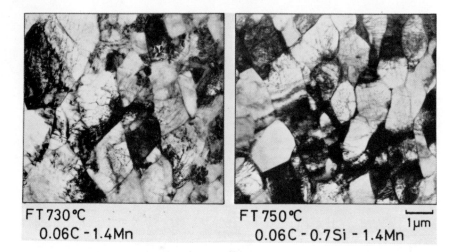

FT 730°C
0.06C - 1.4Mn

FT 750°C
0.06C - 0.7Si - 1.4Mn

1μm

Fig. 17 - Ferrite-substructure found in the steels finished at low
temperatures.

Effects of Cooling Rate after Finish Rolling

So far, the cooling from the FT to the CT was done by oil quenching at
a cooling rate of about 75°C/second (measured in the range from 700 to 300°C).
A series of experiments with various cooling rates after rolling was
conducted according to Figure 10c, using a FT of 760°C for the C-Mn and
800°C for C-Si-Mn steels, and a room temperature CT.

The results are presented in Figure 20. It is clear that the C-Si-Mn
steel allows a milder cooling rate than does the C-Mn steel for obtaining
a low YS/TS.

It is common to both steels that the TS rises and the YS goes through
a minimum as the cooling rate is increased. The Y-El decreases and dis-
appears as the cooling rate is increased to about 45°C/second (for C-Si-Mn)
or about 60°C/second (for C-Mn). These phenomena may be explained as
follows. As the cooling rate increases, the tendency for martensite
formation is increased. On the contrary, a ferrite-pearlite structure is
more prone to be formed as the cooling rate decreases. Since the TS is
dependent on the martensite volume fraction, it increases with increasing
cooling rate. On the other hand, the YS is thought to be reduced by the
martensitic transformation strain and raised by solute carbon in the ferrite.
The effect of the former factor will be largely lost at a low cooling rate
(being lower than about 45°C/second for the C-Si-Mn steel or lower than
about 60°C/second for the C-Mn steel) due to a lack of martensitic transformation, resulting in an increased YS. The effect of the latter factor
may be present at a higher cooling rate; for example, the YS of the C-Mn
steel at 75°C/second is larger than that at 60°C/second, presumably because
of an increased solute carbon content in the ferrite in accordance with
the increased cooling rate. A similar tendency is also observed, though
very slightly, in the C-Si-Mn steel.

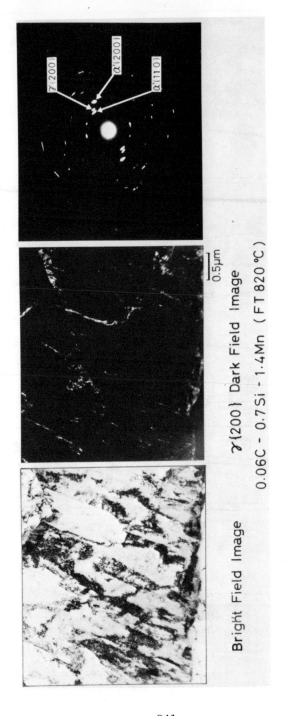

Bright Field Image $\gamma\{200\}$ Dark Field Image

0.06C - 0.7Si - 1.4Mn (FT 820 °C)

Fig. 18 – Transmission electron micrographs showing the presence of retained austenite films between the martensite laths.

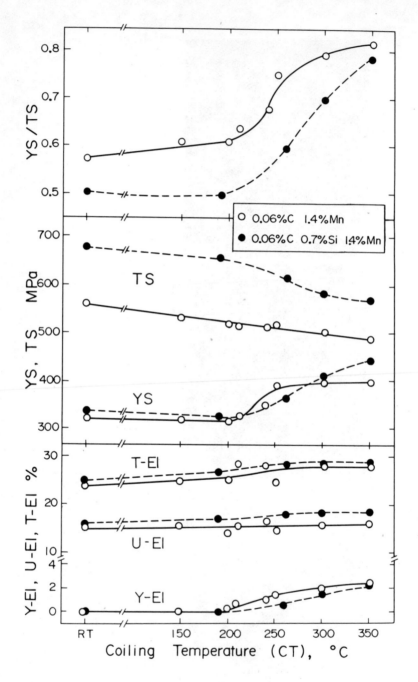

Fig. 19 - Mechanical Properties of DPR processed steels as a function of simulated coiling temperature.

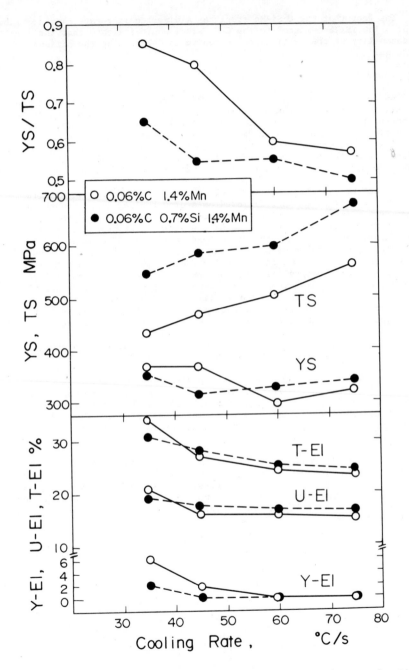

Fig. 20 – Mechanical properties of DPR processed steels as a function of cooling rate after finishing.

The fact that the C-Si-Mn steel has a wider cooling-rate range for a low YS or a low YS/TS compared to the C-Mn steel would indicate that the hardenability of the partitioned austenite is enhanced by the silicon addition.

Importance of Early Phase Separation

As described before, the optimum FT in the DPR process is at or about the Ar_3 temperature of the material. Since the Ar_3 temperature was measured in the unstrained condition, it is very likely that the gamma-to-alpha phase transformation has been fairly advanced when the steel comes out of the final pass, because the steel has just undergone large rolling deformations; the proeutectoid ferrite volume fraction will be larger, being assisted by the strain-induced transformation of gamma-to-alpha.

It may be expected that a FT well above the Ar_3 point, as used in ordinary hot-strip rolling, followed by a slow cooling down into the alpha-gamma temperature range prior to a final rapid cooling would produce a similar result as that obtained using the optimum DPR process. Some experiments have been carried out to test this idea. The steel used was the same as the C-Mn steel used in the previous experiment. The experimental procedure is shown in Figure 10a except that various after-finish delay times (with natural air cooling) prior to the oil quenching were introduced. Two finish temperatures were selected, i.e., 770°C and 900°C. Figure 21 shows the mechanical properties obtained, as a function of the delay time, or of the oil quenching temperature. It is noted that the property-changes with the quenching temperature are much smaller after finishing at 770°C than at 900°C. However, the two curves merge at the same level for a quenching temperature of about 700°C; otherwise inferior mechanical properties are produced by finishing at 900°C.

As shown in Figure 22, the second phase obtained after quenching from 750°C is lath-martensite if finished at 770°C, whereas it becomes bainite if finished at 900°C. When finishing at 900°C, it is necessary to have a delay time (36 seconds, see Figure 21) and quench from 700°C in order to obtain a martensitic phase similar to that obtained after finishing at 770°C and quenching from 750°C.

These results would mean that it is important that the gamma-to-alpha phase transformation starts during or immediately after finishing in order to produce optimum mechanical properties. Taking this into consideration, the finishing temperature should be kept lower than usual, but high enough to ensure that a heavily deformed ferrite structure is not obtained in the final state.

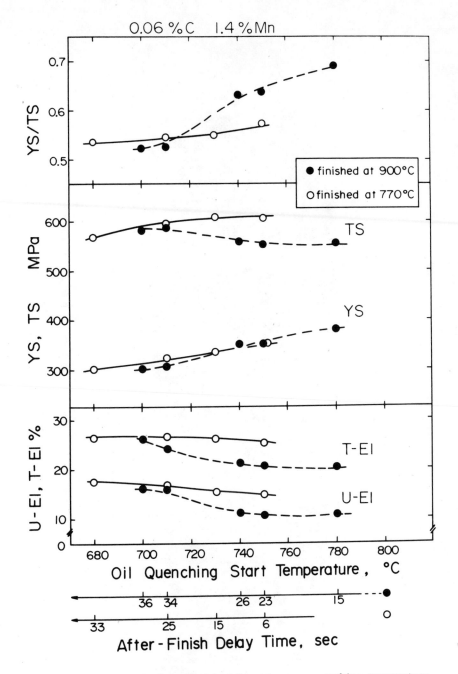

Fig. 21 – Effect of after-finish delay time, or quenching temperature on mechanical properties.

FT 770°C   QT 750°C   FT 900°C   QT 750°C   FT 900°C   QT 700°C

0.06 C - 1.4 Mn

Fig. 22 – Transmission electron micrographs showing the effects of finishing and quenching temperatures on the second phase.

Acknowledgements

The authors wish to thank T. Fujii for helpful discussions and assistance in the Mossbauer spectral analysis. Also thanks are due to M. Endo who helped us in carrying out the experiments on heat treatments, DPR process treatments and mechanical testing.

References

1. S. Hayami, T. Furukawa, H. Gondoh and H. Takechi, "Recent Developments in Formable Hot- and Cold-Rolled HSLA Including Dual-Phase Sheet Steels," pp. 169-182 in Formable HSLA and Dual-Phase Steels, A.T. Davenport, ed.; AIME, New York, N.Y., 1979.

2. P.E. Repas, "Metallurgy, Production Technology, and Properties of Dual-Phase Sheet Steels," SAE paper 790008, 1979.

3. J. Becker and E. Hornbogen, "Microscopic Analysis of the Formation of Dual-Phase Steel," pp. 20-39 in Structure and Properties of Dual-Phase Steels, R.A. Kot and J.W. Morris, ed.; AIME, New York, N.Y., 1979.

4. D.K. Matlock, G. Krauss, L.F. Ramos and G.S. Huppi, "A correlation of Processing Variable with Deformation Behavior of Dual-Phase Steels," ibid., pp. 62-90.

5. T. Tanaka, M. Nishida, K. Hashiguchi and T. Kato, "Formation and Properties of Ferrite plus Martensite Dual-Phase Structures," ibid., pp. 221-241.

6. T. Furukawa, H. Morikawa, H. Takechi and K. Koyama, "Process Factors for Highly Ductile Dual-Phase Sheet Steels," ibid., pp. 281-303.

7. H. Takechi, N. Takahashi, M. Abe, O. Akisue and K. Koyama, "High-Strength Cold-Rolled Steels Produced by Continuous Annealing and Their Applications to Automotive," SAE Paper 810028, 1981.

8. N. Takemoto, T. Furukawa and H. Takechi, "Hot- and Cold-Rolled Dual-Phase Steels - Production Techniques and Mechanical Properties," pp. 32-35 in preprint for the 33rd Annual Conference of the Australasian Institute of Metals, Auckland, New Zealand, May 1980.

9. W. Hume-Rothery, The Structure of Alloys of Iron, p. 287; Pergamon Press, London, 1966.

10. G.R. Speich and R.L. Miller, "Mechanical Properties of Ferrite-Martensite Steels," pp. 145-182 in Structure and Properties of Dual-Phase Steels, R.A. Kot and J.W. Morris, ed.; AIME, New York, N.Y., 1979.

11. A.P. Coldren, G. Tither, A. Cornford and J.R. Hiam, "Development and Mill Trial of As-Rolled Dual-Phase Steel," pp. 205-228 in Formable HSLA and Dual-Phase Steels, A.T. Davenport, ed.; AIME, New York, N.Y., 1979.

12. J.Y. Koo and G. Thomas, "Thermal Cycling Treatments and Microstructures for Improved Properties of Fe-0.12% C-0.5% Mn Steels," Materials Science and Engineering, 24 (1976) pp. 187-198.

247

13. J.M. Rigsbee and P.J. Vander Arend, "Laboratory Studies of Micro-structures and Structure-Property Relationships in Dual-Phase HSLA Steels," pp. 56-86 in Formable HSLA and Dual-Phase Steels, A.T. Davenport, ed.; AIME, New York, N.Y., 1979.

TEMPERING CHARACTERISTICS OF A VANADIUM

CONTAINING DUAL PHASE STEEL

M. S. Rashid
Metallurgy Department
B.V. N. Rao
Analytical Chemistry Department
General Motors Research Laboratories
Warren, Michigan 48090

Dual-phase steels are characterized by a ferrite-"martensite" microstructure, the "martensite" often being a combination of martensite, retained austenite and/or lower bainite. The nature and volume fraction of the "martensite" is governed by steel composition and processing parameters and influences steel mechanical properties. These mechanical properties can be altered by heating or tempering the steel above room temperature.

This paper examines the mechanical property changes produced in a vanadium bearing dual-phase steel upon tempering below 500°C. The phase transformations and microstructural changes produced by tempering were studied by transmission electron microscopy and the observed mechanical property changes were explained on the basis of the changes in microstructure.

Introduction

Dual-phase steels are characterized by a microstructure that consists of 75-85 volume percent fine grained ferrite with the balance being a uniformly distributed mixture of martensite, retained austenite and lower bainite [1-7] in varying proportions. These steels are currently manufactured and used commercially. They are produced by continuously annealing certain low-carbon Si-Mn [4] or microalloyed high strength, low alloy (HSLA) steels [1,2,5], or directly off the hot mill [6]. The continuous annealing generally consists of heating the steel for a short time at temperatures in the ferrite-austenite, $(\alpha+\gamma)$, or austenite, (γ) regions of the Fe-C phase diagram and cooling to room temperature. The cooling rate may be in the range between air cooling and water quenching depending on steel chemistry, available facilities and desired mechanical properties.

The stress-strain behavior of dual-phase steels [2] is characteristically different from that of ferrite-pearlite steels such as plain carbon steel or the HSLA steels (Figure 1). The ferrite-pearlite steels have yield point elongation, a high yield strength to ultimate tensile strength (YS/UTS) ratio and their strength and ductility (uniform elongation) are inversely related (Figure 2). Dual-phase steels have a continuous stress-strain curve with no yield point elongation. They work harden very rapidly at low strains, have a low YS, a high UTS and hence a low YS/UTS ratio. They have better formability than the ferrite-pearlite steels of equivalent tensile strength and their strength-ductility data fall on a separate curve (Figure 2) than that for ferrite-pearlite steels.

The mechanical properties of dual-phase steels can be altered by low temperature heating, or tempering which produces various microstructural changes in the steel. Tempering has been successfully used as a post-continuous annealing step to improve the strength-ductility relationship of certain dual-phase steels. Also, elevated temperatures might be experienced during service use of the steels and knowledge of any mechanical property change resulting from such exposure is desirable. However, data is not publically available on the relationship between mechanical properties and tempering treatments as a function of steel composition and processing parameters, or on the phase transformations and other microstructural changes that can be produced by tempering.

This paper examines the mechanical property changes produced in a dual-phase steel by tempering below 500°C to determine whether mechanical properties can be further improved. The material studied was a laboratory produced, vanadium-bearing dual-phase steel (0.1% C, 1.5% Mn, 0.5% Si, 0.1% V) which meets the GM 980X specification [2]. The microstructural changes accompanying the mechanical property changes were studied by transmission electron microscopy and this information used to explain the observed mechanical property changes. This is the first known investigation of the tempering mechanisms in microalloyed dual-phase steels. Understanding such mechanisms will certainly enable more efficient utilization of these new steels. The response of other dual-phase steels to tempering might be different, being influenced by composition, the ratios of ferrite-martensite-retained austenite and variations in processing parameters and will have to be investigated.

Experimental Procedure

Standard (ASTM E8) tensile specimens were machined from sheet stock with the rolling direction parallel to the specimen axis. Several specimens were heated for 60 minutes in a furnace at either 200, 300, 400 or 500°C and air cooled to room temperature. Duplicate specimens representing each tempering condition were tension tested to failure at room temperature on an Instron testing machine

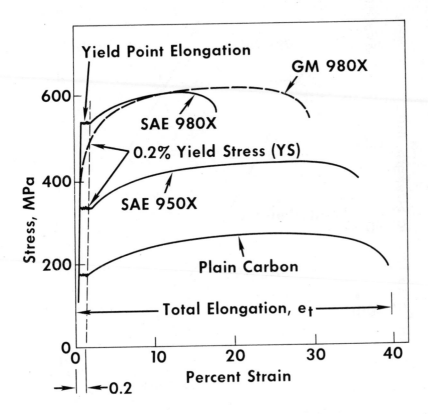

Fig. 1 – Schematic stress-strain curves for plain carbon, HSLA and dual-phase
steels.   SAE 950X and 980X are Society of Automotive Engineers
designations for HSLA steels of different strength levels. GM 980X is a
General Motors developed dual-phase steel. GM 980X is more ductile
than SAE 980X although both steels have similar tensile strength.

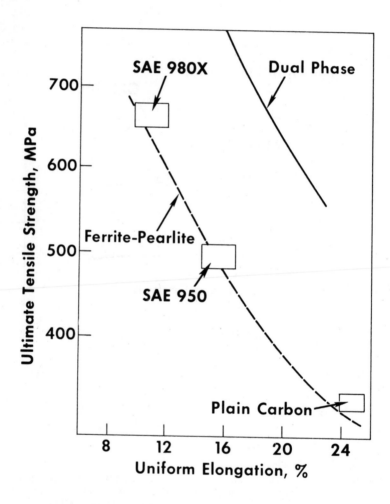

Fig. 2 - Strength-ductility relationship of dual-phase steels compared with that for plain carbon and HSLA steels. The dual-phase steel curve is far above that for ferrite-pearlite steels (Reference 5).

at a crosshead speed of 5 mm/min. Strain was measured with an Instron extensometer over a 50 mm gage length. Strain measurements were made from load-elongation data which was automatically plotted on a strip chart recorder.

Transmission microscopy specimens were obtained from 30 mm x 12.5 mm sections cut from the gage section of tempered, undeformed tensile specimen blanks and then surface ground to 0.625 mm using flood cooling. The ground piece was chemically thinned to 0.125 mm in a stirred and cooled solution containing 5% HF in H_2O_2 (30% concentration). Discs 3 mm in diameter were electro-discharge machined from the chemically thinned sheet, mounted on a flat metal block with Scotch tape, and hand ground on both sides to 0.05 mm on 600 grit SiC paper. The ground discs were cleaned in acetone and alcohol and electropolished at room temperature in a Fischione "Twin-Jet" electropolishing unit in a solution of chromium trioxide, glacial acetic acid and distilled water.

The foils were examined in a JEOL (JEM 200C) transmission electron microscope at an acceleration voltage of 200 kV. In addition to the conventional methods of transmission electron microscopy, such as bright field (BF), dark field (DF), and selected area diffraction (SAD), the weak beam dark field (WBDF) technique was also employed to enable resolution of the very fine precipitates and the interactions between precipitates and dislocations.

<u>Results</u>

The stress-strain curve for the as-received steel was characteristic of dual-phase steels (Figure 1) and the mechanical properties were superior to the minimum specified for GM 980X [2]. On tempering the steel above room temperature the yield point elongation (ype) appeared and various changes in mechanical properties resulted. Mechanical properties of the steel before and after tempering are listed in Table I and also plotted in Figure 3 as a function of tempering temperature. The average absolute changes in strength and ductility were calculated from the preceding data and are also listed in Table I.

Minimal strength and ductility changes were observed on heating the steel at 200°C. The yield strength (YS) increased slightly but the opposite was observed in the ultimate tensile strength (UTS); the total elongation (e_T) and uniform elongation (e_u) did not change. However, larger changes were observed on heating above 200°C. Yield strength increased with increasing heating temperature and reached a maximum value at about 400°C. The UTS decreased continually with increasing heating temperature and tended to level off at the higher temperatures.

The e_T and e_u were constant with tempering temperature up to 300°C and dropped rapidly with higher tempering temperatures. The combined reduction in UTS and elongation after the 400°C tempering suggests a reduction in the toughness of the steel.

The post uniform elongation (e_T-e_u), and ype are also plotted in Figure 3 as a function of tempering temperature. The (e_T-e_u) was only minimally higher at tempering temperatures above 300°C compared to the values in the as-received steel. The ype appeared on heating at 200°C and increased in direct relation to heating temperature, it was 1.4% on heating at 300°C and reached a maximum of about 2.4% on heating at higher temperatures, suggesting a rate dependent behavior. These maximum changes in ype correspond to similar limiting changes observed in strength and ductility on heating above 400°C.

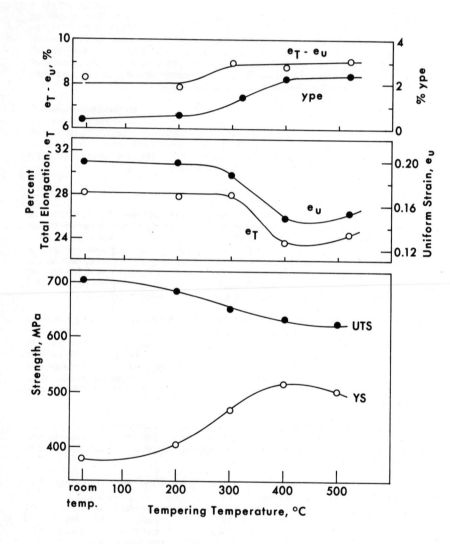

Fig. 3 -   Effect of tempering temperature on the mechanical properties.

Table I: Effect of Tempering for One Hour on Mechanical Properties

Tempering Temperature °C	Tensile Properties						Average Change in Properties					
	YS MPa	UTS MPa	e_T %	e_u %	e_T-e_u %	ype %	ΔYS MPa	ΔUTS MPa	$\Delta(\% e_T)$	$\Delta(\% e_u)$	$\Delta(\% e_T-e_u)$	$\Delta(\% ype)$
AR	390	699	28.1	19.9	8.2	0.0						
	386	707	28.3	19.9	8.4	0.0						
200	400	683	28.5	20.0	8.5	0.5	+16	-19	-0.4	0.0	-0.4	+0.4
	409	685	27.1	19.8	7.3	0.7						
300	474	660	28.9	19.5	9.9	1.5	+80	-50	-0.2	-0.9	+0.7	+1.2
	462	647	27.0	18.5	8.5	1.3						
400	521	637	22.8	14.8	8.0	2.2	+128	-70	-4.5	-5.0	+0.5	+2.1
	512	629	24.6	15.0	9.6	2.3						
500	499	612	24.8	15.8	9.0	2.5	+115	-86	-3.8	-4.6	+0.8	+2.2
	508	622	24.0	14.8	9.2	2.3						

255

Earlier transmission electron microscopy of the same steel [3] showed that the as-received steel consisted of transformed and retained ferrite, martensite and retained austenite. Transformed ferrite is ferrite that nucleates out of the austenite that forms during continuous annealing. The untransformed or retained ferrite was essentially free of fine vanadium carbonitride precipitates and contained a high density of accommodation dislocations, while the transformed ferrite contained banded precipitates and a lower density of dislocations, some of which were pinned by the precipitates. Substantial substructural twinning was found in the predominantly plate martensite. The retained austenite was principally in the form of 2-6 μm diameter particles although some submicron size particles were also present.

In conjunction with the minimal changes produced in the mechanical properties after tempering at 200°C, minimal changes were also observed in the ferrite except for relaxation of residual stresses at the retained ferrite-martensite interfaces and rearrangement of the highly heterogeneous dislocation distributions into uniform low energy configurations. However, extensive twin boundary cementite particles were found in the martensite (Figures 4a and b). The cementite was in the form of thin platelets or films and is indicated by the arrows in Figure 4c. Figure 4c was obtained from a superimposed cementite and double diffracted spot as revealed by the selected area diffraction pattern (Figure 4d). Diffraction patterns obtained from other martensite particles showed effects indicative of carbon clustering prior to precipitation. Minimal microstructural changes were observed in the retained austenite except for some very fine precipitates which were suspected to be carbides.

On tempering at 300°C, a larger density of fine precipitates (<2.5 nm in diameter) was found in both ferrites (for example, as seen in the retained ferrite in Figures 5a and b) compared to that found by the lower tempering treatment. Weak beam dark field analyses of some areas indicated that the precipitates were dislocation nucleated. These precipitates were suspected to be vanadium carbonitrides but strong diffraction evidence could not be obtained. Also, some coarsening of the banded carbonitrides was observed in the transformed ferrite.

Coarser cementite platelets (about 50 nm thick by 200 nm long) were found in the martensite, M in Figure 5a, compared to the thin films observed after the 200°C tempering. The substructural twinning in the martensite was no longer evident indicating migration of the twin boundaries. Most of the retained austenite (about 90 volume percent by visual estimate) remained unchanged (Figure 5c) and was similar to that in the as-received steel while the remainder seemed to transform to a bainitic product.

Corresponding to the observed maximum changes in mechanical properties, gross microstructural changes were observed in specimens tempered at 400°C and above. A substantial increase was observed in the density of fine precipitates (<5 nm diameter), and in the amount of grain boundary and sub-boundary nucleated cementite and the intergranular cementite platelets in the ferrite (Figures 6a and b). Some dislocation pinning of the precipitates suggested that secondary hardening might result from such interactions. Extensive coarse cementite particles were found in the martenite similar to and more pronounced than those observed after tempering at 300°C.

Most of the retained austenite (more than 90% by visual estimate) was decomposed and was replaced by an upper bainitic product consisting of ferrite laths and lath boundary stringers of continuous cementite platelets. The existence of upper bainite was confirmed by morphological data and orientation

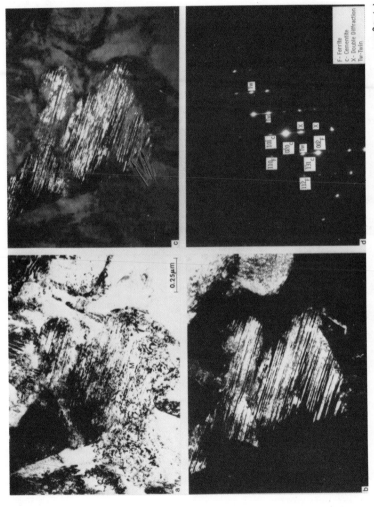

Fig. 4 – Bright field transmission electron micrograph of a martensite particle in steel tempered at 200°C (a), and (b) dark field micrograph of the same area. The dark field micrograph from superimposed cementite, double diffracted and streak from twinning reflections is shown in (c) and the corresponding indexed selected area diffraction pattern in (d) revealing substructural twinning and twin boundary cementite precipitation. Arrows in (c) point to the twin boundary cementite films and the reflection used to form (c) is marked by aperture in (d).

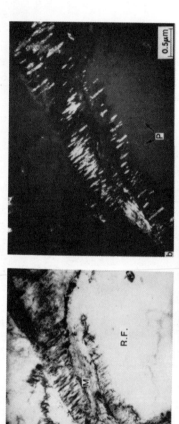

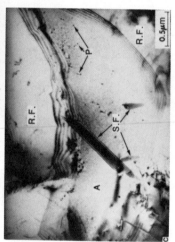

Fig. 5 - Bright field (a) and dark field (b) micrographs revealing extensive cementite precipitation in a specimen tempered at 300°C. 'P' in (b) shows extremely fine precipitation in the retained ferrite, R.F. (c) is the bright field micrograph from a different area showing the undecomposed retained austenite (A) following 300°C tempering. S.F. reveals stacking faults and 'P', fine precipitation.

258

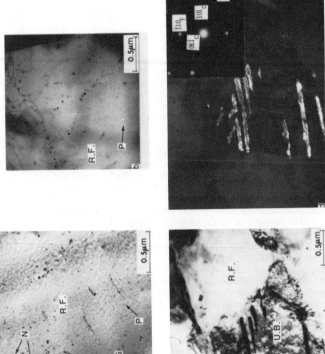

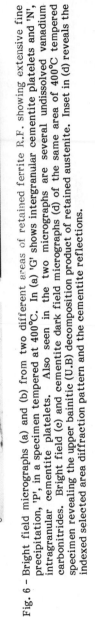

Fig. 6 – Bright field micrographs (a) and (b) from two different areas of retained ferrite R.F. showing extensive fine precipitation, 'P', in a specimen tempered at 400°C. In (a) 'G' shows intergranular cementite platelets and 'N', intragranular cementite platelets. Also seen in the two micrographs are several undissolved vanadium carbonitrides. Bright field (c) and cementite dark field micrographs (d) of the same area of 400°C tempered specimen revealing the upper bainitic (U.B) decomposition product of retained austenite. Inset in (d) reveals the indexed selected area diffraction pattern and the cementite reflections.

relationships. Figures 6c and d show a completely decomposed retained austenite particle with cementite platelets while the inset in Figure 6d shows the indexed diffraction pattern for the cementite.

Yet more changes were observed after tempering at 500°C. The vanadium carbonitride precipitates in both ferrites were coarser than observed after lower temperature tempering and the banded nature of precipitates in the transformed ferrite was no longer evident (Figure 7a). Some cementite platelets were observed at grain boundaries. Coarse, partially spheroidized cementite particles were observed in the martensite (Figure 7b), and all the retained austenite was decomposed. About half decomposed to an upper bainitic product (Figures 8a and b) similar to that observed after tempering at 400°C. The remainder decomposed by an interphase precipitation mechanism producing several ferrite-carbide morphologies including banded precipitation (Figure 8c).

Discussion

The mechanical properties of the steel were altered minimally on heating at 300°C or below except for a small increase in YS and a similar decrease in UTS, the two latter changes indicating a corresponding decrease in the strain hardening rate of the steel. Heating above 300°C decreased the UTS and ductility, however. These data suggest that the mechanical properties of this steel cannot be improved further by tempering below 500°C.

The mechanical property changes produced by tempering seemed to be explainable in terms of the microstrutural changes observed in the ferrite, martensite and retained austenite. Tempering at 200°C resulted in a slight increase in yield strength and a slight decrease in tensile strength. The yield strength increase is evidently caused by the observed relaxation of residual stress at the ferrite-martensite interfaces and rearrangement of dislocations in the ferrite. A part of the strength increase might be caused by the fine precipitation that was observed in the retained austenite. the slight decrease in tensile strength must be related to the twin boundary precipitation of cementite particles in the martensite (Figures 4a and b).

The larger observed increase in yield strength and decrease in tensile strength after tempering at 300°C compared to the 200°C temper, suggests an increase in the same microstructural changes noted at 200°C, as was observed. The density of fine precipitates in both ferrites was higher as was the precipitation in the retained austenite resulting in the observed higher yield strength. The latter would cause increased mechanical stability of the austenite [8,9] and contribute to the increased YS. The cementite precipitates in the martensite were coarser than at 200°C and (Figures 5a and b) explain the larger decrease in UTS. These changes evidently effect e_T and e_u minimally. However, there was a slight increase in the ype suggesting an increase of dislocation-interstitial coupling which was also observed by transmission microscopy.

Tempering at 400°C resulted in a decrease in tensile strength and ductility, indicating an overall decrease in the toughness. It must be emphasized, however, that even in this condition the steel has ample toughness for most practical applications.

Extensive coarse cementite particles were found in the martensite indicating continued progression of the tempering observed at lower temperatures. Such changes are known to decrease strength but increase ductility. However, since both strength and ductility decreased in this steel, the decrease in ductility was attributed principally to the observed decomposition of retained austenite to upper bainite (Figures 6c and d) and to the morphology of the upper bainite.

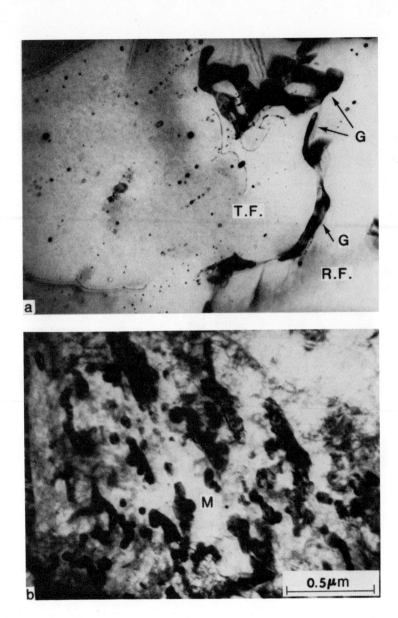

Fig. 7 - Bright field micrographs from two different areas of the 500°C tempered specimen revealing general coarsening of the alloy carbonitrides in the transformed ferrite, T.F. (a) and the early stages of spheroidization in the martensite, M, (b). 'G' in (a) shows intergranular cementite precipitation.

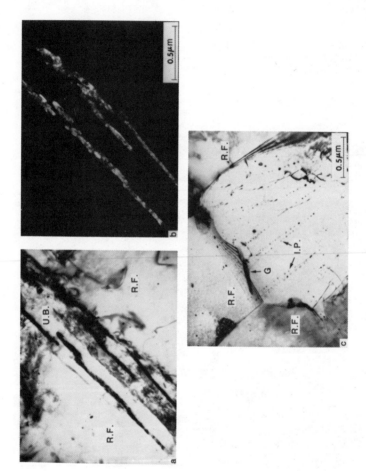

Fig. 8 – Bright field (a) and cementite dark field micrographs (b) of the same area of 500°C tempered specimen revealing upper bainitic (U.B.) decomposition product of retained austenite. (c) is the bright field micrograph from a different area of the same specimen revealing the interphase (I.P.) precipitation resulting from the decomposition of retained austenite. 'G' shows a intergranular cementite platelet.

Retained austenite is reported to influence dual-phase steel strength and ductility. Upon deformation of the as-received steel the retained austenite transforms to martensite [11]. The transformation induced plasticity or TRIP enhances steel ductility while the martensite that forms upon deformation increases tensile strength. The decomposition of retained austenite upon tempering eliminates the TRIP mechanism in the tempered steel. This will result in a decrease in both strength and ductility, as was observed. Furthermore, the upper bainite was comprised of continuous interlath cementite stringers. These particles provide less fracture resistance and reduce ductility [10]. Similar tempering behavior has been reported previously in various structural steels [9,10].

Transmission microscopy also produced evidence of dislocation nucleated precipitation of fine vanadium carbonitrides in the ferrite. This could cause "secondary hardening" and would explain some of the yield strength increase and the large ype observed.

The kinetics of martensite and austenite decomposition are faster at 500°C and resulted in precipitate coarsening in the ferrite and coarsening of the cementite particles in the martensite. This was commensurate with the observed decrease in tensile strength. However, ductility increased and yield strength decreased slightly contrary to that observed after tempering at 400°C. This seemed to be related to the unique decomposition observed in the retained austenite after this tempering treatment.

About half the retained austenite transformed to upper bainite similar to that observed at 400°C but the balance transformed to ferrite accompanied by the interphase precipitation of carbides, the latter microstructure being more ductile and with greater resistance to fracture than the former. Consequently, somewhat higher ductility was observed on tempering at 500°C.

In summary, tempering of this dual-phase steel below 500°C did not improve mechanical properties. However, the response of other dual-phase steels might be different. In this study the observed mechanical property changes produced by tempering below 500°C could be explained reasonably well by the various microstructural changes that are produced by tempering. The operative mechanisms were precipitation in the ferrite and austenite, tempering of the martensite and decomposition of the retained austenite. Different mechanisms dominated at different tempering temperatures and dictated the mechanical property changes. The largest decrease in ductility occurred on tempering above 400°C and seemed to be related primarily to the decomposition of the retained austenite to upper bainite.

Conclusions

1. The mechanical properties of the as-received steel could not be improved by tempering below 500°C.

2. The microstructure appears quite stable to heating at 200°C and minimal changes in mechanical properties were observed. At higher tempering temperatures, tensile strength and ductility decreased with increasing temperature while yield strength increased and exhibited maximum values on heating above 400°C.

3.   At least four distinct microstructural changes seemed to occur during tempering, one or more being dominant in different tempering ranges. Precipitation of fine carbonitrides in the ferrite and austenite, and precipitation of cementite particles in the martensite seemed to govern mechanical properties for tempering treatments below 300°C. Upon tempering above 400°C, the decomposition of retained austenite appeared to have the principal influence on mechanical properties.

4.   The largest loss in ductility occurred on heating at 400°C and was attributed primarily to the nearly complete decomposition of retained austenite to an upper bainitic product and the absence of the transformation induced plasticity or TRIP phenomenon.

5.   Two competing mechanisms were identified for the decomposition of retained austenite on tempering at 500°C, an upper bainitic transformation similar to that observed at 400°C, and an interphase precipitation mechanism leading to ferrite with banded carbonitrides.

Acknowledgment

The expert assistance received from Messrs. H. Sturner, C. Ligotti and H. G. Lemmer in the experimental aspects of this work is gratefully acknowledged.

References

1.   M. S. Rashid, SAE Trans., (1976) Vol 85, Sec 2, p 938.

2.   M. S. Rashid, SAE Trans., (1977) Vol 86, Sec 2, p 935.

3.   B. V. N. Rao and A. K. Sachdev, 1980 to be published.

4.   S. Hayami and F. Furukawa, Proc. Microalloying '75, Union Carbide, New York, 1975, p 311.

5.   J. H. Bucher and E. G. Hamburg, SAE Trans., (1977) Vol 86, Sec 1, p 730.

6.   A. P. Coldren and G. Tither, J. Metals, (1978) Vol 30 (4), p 6.

7.   J. M. Righbee and P. J. VanderAhrend, Formable HSLA and Dual-Phase Steels, Conference Proceedings, The Met. Soc. of AIME, Ed. A. T. Davenport, 1977.

8.   S. Jin and D. Huang, Met. Trans. A (1976), Vol 7A, p 745.

9.   R. M. Horn and R. O. Ritchie, Met. Trans. A (1978), Vol 9A, p 1039.

10. B. V. N. Rao and G. Thomas, Met. Trans. A (1980), Vol 11A, p 441.

11. B. V. N. Rao and M. S. Rashid, to be published.

TEMPERING OF DUAL-PHASE STEELS

R. G. Davies

Research
Ford Motor Company
Dearborn, MI 48121

A study has been made of the influence of tempering at temperatures up to 550°C upon the mechanical properties of dual-phase steels. These steels, which were representative of both commercially available and experimental steels, contained various amounts of Mn, V, Cr and Mo and were either water-quenched or air-cooled from the intercritical annealing temperature. It was observed that water-quenching led to embrittlement, however, tempering at about 130°C restored the ductility. The bake-hardening (20 mins. at 170°C) of a lean alloy (LA) water-quenched steel was influenced by the initial tempering temperature and the amount of prestrain; the best combination of tensile strength and total elongation was obtained after an initial temper at 100°C. The loss in strength upon tempering at up to 550°C was a minimum for all the air-cooled dual-phase steels, independent of alloy content; this is thought to be due to auto-tempering of the higher carbon martensite and precipitation in the ferrite. Tempering is also shown to improve the hydrogen embrittlement resistance and the percentage hole expansion of both water-quenched and air-cooled dual-phase steels. As generally supplied the water-quenched steels are superior to the air-cooled steels in edge-cracking resistance; this is probably due to the presence of lower carbon more ductile martensite in the LA water-quenched steel.

Introduction

Dual-phase steels, which consist of a mixture of high carbon martensite (>0.5%C) in a fine grained ferrite (<5 μm diameter), are produced by cooling, at a rate appropriate to the alloy content, from the α + γ region of the phase diagram. Highly alloyed (H.A.) steels, which may contain 1.5% Mn and 0.1 - 0.3% of V, Cr and/or Mo to promote hardenability, are often air cooled ($\sim 20^{\circ}$C/sec); lean alloy (L.A.) steels containing only 0.6% Mn for hardenability are water-quenched ($\sim 2000^{\circ}$C/sec) to produce the dual-phase structure. These steels are sometimes subjected to a tempering treatment prior to or during fabrication; for example, the material may be hot-dip galvanized ($\sim 420^{\circ}$C for a few seconds) and/or put through the paint bake oven ($\sim 170^{\circ}$C for 20 mins.). Galvanizing, which is important for corrosion protection, is carried out prior to forming any components, while finished components are paint-baked.

Tempering will, by affecting the carbon distribution, change the properties of both the martensite and the ferrite which could lead to a degradation of the mechanical properties and formability of the dual-phase steel. As a result of the differences in carbon content in the ferrite and martensite, it is to be expected that the lean and highly alloyed dual-phase steels will respond differently to tempering. The amount of carbon super-saturation in the ferrite and martensite will be a function of the cooling rate from the intercritical annealing temperature; the faster the cooling rate the greater will be the supersaturation. The mechanical properties of both lean and high alloy dual-phase steels have been investigated for both the as-quenched and tempered conditions; the interaction of prestrain and tempering was extensively studied in the LA dual-phase steel.

Experimental Procedure

The composition of the steels investigated are given in Table I; the steels were all approximately 0.9mm (0.036 ins.) thick. All the alloys were intercritically annealed at 760°C for 10 mins. in a salt pot and then either forced air cooled or quenched into brine. Tempering of the individual samples was done in an air furnace between preheated steel plates. Standard tensile specimens with a 50mm (2 ins.) by 12.5mm (0.5 in.) gage section were tested in an Instron machine at room temperature at a crosshead rate of approximately 0.04mm/s.

Table I. Compositions in Wt.% of Alloys Studied, Balance Iron

Alloy Designation	C	MN	Si	V	Mo	Cr
LA	0.08	0.64	0.25	-	-	-
V-HA	0.12	1.45	0.53	0.09	-	-
Cr-HA	0.09	1.51	0.56	-	-	0.21
Mo-HA	0.09	1.54	0.54	-	0.10	-
Mn-HA	0.08	1.90	0.50	-	-	-

Results

As-Quenched

It was found that both low and high alloy dual-phase steels exhibited some brittleness in the as-quenched condition; this was manifest as low uniform elongation with no localized necking prior to fracture. However, tempering at a low temperature (<150°C) for a few minutes increased the elongation and led to necking; this is illustrated in Fig. 1 for the V - HA steel. In general, it was found that the HA steels should be tempered in excess of 120°C, while for the LA steels two hours at room temperature (23°C) was sufficient to increase the total elongation from 15 to 22%.

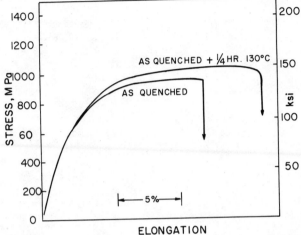

Fig. 1 Stress-strain curves for the V-HA steel showing as-quenched brittleness and necking after low temperature anneal.

Scanning electron micrographs revealed patches of cleavage in some of the freshly quenched samples, Fig. 2(a), while only ductile dimpling was found in the tempered, fully ductile, specimens, Fig. 2(b). The as-quenched brittleness could be related to the delayed fracture phenomena that is observed in fully martensitic steels (1,2). It has been proposed (2) that in these steels dislocations move under the action of the applied stress to interact with grain boundary defects to form cracks which then propagate; tempering reduces the dislocation mobility and thus ameliorates the brittleness. The difference in tempering response, i.e. temperature to restore full ductility, between the HA and LA types of dual-phase steels may be a consequence of: 1. the larger volume expansion of martensite in the HA steel (higher carbon content) which may lead to higher internal stresses, 2. more auto tempering of the martensite in the LA steel as the higher M_s temperature could lead to relaxation of internal stresses, and/or 3. the presence of $\overset{\circ}{V}$ in the HA steel could reduce the strain-aging response of the ferrite (3,4). Whatever the actual reason for the brittleness, it is clear that dual-phase steels should not be used in the as-quenched condition.

Bake-Hardening

The yield strength of both air-cooled (5,6) and water-quenched (7,8) dual-phase steels increases upon aging for 20-30 mins. at about 170°C; this time and temperature are typical of present paint oven conditions and this aging is often referred to as "bake-hardening. The strengthening is a result of the excess carbon in solution diffusing so as to pin the free dislocations in the ferrite and/or form fine iron carbides. From the above observations of brittleness in as-quenched dual-phase steels, it is clear that these steels should be given a low temperature tempering treatment prior to use so as to restore the ductility. However, this low temperature tempering treatment will remove some of the carbon from solution in the form of iron carbides and so change the bake-hardening characteristics. In addition, the amount of cold-work prior to bake-hardening will, by changing the number of nucleation sites for carbide precipitation, also influence the final strength. Thus, this study was undertaken to investigate the influence of initial tempering temperature and the amount of subsequent tensile deformation on the bake-hardening of the lean alloy dual-phase steel that had been quenched into brine from 760°C; all the specimens had a final heat treatment of 20 mins. at 170°C. The initial tempering was for 5 mins. at temperatures up to 350°C.

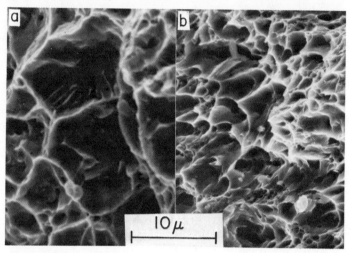

Fig. 2 Scanning electron micrographs:
a) as-quenched LA steel showing patches   b) ductile dimpling in the tempered V-HA
   of cleavage, and   steel.

The increase in yield stress after bake-hardening as a function of initial tempering temperature, for specimens subsequently deformed 5% in tension, is shown in Fig. 3. It can be seen that there is a broad maximum in the increase in yield stress at around 200°C and that even no deliberate initial tempering (25°C data point) results in a substantial increase in the yield stress. In Fig. 4 are shown the increases in yield stress as a function of prestrain following one of three initial tempering temperatures (100, 175 and 250°C). Although there is a minimum in the increase in yield stress at about 8 - 10% prestrain, the increase is greater than 83 MPa (12 ksi) for all prestrains and tempering temperatures.

While the increase in yield stress is greatest for the specimens initially tempered at 175°C, the absolute value of the flow stress is the greatest, as shown in Fig. 5, after initially tempering at 100°C. This sequence of strength values is a result of the decrease in the initial strain-hardening rate with increasing tempering temperature (9). The change in strain-hardening rate is no doubt influenced by the size and distribution of carbide precipitates and by the dislocation pinning and rearrangement that takes place in the ferrite during tempering.

In addition, as shown in Fig. 6, the lower initial tempering temperature gives a better trade-off between tensile strength and total elongation; the results for the 175°C temper were very similar to those obtained after the 100°C temper. Thus, there is a higher strength at a given elongation for specimens tempered at the lower temperatures. As has been previously demonstrated (10), the improved strength-elongation trade-off is a reflection of a higher strain-hardening rate at any given strain.

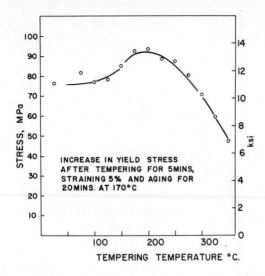

Fig. 3 Increase in yield stress as a function of initial tempering temperature for LA steel specimens that were subsequently deformed 5% and bake-hardened.

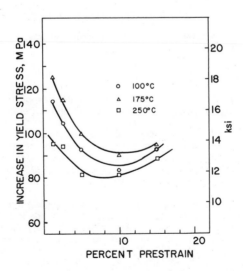

Fig. 4 Increase in yield stress due to bake-hardening as a function of prestrain for LA steel specimens initially tempered at 100, 175 and 250°C.

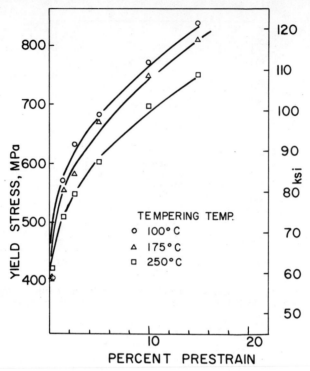

Fig. 5   Flow stress as a function of prestrain for specimens initially tempered at 100, 175 and 250°C.

Higher Temperature Tempering

In order to investigate the changes in mechanical properties following higher temperature tempering, a series of both air-cooled and water-quenched dual-phase steels were heated for 5 mins. at temperatures up to 550°C. Figure 7 shows the mechanical properties of an air-cooled V-dual-phase steel; it can be seen that the tensile strength decreases slowly with increasing temperature while the yield stress increases rapidly at temperatures above 300°C. The elongation, both total and uniform, are relatively unchanged for temperatures up to 400°C but then up to 550°C there is a decrease in the elongation of about 3-4%.

A comparison of the changes in strength with tempering temperature for an HA air-cooled and a LA water-quenched dual-phase steel is made in Fig. 8. Although the tensile strength of these two steels differ initially by only 28 MPa (4 ksi), after tempering at 450°C the LA steel has softened considerably so that the tensile strengths now differ by 117 MPa (17 ksi). By tempering at 450°C the tensile strength of the HA steel is decreased by 69 MPa (10 ksi), while for the LA steel the decrease is 152 MPa (22 ksi). Figure 9 shows the tensile strength as a function of tempering temperature for all the air-cooled HA steels; again the decrease in strength by tempering at 450°C is about 69 MPa (10 ksi). However, when these same HA materials are water-quenched to obtain the dual-phase structure, the decrease in tensile strength on tempering at 450°C is approximately 138 MPa (20 ksi), Fig. 10; this decrease in strength is similar to that observed in the water-quenched LA steels. The HA dual-phase steels, because of their higher carbon content, contain a larger volume fraction of martensite and therefore are stronger after quenching; thus, even when tempered at 450°C these steels have tensile strengths in excess of 620 MPa (90 ksi).

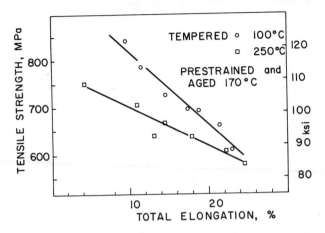

Fig. 6   Tensile strength as a function of total elongation for specimens tempered at 100 and 250°C, strained various amounts and bake-hardened.

The softening of both the quenched and air-cooled dural-phase steels is essentially independent of the presence of such carbide forming elements as Mo, V or Cr. This is maybe not surprising since the short times and low temperatures studied would not allow for much substitutional element diffusion. There are several factors that may contribute to the differences in tempering behavior between the air-cooled and water-quenched dual-phase steels: 1. Auto-tempering of the martensite during the air-cooling operation (11). This will result in less softening of the martensite upon deliberate subsequent tempering. 2. Higher carbon martensite in the air-cooled dual-phase steels arising from some of the austenite transforming to ferrite during the cooling (6,12), and so increasing the carbon content of the remaining austenite; this austenite eventually transforms to martensite. The higher the carbon content of the martensite the stronger it will remain after tempering (13); the stronger the martensite the better will be elongation at a given tensile strength (6). 3. There is the possibility of precipitation in the ferrite during the air-cooling. Again this will make the material less susceptible to softening on further tempering by providing homogeneous nucleation sites. The precipitation in the ferrite may take the form of general precipitates in the ferrite matrix and row-precipitates in the ferrite formed from the austenite during the cooling (14). If the auto-tempering of both the martensite and ferrite are important factors in making the air-cooled dual-phase resistant to tempering, then it is unlikely that quenched dual-phase steels can be hot-dip galvanized and retain an acceptable combination of strength and ductility.

The ductility, as measured by total elongation, of the water-quenched dual-phase steels never approached that of the air-cooled steels even when tempering resulted in similar tensile strengths; this is shown in Fig. 11. For the air-cooled steels tempering resulted in a slight loss in ductility (and strength) while the water-quenched materials showed a slight gain in ductility along with a considerable loss in strength. Thus, it appears that the ductility of a dual-phase steel is essentially determined by the initial structure. This confirms that the percentage of ferrite in the structure is the dominant factor in controlling the ductility of dual-phase steels (6); the greater the amount of ferrite the greater is the ductility. The strength of

271

these steels is very dependent upon the strength of the martensite phase. Thus, softening the martensite by tempering will reduce the strength of the dual-phase steel, but not necessarily increase the ductility.

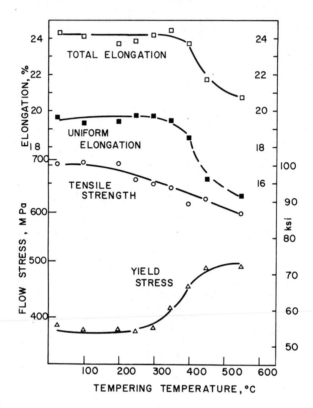

Fig. 7   Mechanical properties as a function of tempering temperature for an air-cooled V-HA steel.

In addition to changing tensile strength and elongation, tempering also influences more localized fracture phenomena as found in hydrogen embrittlement (15) and hole expansion (9). Figure 12 shows the loss in ductility as a function of tempering temperature (20 mins. at temperature) for both air-cooled and water-quenched dual-phase steel hydrogen changed at 6ma/cm^2 in a 4% H_2SO_4 solution (15); the loss in ductility is defined as the decrease in uniform elongation due to the presence of hydrogen. It can be seen that both the alloys respond in a similar manner to the tempering. Over this temperature range there is little change in elongation of the base material, see Fig. 7.   Thus, the changes in ductility with hydrogen embrittlement are probably controlled solely by the strength of the martensite, especially as the fracture appears to initiate in these regions.

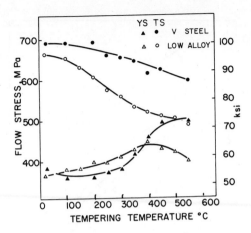

Fig. 8   Yield and tensile strength as a function of tempering temperature for an air-cooled V-HA and a water-quenched LA steel.

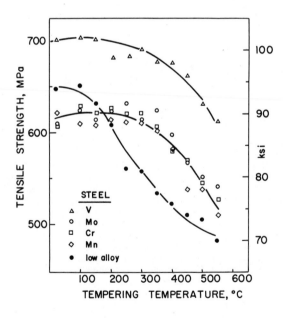

Fig. 9   Tensile strength as a function of tempering temperature for the air-cooled HA steels; the data for the LA water-quenched steel is added for comparison.

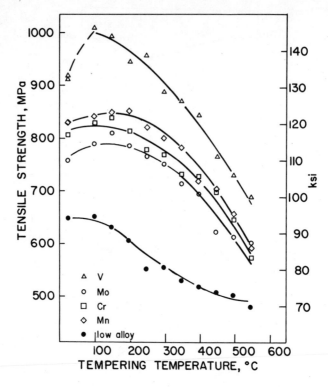

Fig. 10   Tensile strength as a function of tempering temperature for the HA and LA water-quenched steels.

The hole expansion as a function of tempering temperature, Fig. 13, shows quite different behavior between the air-cooled V-HA and water-quenched dual-phase steels. These steels were similar, but not identical to those listed in Table I, and the differences in behavior cannot be ascribed to changes in strength level alone (9). The water-quenched material exhibits a large increase in percent expansion after the air-cooled steel has to be tempered at 450°C for a comparable expansion. The expansion ductility of the water-quenched steel appears to be related to the as-quenched brittleness noted above. The differences in behavior between the two types of dual-phase steels is believed to be a consequence of the differences in carbon level in the martensite phase. The higher carbon martensite in the air-cooled steel will be stronger, less ductile and more susceptible to local fracture during the hole punching operation than the lower carbon martensite in the water-quenched steel. Thus, the improvement in percent hole expansion with tempering temperature is a reflection of the softening of the higher carbon martensite.

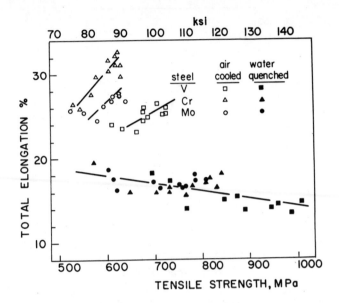

Fig. 11   Tensile strength as a function of total elongation for both the air-cooled
and water-quenched steels following tempering (see Figs. 9 and 10).

Summary

It has been observed that dual-phase steels when water-quenched exhibit
reduced elongations and no necking prior to fracture; a low temperature (~130°C)
temper will restore the ductility. This brittleness is thought to be related to the
delayed failure phenomena noted in quenched fully martensitic steels. The influence
of initial tempering temperature and percent prestrain on the bake-hardening (20
mins. at 170°C) of a LA water-quenched dual-phase steel was studied. It was found
that the increase in the yield stress was a maximum after tempering at about 200°C.
However, the maximum absolute value of the yield stress and the best combination of
tensile strength and total elongation was obtained by tempering at 100°C. The
increase in the yield stress is a consequence of the diffusion of carbon to pin the free
dislocations in the ferrite and/or to form very fine iron carbides. After higher
temperature (up to 550°C for 5 min.) tempering it was observed that the air-cooled
dual-phase steels retained their strength much better than the water-quenched steels;
this was independent of the presence of carbide forming elements such as V, Mo or
Cr. The better retention of strength is thought to be due to auto-tempering of the
higher carbon martensite and the presence of a very fine precipitate in the ferrite.
Tempering is also shown to improve the hydrogen embrittlement resistance and the
percentage hole expansion in both water-quenched and air-cooled dual-phase steels.

Acknowledgements

The author is grateful to W. Stewart for technical assistance in the many facets
of this investigation.

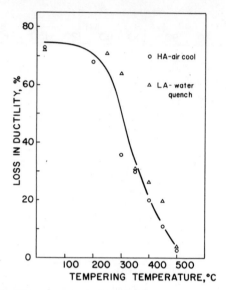

Fig. 12   Loss in ductility due to hydrogen charging as a function of tempering temperature for air-cooled V-HA and the water-quenched LA steel (14).

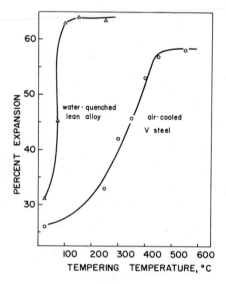

Fig. 13   Percent hole expansion as a function of tempering temperature for air-cooled V-HA and water-quenched LA steel (9).

276

References

1. V. I. Sarrak and G. A. Filippov, "The Nature of Delayed Fracture of Quenched Steel," Met. Sci. Heat Treat., 18 (1976) pp. 1050-54.

2. K. Mazanec and R. Sejnoka, "Delayed Fractures in Martensite," Trans. AIME., 233 (1965) pp. 1602-08.

3. S. Epstein, H. J. Cutler and W. Frame, "Vanadium Treated, Non-Aging, Rimming Steel," J. Metals, 2 (1950) pp. 830-34.

4. W. R. D. Jones and G. Combes, "Effects of Vanadium or Chromium on the Strain Ageing of Rimming Steels, "J. Iron Steel Inst. 174 (1953) pp. 9-15.

5. S. Hayami and T. Furukawa, "A Family of High-Strength Cold-Rolled Steels," Microalloying 75 proceedings, pp. 311-320; Union Carbided Corp., New York, N.Y. 1977.

6. R. G. Davies, "The Deformation Behavior of a Vanadium-Strengthened Dual Phase Steel," Met. Trans., 9A (1978) pp. 41-52.

7. K. Araki, S. Fukunaka and K. Uchida, "Development of Continuously Annealed High Strength Cold Rolled Sheet Steels," Trans. Iron Steel Inst. Japan, 17 (1977) pp. 701-09.

8. R. G. Davies, "Early Stages of Yielding and Strain-Aging of a Vanadium-Containing Dual-Phase Steel," Met. trans. 10A (1979) pp. 1549-55.

9. R. G. Davies, unpublished research.

10. R. G. Davies and C. L. Magee, "Physical Metallurgy of Automotive High Strength Steels," J. Metals, 31, No. 11 (1979) pp. 17-23.

11. G. R. Speich, "Tempering of Low-Carbon Martensite," Trans. Met Soc. AIME, 245 (1969) pp. 2553-64.

12. G. S. Huppi, D. K. Matlock and G. Krauss, "A Evaluation of the Importance of Epitaxial Ferrite in Dual-Phase Steel Microstructures," Scripta Met., 14 (1980) pp. 1239-43.

13. R. A. Grange, C. R. Hribal and L. F. Porter, "Hardness of Tempered Martensite in Carbon and Low-Alloy Steels," Met. trans., 8A (1977) pp. 1775-85.

14. A. T. Davenport, F. G. Berry and R. W. K. Honeycombe, "Interphase Precipitation in Iron Alloys," Metal Sci. J., 2 (1968) pp. 104-06.

15. R. G. Davies, "Hydrogen Embrittlement of Dual-Phase Steels," Met. Trans., 12A (1981), in press.

TEMPERING OF FERRITE-MARTENSITE STEELS

G. R. Speich and R. L. Miller

United States Steel Corporation
Research Laboratory
Monroeville, PA   15146

Abstract

The effect of tempering at low (200°C, 392°F) and high (650°C, 1192°F) temperatures on the strength, yield behavior, and ductility of ferrite-martensite mixtures was investigated in a series of 1.5 percent manganese steels with various carbon contents which had been inter-critically annealed at 760°C (1400°F) and water-quenched. The change in yield strength upon tempering was complex because of the relief of residual stresses, carbon segregation to dislocations, and the return of discontinuous yielding. After tempering at low-temperature, the yield strength increased in all the steels, but discontinuous yielding returned only in those steels containing low-volume fractions of martensite. After tempering at high temperature, the yield strength decreased and discontinuous yielding returned in all the steels. In contrast, the tensile strength decreased after tempering at either low or high temperature in a simple manner that could be predicted from the measured volume fraction and change in hardness of the martensite phase.

Although both uniform and postuniform elongation increased after tempering, as expected from the decrease in tensile strength, the postuniform elongation increased much more than the uniform elongation. As a result, when the steels were compared at the same tensile-strength level, uniform elongation decreased after tempering, whereas postuniform elongation, reduction of area, and total elongation increased. As-quenched ferrite-martensite steels had the highest ratio of uniform to total elongation, which is probably a factor contributing to their better formability.

279

Introduction

Dual-phase (ferrite-martensite) sheet steels are of considerable interest, especially to automobile manufacturers, because they offer attractive combinations of strength, ductility, and formability at the 550 to 690 MPa (80 to 100 ksi) tensile-strength level.[1-3]* Although the heat treatment of these steels usually involves only an intercritical anneal followed by accelerated cooling,[3] tempering at low temperatures may be used in some production practices such as paint baking. As a result, an understanding of the tempering behavior of these steels is of considerable importance.

In the present work, the effect of tempering at low (200°C, 392°F) and high (650°C, 1192°F) temperatures on the strength, yielding behavior, and ductility of a series of low-carbon, 1.5 percent manganese steels with a ferrite-martensite microstructure was determined. The properties of the same steels in the as-quenched condition had been studied in earlier work.[4] Carbon segregation, carbide distribution, and the dislocation substructure of both the ferrite and martensite phases after tempering were determined by internal friction, scanning electron microscope, and transmission electron microscope techniques, respectively. The changes in yield strength were complex and are discussed in terms of the residual stress patterns, carbon segregation to dislocations, and the return of discontinuous yielding. The changes in tensile strength were simpler and are explained in terms of a simple composite strengthening model based on the volume fraction and change in strength of the martensite phase.

Materials and Experimental Procedures

The steels selected for study consisted of seven 1.5 percent Mn steels containing 0.005, 0.06, 0.12, 0.16, 0.20, 0.29, and 0.40 percent carbon. The manganese content of 1.5 percent was chosen for all the steels because this is similar to the manganese content of most dual-phase steels.[4] The carbon-content range of most dual-phase steels is 0.06-0.12 percent,[4] and the carbon content of two of the steels were selected to include this range. However, in addition, a very low-carbon steel and several higher carbon steels were included to study the effect of tempering on the ferrite matrix, and to examine the effect of tempering on a wider volume fraction range of ferrite-martensite mixtures than those normally encountered in dual-phase steels. All steels were prepared as aluminum-killed vacuum-induction-melted 45 kg (100 lb) heats. Their chemical compositions are given in Table I.

---

* See References.

Table I. Chemical Composition of Steels, Weight Percent

Steel	C	Mn	P	S	Si	Al	N
0.005C-1.5Mn	0.005	1.51	0.009	0.010	0.24	0.010	0.006
0.06C-1.5Mn	0.060	1.47	0.009	0.010	0.23	0.009	0.007
0.12C-1.5Mn	0.120	1.47	0.009	0.010	0.24	0.008	0.007
0.16C-1.5Mn	0.160	1.53	0.009	0.009	0.24	0.024	0.005
0.20C-1.5Mn	0.200	1.53	0.009	0.010	0.25	0.023	0.006
0.29C-1.5Mn	0.290	1.51	0.009	0.010	0.26	0.021	0.007
0.40C-1.5Mn	0.400	1.53	0.009	0.010	0.25	0.026	0.005

The steels were hot-rolled straightaway to 13-mm-thick (1/2-in.) plate, and specimen blanks were cut in the transverse direction from the hot-rolled plate. To reduce the amount of pearlite banding,[1] these blanks were first austenitized 30 minutes at 870°C (1598°F), followed by isothermal transformation for 2 hours at 550°C (1020°F). These specimens were then intercritically annealed 1 hour at 760°C (1400°F) and water-quenched. This intercritical heat treatment produced a series of specimens with various volume fractions of martensite, depending on the carbon content of the steel.[4] These specimens were then tempered 1 hour at 200°C (392°F) or at 650°C (1192°F), and air-cooled.

Axisymmetric tension-test specimens with a diameter of 6.35 mm (0.250 in.) and a gage length of 25.4 mm (1 in.) were machined from the tempered blanks. The yield strength, tensile strength, work-hardening behavior, uniform elongation,* total elongation, and reduction of area were then determined.

The carbide distribution in the tempered specimens was studied by scanning electron microscopy after the specimens were polished and etched in two percent nital. A number of specimens were also electrolytically thinned and examined by transmission electron microscopy to determine the dislocation substructure. The segregation of carbon to dislocations in the ferrite was studied by internal friction. Retained-austenite content was measured by x-ray diffraction analysis. The percent martensite in the as-quenched specimens had been determined by quantitative metallography in the earlier work.[4]

---

* Uniform elongation was determined from computer plotting of the data and application of the Considere' condition near maximum load.[4]

Results

Metallography

The percent martensite,, percent retained austenite, and carbon content of the martensite phase (determined from a simple carbon mass balance) for the seven steels are given in Table II. These data are taken from the earlier work.[4] When the steels were tempered at either 200 or 650°C, the amount of martensite and its overall carbon content were assumed to remain constant. The retained-austenite content did not change upon tempering at 200°C but decreased to zero upon tempering at 650°C, as shown in Table II.

The microstructures of the 0.06C-1.5Mn steel in the as-quenched condition and after tempering are shown in Figures 1A, B, and C. Scanning electron micrographs (upper) indicate the nature of the carbide precipitation in the martensite phase, and transmission electron micrographs (lower) indicate the dislocation density and carbide precipitation in the ferrite phase. The microstructures of the other steels were very similar except that the percent martensite increased in the steels with higher carbon contents.

In the as-quenched condition, Figure 1A, the martensite phase was nearly featureless and was distributed as small particles at the ferrite grain boundaries. The dislocation density in the ferrite phase was very high, especially in the vicinity of the martensite particles (marked with M), because of plastic deformation needed to accommodate the volume change accompanying the austenite-to-martensite transformation.[5]

Tempering at 200°C, Figure 1B, resulted in carbide precipitation in both the martensite and ferrite phases. Most of the carbide precipitation in the martensite phase occurred at interlath or interplate boundaries, causing the martensite structure to become much more clearly delineated than in the as-quenched condition.

Tempering at 650°C, Figure 1C, resulted in a marked lowering of the dislocation density in the ferrite phase and the precipitation of coarse carbide particles in the martensite phase.

Internal Friction

Internal-friction results for the 0.06C-1.5Mn steel in the as-quenched and tempered conditions are shown in Figure 2. The low-temperature Snoek peak (40°C, 1 cycle/s) is associated with the amount of carbon in solution in the ferrite.* However, exact estimates of the amount of carbon in solution are difficult because of possible contributions from nitrogen or from manganese-nitrogen interactions to the Snoek peak,[8] since all the steels contained 0.007 percent nitrogen. However, from the Fe-Mn-C phase diagram,[9] the carbon-solubility limit of the ferrite at 760°C is much lower in a 1.5Mn steel than in an Fe-C alloy. The small amount of martensite observed in the 0.005C-1.5Mn steel (Table II) is evidence of the prior existence of austenite at the

---

* Carbon in the martensite phase does not contribute to the overall Snoek peak because the carbon atoms tend to remain in an ordered array as a result of elastic energy considerations.[6,7]

Table II. Metallographic Parameters for As-Quenched and Tempered Ferrite-Martensite Steels

Steel	Percent Martensite	Percent Retained Austenite	Carbon Content* of Martensite Phase, Percent
		As-Quenched	
0.005C-1.5Mn	0.3	0.0	-
0.06C-1.5Mn	24.8	1.0	0.25
0.12C-1.5Mn	35.5	1.4	0.37
0.16C-1.5Mn	41.1	1.3	0.40
0.20C-1.5Mn	48.9	2.0	0.41
0.29C-1.5Mn	60.6	3.1	0.48
0.40C-1.5Mn	89.7	4.1	0.44
		1 hr 200°C	
0.005C-1.5Mn		0.0	
0.06C-1.5Mn		1.0	
0.12C-1.5Mn		1.4	
0.16C-1.5Mn		1.8	
0.20C-1.5Mn		2.3	
0.29C-1.5Mn		3.4	
0.40C-1.5Mn		4.4	
		1 hr 650°C	
0.005C-1.5Mn		-	
0.06C-1.5Mn		0.0	
0.12C-1.5Mn		-	
0.16C-1.5Mn		0.0	
0.20C-1.5Mn		-	
0.29C-1.5Mn		-	
0.40C-1.5Mn		-	

* Calculated from carbon mass balance.[4]

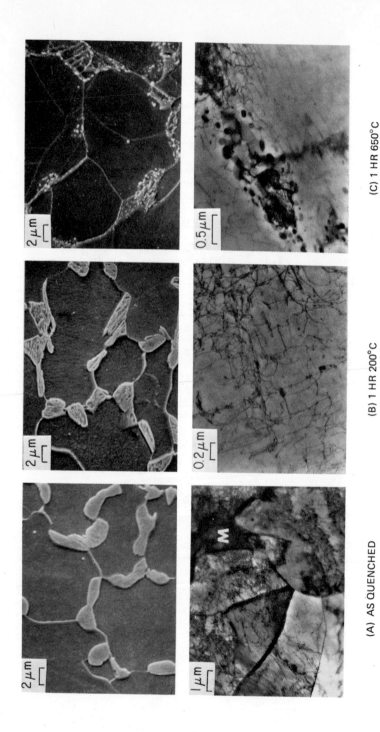

(A) AS QUENCHED   (B) 1 HR 200°C   (C) 1 HR 650°C

Fig. 1 — Microstructure of 0.06C-1.5Mn steel intercritically annealed 1 hour at 760°C, water-quenched, and tempered 1 hour at 200°C (392°F) or 650°C (1192°F). [Scanning electron micrographs (upper), 2% nital etch; transmission electron micrographs (lower).]

intercritical annealing temperature. Clearly, then, the carbon-solubility limit of ferrite at 760°C must be less than 0.005 weight percent, and the carbon content of the ferrite at room temperature would be less than this value.

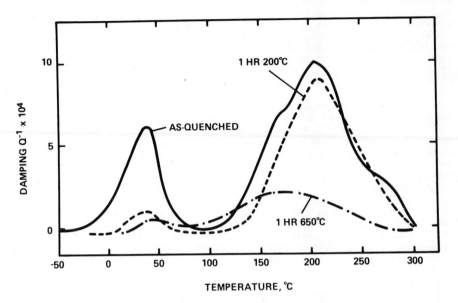

Fig. 2 — Internal friction peaks of 0.06C-1.5Mn steel in as-quenched condition and after tempering 1 hour at 200°C (392°F) or 650°C (1192°F).

The high-temperature Köster peak at 200°C is associated with dislocation-carbon interactions. The Köster peak is expected to increase when the dislocation density is increased and when there is significant segregation of carbon to dislocation sites.

For specimens in the as-quenched condition, both substantial Snoek and Köster peaks are observed, indicating that both a significant amount of free carbon and a high dislocation density are present. After the specimens were tempered at 200°C, the Snoek peak was drastically lowered, consistent with the lowering of the dissolved carbon in the ferrite as a result of carbide precipitation in the ferrite (Figure 1B). However, the Köster peak was only slightly lowered because the dislocation density in the ferrite was still high (Figure 1B).

After the specimens were tempered at 650°C, both the Snoek and Köster peaks were drastically lowered. Although substantial carbon could have been dissolved in the ferrite by heating to 650°C, the precipitation of carbon during air cooling from 650°C[6] still results in a lowering of the Snoek peak. The lowering of the Köster peak is associated with the drastically lowered dislocation density caused by recovery and recrystallization of the ferrite (Figure 1C).

285

Yield and Tensile Strength

The yield strength (0.2% offset), tensile strength, percent uniform and total elongations, and percent reduction of area of the various steels in the as-quenched[4] and in the tempered conditions are given in Table III. The yield strengths of the as-quenched and the tempered steels are plotted versus percent martensite in Figure 3 and are seen to increase with increasing percent martensite in a nonlinear manner. For any given percent martensite, the yield strength increased slightly upon tempering at 200°C, but decreased slightly upon tempering at 650°C.

The tensile strengths of the as-quenched and the tempered steels are plotted versus percent martensite in Figure 4 (Data for the 0.29C-1.5Mn, and 0.40C-1.5Mn steels are not shown because the specimens broke before maximum load was reached.) The tensile strengths of both the as-quenched and the tempered steels increased smoothly with increasing percent martensite. For any given percent martensite, the tensile strength decreased slightly upon tempering at 200°C and greatly upon tempering at 650°C.

The changes in yield and tensile strengths for the seven steels upon tempering at 200°C are shown more clearly in Figure 5. The changes in yield strengths were positive but showed a complex variation with percent martensite, first increasing, then decreasing, and again increasing as the percent martensite increased. The initial maximum occurred at about 25 percent martensite, just before discontinuous yielding was replaced by continuous yielding, Table III. A second maximum occurred near 100 percent martensite. In contrast, the change in tensile strength was negative and decreased smoothly with increasing percent martensite. (Again, changes in tensile strength for the 0.29C-1.5Mn and 0.40C-1.5Mn could not be obtained because the specimens broke before maximum load was reached.

The changes in yield and tensile strengths of the steels upon tempering at 650°C are shown in Figure 6. The changes in yield and tensile strengths were negative and both decreased smoothly with increasing percent martensite, in contrast to the more complex behavior observed after tempering at 200°C.

Uniform Elongation, Total
Elongation, and Reduction of Area

Uniform and total elongations of the as-quenched and the tempered steels are plotted versus percent martensite in Figures 7 and 8, respectively, for all but the 0.40C-1.5Mn steel. The linear variation of uniform and total elongations of the tempered ferrite-martensite steels with percent martensite is similar to results reported earlier for as-quenched ferrite-martensite steels.[1] Tempering at 200°C did not affect uniform elongation but slightly increased total elongation for any given percent martensite. Tempering at 650°C led to a large increase in both uniform and total elongation for any given percent martensite.

Table III. Mechanical Properties of Tempered Ferrite-Martensite Steels

Steel	Heat Treatment	Yield Strength, MPa (ksi)*	Tensile Strength, MPa (ksi)	Uniform Elongation, %	Total Elongation, %	Reduction of Area, %
0.005C-1.5Mn	As-quenched	257 (37.3)C	401 (58.2)	21.1	37.5	81.8
0.06C-1.5Mn	"	329 (47.8)C	643 (93.4)	15.2	25.0	54.8
0.12C-1.5Mn	"	400 (58.1)C	824 (119.5)	10.5	17.0	33.2
0.16C-1.5Mn	"	448 (65.1)C	924 (134.0)	9.2	14.5	27.8
0.20C-1.5Mn	"	569 (82.5)C	1159 (168.1)	7.2	10.0	17.9
0.29C-1.5Mn	"	769 (110.4)C	–	–	4.0	2.6
0.40C-1.3Mn	"	954 (138.5)C	–	–	1.0	0.3
0.005C-1.5Mn	1 hr 200°C (392°F)	266 (38.6)D	388 (56.4)	21.5	38.0	83.5
0.06C-1.5Mn	"	392 (56.9)D	613 (88.9)	14.9	27.0	65.9
0.12C-1.5Mn	"	446 (64.7)D	786 (114.0)	9.9	18.0	41.5
0.16C-1.5Mn	"	490 (71.2)C	905 (131.2)	9.3	17.0	39.6
0.20C-1.5Mn	"	594 (86.2)C	1095 (158.9)	8.5	13.0	27.9
0.29C-1.5Mn	"	767 (111.3)C	1385 (200.9)	–	5.0	18.8
0.40C-1.5Mn	"	1126 (163.4)C	–	–	4.0	6.2
0.005C-1.5Mn	1 hr 650°C (1192°F)	272 (39.5)D	367 (53.3)	22.8	44.0	84.2
0.06C-1.5Mn	"	302 (43.9)D	424 (61.5)	20.9	39.0	80.1
0.12C-1.5Mn	"	366 (53.1)D	482 (70.0)	17.7	35.0	74.2
0.16C-1.5Mn	"	393 (57.1)D	512 (74.3)	16.5	34.0	69.9
0.20C-1.5Mn	"	437 (63.4)D	562 (81.5)	14.8	29.0	69.0
0.29C-1.5Mn	"	514 (74.5)D	633 (91.9)	13.3	26.0	62.5
0.40C-1.5Mn	"	623 (90.4)D	722 (104.7)	10.9	24.0	57.0

*D = Discontinuous Yielding (lower yield)
C = Continuous Yielding (0.2% offset)

287

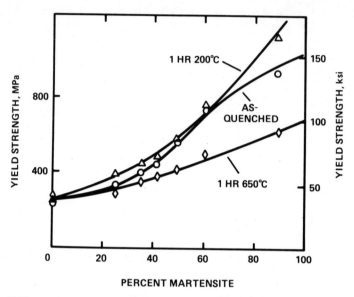

Fig. 3 — Yield strength of ferrite-martensite steels in as-quenched condition and after tempering 1 hour at 200°C (392°F) or 650°C (1192°F).

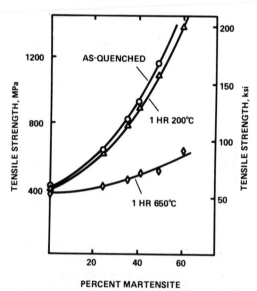

Fig. 4 — Tensile strength of ferrite-martensite steels in as-quenched condition and after tempering 1 hour at 200°C (392°F) or 650°C (1192°F).

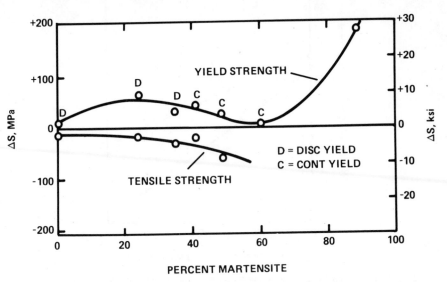

Fig. 5 — Changes in yield strength (0.2% offset) and tensile strength of ferrite-martensite steels after tempering 1 hour at 200°C (392°F).

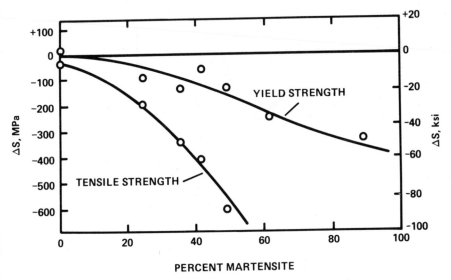

Fig. 6 — Changes in yield (0.2% offset) and tensile strengths of ferrite-martensite steels after tempering 1 hour at 650°C (1192°F).

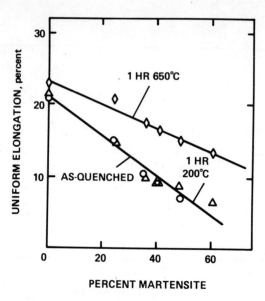

Fig. 7 — Uniform elongation of ferrite-martensite steels in as-quenched condition and after tempering 1 hour at 200°C (392°F) or 650°C (1192°F).

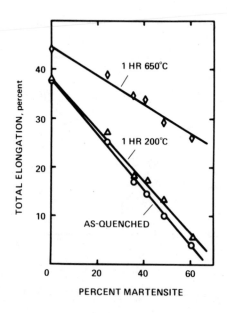

Fig. 8 — Total elongation of ferrite-martensite steels in as-quenched condition and after tempering 1 hour at 200°C (392°F) or 650°C (1192°F).

The same data are plotted versus tensile strength in Figure 9. Both uniform, e_u, and total elongation, e_t, increased as the tensile strength decreased.* However, when the steels are compared at the same tensile strength, the effect of tempering was to decrease the uniform elongation and to increase the total elongation. As a result, the ratio of uniform elongation to total elongation, e_u/e_t, was much higher in the as-quenched steels than in the tempered steels, and somewhat higher for steels tempered at 200°C than for those tempered at 650°C, at all strength levels, as shown in Figure 10. The postuniform elongation (difference between total and uniform elongation) is also plotted versus tensile strength in Figure 11. The effect of tempering was to significantly increase postuniform elongation for any given tensile-strength level, with highest values being exhibited by specimens tempered at 650°C.

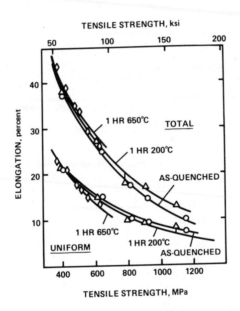

Fig. 9 — Uniform and total elongation of ferrite-martensite steels in as-quenched condition and after tempering 1 hour at 200°C (392°F) or 650°C (1192°F).

---

* Both uniform and total elongation are reported in terms of engineering strain, Table III. The symbol ε_u will be used for true uniform strain.

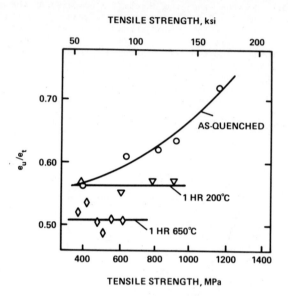

Fig. 10 — Effect of tempering on ratio of uniform to total elongation of ferrite-martensite steels.

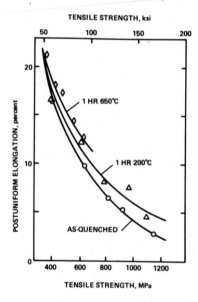

Fig. 11 — Effect of tempering on postuniform elongation of ferrite-martensite steels.

Reduction of area of the as-quenched and the tempered steels is plotted versus percent martensite in Figure 12. The linear decrease in reduction of area with increase in percent martensite is similar to results reported earlier for as-quenched ferrite-martensite steels.[4] Tempering at 200°C slightly increased reduction of area for any given percent martensite. Tempering at 650°C led to a much larger increase in reduction of area. The same data are plotted versus tensile strength in Figure 13. At the same tensile-strength level, tempering increased the reduction of area. These results are consistent with those obtained for postuniform elongation (Figure 11).

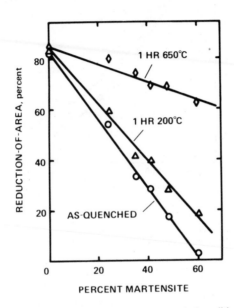

Fig. 12 — Reduction of area of ferrite-martensite steels in as-quenched condition and after tempering 1 hour at 200°C (392°F) or 650°C (1192°F).

Stress-Strain Behavior

Because of the complex response of these steels to tempering, the complete true-stress—true-strain curves for the steels in the as-quenched and the tempered conditions were calculated and plotted automatically by using a computer program[10]. The results for the 0.06C-1.5Mn, 0.20C-1.5Mn, and 0.40C-1.5Mn steels in the as-quenched condition and after tempering at 200°C are shown in Figure 14. The results for the same steels after tempering at 650°C are shown in Figure 15.

Tempering at 200°C had a complex effect on the stress-strain behavior, which varied with the amount of martensite in the steels (Figure 14). For the 0.06C-1.5Mn steel (containing 24.8% martensite), tempering at 200°C resulted in a return of discontinuous yielding, an increase in flow stress up to 0.8 percent strain, and a decrease in flow

293

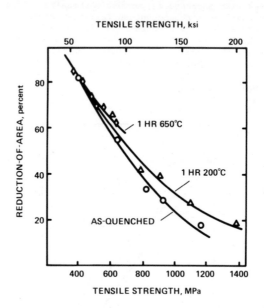

TENSILE STRENGTH, ksi

1 HR 650°C

1 HR 200°C

AS-QUENCHED

TENSILE STRENGTH, MPa

Fig. 13 — Reduction-of-area of ferrite-martensite steels in as-quenched condition and after tempering 1 hour at 200°C (392°F) or 650°C (1192°F).

stress from 0.8 percent strain up to the necking strain. For the 0.20C-1.5Mn steel (containing 48.9% martensite), tempering at 200°C did not change the continuous-yielding behavior of the as-quenched steels nor did it change the flow stress up to 1.5 percent strain, but it did result in a decrease in flow stress at higher strains. For the 0.40C-1.5Mn steel (containing 89.7% martensite), tempering at 200°C did not change the continuous yielding behavior of the as-quenched steels but resulted in an increase in flow stress up to 1 percent strain. (The as-quenched specimens broke in a brittle manner at higher strains.) As a result of this complex behavior, the yield strength upon tempering at 200°C first increased, then decreased, then again increased as the percent martensite increased, as shown in Figure 5. In contrast, the tensile strength decreased after tempering at 200°C for all percentages of martensite, as shown in Figure 5.

After tempering at 650°C, discontinuous yielding returned in all the steels. Typical results for the 0.06C-1.5Mn, 0.20C-1.5Mn, and 0.40C-1.5Mn steels are shown in Figure 15. Also, both the yield and tensile strengths decreased after tempering at 650°C (compare with curves for as-quenched specimens of Figure 14).

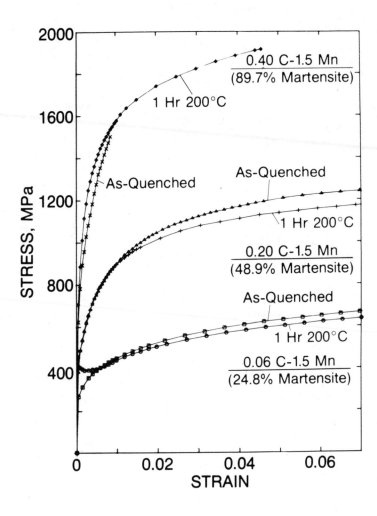

Fig. 14 — True-stress—true-strain curves for ferrite-martensite steels in as-quenched condition and after tempering 1 hour at 200°C.

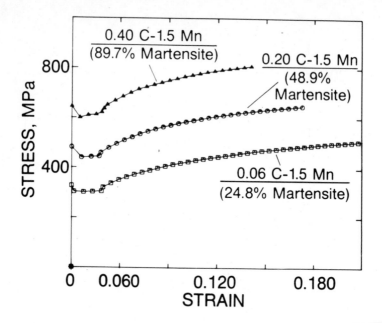

Fig. 15 — True-stress–true-strain curves for ferrite-martensite steels after tempering 1 hour at 650°C.

Tensile Strength

      The variation in tensile strength of ferrite-martensite steels with the amount and carbon content of the martensite phase has been previously analyzed by using a simple composite-strengthening model.[4] Increasing the carbon content of the martensite phase was shown to increase the tensile strength of the composite because of the increased strength of the martensite phase. In a similar manner, tempering will decrease the strength of the martensite phase, and thus should result in a decrease in the strength of the composite.

      From previous work,[4] let S_T, $S_{T,\alpha}$, and $S_{Y,m}$ be the tensile strengths of the composite and of the ferrite and the yield strength of martensite, respectively, and let P_α and P_m be the volume percent ferrite and martensite, respectively. Then,

$$S_T = S_{T,\alpha} P_\alpha/100 + S_{Y,m} P_m/100 \qquad (1)$$

The yield strength of the martensite rather than the tensile strength is used because most of the plastic strain is concentrated in the softer ferrite phase.[4]

The change in tensile strength upon tempering, ΔS_T, is then given by

$$\Delta S_T = \Delta S_{T,\alpha} + (\Delta S_{Y,m} - \Delta S_{T,\alpha})\ P_m/100 \qquad (2)$$

where $\Delta S_{T,\alpha}$ and $\Delta S_{Y,m}$ are the changes upon tempering in the tensile strength of the ferrite and the yield strength of the martensite, respectively, and where the relationship $P_m + P_\alpha = 100$ has been used. Because the change in hardness (and thus the tensile strength) can be obtained for different carbon contents from the work of Grange et al.[11] whereas the change in yield strength cannot, it is simpler to rewrite Equation 2 as

$$\Delta S_T = \Delta S_{T,\alpha} + (\beta\ \Delta S_{T,m} - \Delta S_{T,\alpha})\ P_m/100 \qquad (3)$$

where β is the yield/tensile strength ratio and $\Delta S_{T,m}$ is the change in tensile strength of the martensite upon tempering.

From the work of Grange et al.,[11] the change in hardness of the martensite upon tempering was determined for martensite with the carbon contents given in Table II. Then, from an experimentally determined conversion of hardness to tensile strength,[12] the value of $\Delta S_{T,m}$ was determined. The value of $\Delta S_{T,\alpha}$ was obtained from the tensile-strength data for the 0.005C-1.5Mn steel, Table III.

From the measured values of $\Delta S_{T,m}$ and $\Delta S_{T,\alpha}$, a value of 0.80 for β, (characteristic of tempered martensite*) and the measured values of P_m (Table II), values were calculated for ΔS_T by using Equation 3. The calculated values are compared with the experimental values in Figure 16. The correct dependence of ΔS_T on percent martensite is predicted by Equation 3, although the calculated values are slightly greater than the observed values.

Yield Strength and Yielding Behavior

The application of simple composite-strengthening models to explain the changes in yield strength upon tempering at low temperatures is much more difficult because the yield behavior at small plastic strains in these steels is affected by several complex variables including residual stresses, mobile dislocation density, transformation of retained austenite, and plastic incompatibility of the hard and soft phases.[4,13-15]

---

* Values of β range from 0.68 to 0.89 depending on tempering temperature. We have taken an average value.

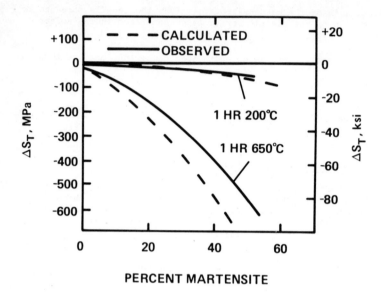

Fig. 16 — Comparison of calculated and observed changes in tensile strength after tempering 1 hour at 200°C (392°F) or 650°C (1192°F).

Of these factors, the changes in either residual stresses or in mobile dislocation density are believed to be primarily responsible for changes in the yield strength upon tempering at low temperatures. The other two factors are not considered here because the retained-austenite content was low (Table I) and changes in plastic incompatibility of the two phases would be expected to lower, not raise, the yield strength as experimentally observed after tempering at low temperatures (Figure 5).

The residual-stress pattern about the martensite particles is similar to that developed for the case of a spherical hole in an infinite matrix when the hole is expanded by a uniform internal pressure.[16] Analysis of the stresses for this case shows there is a zone immediately adjacent to the particle in which the yield stress of the ferrite is exceeded. As a result, a plastic zone is generated about each particle. Further out from this zone, the stresses are in the elastic range and decay rapidly with increasing distance from the particle. When an applied stress is added to these residual elastic stresses, the net effect is to produce plastic deformation, even when the applied stresses are below the normal yield stress.[17]

Gerbase et al.[15] have analyzed the effect of residual stresses on initial work-hardening behavior of dual-phase steels. By using a simple distribution of residual stresses, they predict that the flow stress is dependent on the magnitude of the residual stress and the volume fraction of the ferrite subject to the stress. If the volume percent of ferrite subject to the residual stress is assumed to decrease exponentially with strain, then the initial high work-hardening behavior of dual-phase steels can be explained.

Tempering is known to lower the magnitude of the residual stresses in quenched steels because of the volume contraction accompanying carbide precipitation in the martensite phase, even at temperatures as low as 200°C.[18] According to the residual-stress model, this would increase the yield strength as experimentally observed.

The complex variation of the increase in yield strength with percent martensite after tempering at 200°C (Figure 5) can be explained by a complicated interplay between the magnitude of the residual stresses in the ferrite and martensite phases and the percent martensite. At low percentages of martensite the residual stresses in the ferrite are low and can be relieved by tempering; as a result, discontinuous yielding returns and the increase in yield strength is large. At intermediate percentages of martensite, the residual stresses in the ferrite are much larger and are only partly relieved by tempering; as a result, the continuous-yielding behavior is maintained and the increase in yield strength is small. At high percentages of martensite, where martensite is the matrix phase, residual stresses in the martensite phase become important. These residual stresses are more easily relieved by tempering;[18] as a result, the increase in yield strength again becomes large. Of course, exact calculations cannot be made because of the complex nature of the residual-stress patterns.

Tanaka et al.[19] and Matlock et al.[14] have used a mobile-dislocation-density model to explain the initial yielding behavior of dual-phase steels. Tanaka et al. have also used this model to explain the strain-aging response of dual-phase steels. In this model, tempering at 200°C would cause carbon to segregate to dislocations in the ferrite and lower the mobile dislocation density, as shown by the internal-friction results (Figure 2). This would result in an increase in the flow stress, as experimentally observed. The segregation of carbon to dislocations in the ferrite can also explain the return of the yield point in those steels with lower volume fractions of martensite.[20] The lack of return of yield point at higher volume fractions of martensite (Figure 12) would have to be explained by assuming that the dislocation density generated in the ferrite was so high that not all the dislocations were pinned by the available carbon. As in the case of the residual-stress model, quantitative prediction is not possible because there is no way of quantitatively measuring the mobile dislocation density.

After tempering at high temperatures, residual stresses are absent, and the overall dislocation density has been greatly lowered (Figure 1). As a result, discontinuous yielding returns in all the steels (Figure 8). Also, because of the large decrease in the hardness of the martensite phase, the yield strength of these steels is greatly reduced (Figure 6).

Ductility

Ductility, as measured by total elongation, consists of uniform and postuniform elongation. The uniform elongation is related to the percentage of martensite and to its hardness because of their influence on the strength level and the work-hardening rate.[4] The strength level and work-hardening rate must be such that the Considere condition,

299

$$\sigma = d\sigma/d\varepsilon \qquad\qquad (4)$$

is satisfied at a strain value equal to the uniform elongation, where σ and ε are true stress and true strain. Because $d\sigma/d\varepsilon$ decreases with increasing strain, and σ increases with increasing strain, at some strain these values become equal, as shown schematically in Figure 17(A). The strain at which this occurs is the true uniform strain, ε_u.

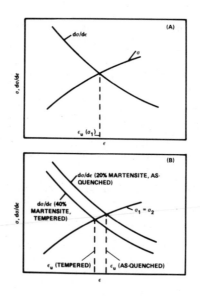

Fig. 17 — Considere condition for uniform strain.

For higher percentages of martensite, both the strength level and the work-hardening rate increase.[4] The uniform elongation then decreases, because the relative shift of the σ and $d\sigma/d\varepsilon$ curves results in an intersection at a lower value of plastic strain. Upon tempering, the hardness of the martensite phase decreases and the tensile strength and work-hardening rate decrease. It is expected that the uniform elongation will then increase, as shown in Figure 7, because the intersection point now moves to higher plastic strains.

However if the steels are compared at the same tensile strength, the uniform elongation actually decreases upon tempering, as shown in Figure 9. This is caused by the work-hardening rate of a steel with a higher volume fraction of softer martensite being lower than for a steel with a lower volume fraction of harder martensite, as shown schematically in Figure 17(B). (The volume fraction of martensite and the hardness of the martensite must be balanced to achieve a given tensile strength.) Because of this effect, the ratio of uniform to total elongation for any .

4) Although both uniform and postuniform elongation increased upon tempering, as expected from the decrease in tensile strength, the increase in postuniform elongation was much greater than the increase in uniform elongation. As a result, when the steels were compared at the same tensile-strength level, uniform elongation decreased upon tempering whereas postuniform elongation, reduction of area, and total elongation increased.

5) As-quenched ferrite-martensite steels had the highest ratio of uniform to total elongation at all strength levels, which is probably a factor contributing to their better formability.

Acknowledgment

The authors would like to express their appreciation to their colleagues, J. T. Michalak and J. S. Lally for performing the internal friction and transmission electron microscope studies.

References

1. Rashid, M. S., "GM980X - Potential Applications and Review," SAE Reprint No. 770211, March, 1977.

2. Davies, R. G., "The Deformation Behavior of the Vanadium-Strengthened Dual-Phase Steel," Met. Trans, 9A (1978), pp. 41-52.

3. Repas, P. E., "Metallurgy, Production, Technology, and Properties of Dual-Phase Sheet Steels," pp. 13-22, in Dual-Phase and Cold-Pressing Vanadium Steels in the Automobile Industry, Vanitech, London, 1978.

4. Speich, G. R., and Miller, R. L., "Mechanical Properties of Ferrite-Martensite Steels," pp. 145-82, in Structure and Properties of Dual-Phase Steels, AIME, New York, NY, 1979.

5. Moyer, J. M., and Ansell, G. S., "The Volume Expansion Accompanying the Martensite Transformation in Iron-Carbon Alloys," Met. Trans., 6A (1975), pp., 178-91.

6. Stark, P., Averbach, B. L., and Cohen, M., "Influence of Microstructure on Carbon Damping in Iron-Carbon Alloys," Acta Metall., 6, (1956), pp. 149-55.

7. Stark, P., Averbach, B. L., and Cohen, M., "Internal Friction Measurements on Tempered Martensite," Acta Metall. 3, (1956), p. 91.

8. Enrietto, J. F., "Complex Damping Effects in Fe-Mn-N Alloys," Trans. AIME, 224 (1962), pp. 1119-23.

9. Hillert, M., and Waldenstrom, M., "Isothermal Sections of the Fe-Mn-C System in the Temperature Range 873K - 1373K," Calphad, 1 (1977), pp. 97-132.

10. Wray, P. J., private communication, U. S. Steel Research Laboratory, Monroeville, PA 15146, 1979.

11. Grange, R. A., Hribal, C.K., and Porter, L. F., "Hardness of Tempered Martensite in Carbon and Low-Alloy Steels," Met. Trans, 9A, (1977), pp. 1175-85.

12. Speich, G. R., and Schwoeble, A. J., private communication, U. S. Steel Research Laboratory, Monroeville, PA 15146, 1979.

13. Rigsbee, J. M., and VanderArend, P. J., Laboratory Studies of Micro-structures and Structure-Property Relationships in Dual-Phase Steels," pp. 58-88, in "Formable HSLA and Dual-Phase Steels," AIME, New York, NY, 1979.

14. Matlock, D. K. Krauss, G., Ramos, L. F., and Huppi, G. S., A Correlation of Processing Variables with Deformation Behavior of Dual-Phase Steels," pp. 62-90, in Structure and Properties of Dual-Phase Steels, AIME, New York, NY, 1979.

15. Gerbase, J., Embury, J. D., and Hobbs, R. M., "The Mechanical Behavior of Some Dual-Phase Steels and Their Relation to Theoretical Models," pp. 118-44, in Structure and Properties of Dual-Phase Steels, AIME, New York, NY, 1979.

16. Hillert, M., "Pressure-Induced Diffusion and Deformation During Precipitation, Especially Graphitization," Jerkonterets Annaler, 41, (1957), pp. 67-81.

17. Muir, H., Averbach, B. L., Cohen, M., "The Elastic Limit and Yield Behavior of Hardened Steels," ASM Trans., 47, (1955), pp. 380-407.

18. Brown, R. L., Rack, H. J., and Cohen, M., "Stress Relaxation during the Tempering of Hardened Steel," Mat. Sci. and Eng., 21, (1975), pp. 25-34.

19. Tanaka, T., Nishida, M., Hasiguchi, K., and Kato, T., "Formation and Properties of Ferrite Plus Martensite Dual-Phase Structures," pp. 221-41 in Structure and Properties of Dual-Phase Steels," AIME, New York, NY, 1979.

20. Baird, J. D., "Strain-Aging of Steel - A Critical Review," Iron and Steel, 36, (1963), Parts I-III, pp. 186-450.

21. Brown, L. M., and Embury, J. D., "The Initiation and Growth of Voids at Second-Phase Particles," pp. 164-169, in Microstructure and Design of Alloys, 1, (1973), Cambridge.

---

* It is understood that the material in this paper is intended for general information only and should not be used in relation to any specific application without independent examination and verification of its applicability and suitability by professionally qualified personnel. Those making use thereof or relying thereon assume all risk and liability arising from such use or reliance.

Fig. 1 — Microstructure of 0.06C-1.5Mn steel intercritically annealed 1 hour at 760°C, water-quenched, and tempered 1 hour at 200°C (392°F) or 650°C (1192°F). [Scanning electron micrographs (upper), 2% nital etch; transmission electron micrographs (lower).]

Fig. 2 — Internal friction peaks of 0.06C-1.5Mn steel in as-quenched condition and after tempering 1 hour at 200°C (392°F) or 650°C (1192°F).

Fig. 3 — Yield strength of ferrite-martensite steels in as-quenched condition and after tempering 1 hour at 200°C (392°F) or 650°C (1192°F).

Fig. 4 — Tensile strength of ferrite-martensite steels in as-quenched condition and after tempering 1 hour at 200°C (392°F) or 650°C (1192°F).

Fig. 5 — Changes in yield strength (0.2% offset) and tensile strength of ferrite-martensite steels after tempering 1 hour at 200°C (392°F).

Fig. 6 — Changes in yield (0.2% offset) and tensile strengths of ferrite-martensite steels after tempering 1 hour at 650°C (1192°F).

Fig. 7 — Uniform elongation of ferrite-martensite steels in as-quenched condition and after tempering 1 hour at 200°C (392°F) or 650°C (1192°F).

Fig. 8 — Total elongation of ferrite-martensite steels in as-quenched condition and after tempering 1 hour at 200°C (392°F) or 650°C (1192°F).

Fig. 9 — Uniform and total elongation of ferrite-martensite steels in as-quenched condition and after tempering 1 hour at 200°C (392°F) or 650°C (1192°F).

Fig. 10 — Effect of tempering on ratio of uniform to total elongation of ferrite-martensite steels.

Fig. 11 — Effect of tempering on postuniform elongation of ferrite-martensite steels.

Fig. 12 — Reduction of area of ferrite-martensite steels in as-quenched condition and after tempering 1 hour at 200°C (392°F) or 650°C (1192°F).

Fig. 13 — Reduction-of-area of ferrite-martensite steels in as-quenched condition and after tempering 1 hour at 200°C (392°F) or 650°C (1192°F).

Fig. 14 — True-stress—true-strain curves for ferrite-martensite steels in as-quenched condition and after tempering 1 hour at 200°C.

Fig. 15 — True-stress—true-strain curves for ferrite-martensite steels after tempering 1 hour at 650°C.

Fig. 16 — Comparison of calculated and observed changes in tensile strength after tempering 1 hour at 200°C (392°F) or 650°C (1192°F).

Fig. 17 — Considere condition for uniform strain.

given tensile-strength level is greatest for the as-quenched ferrite-martensite steels (Figure 10). This observation may offer a rationale for the superior formability of these steels.[1]

The postuniform elongation arises from the additional elongation that occurs because of necking of the specimen and is equal to the difference between the total and uniform elongation. For any given tensile-strength level, postuniform elongation increases substantially when the specimens are tempered as shown in Figure 11. A similar result occurs for reduction of area, as shown in Figure 13. This, of course, is expected since both of these ductility parameters are related to the necking of the tension-test specimen.

Reduction of area and postuniform elongation are related to the micro-void coalescence and growth processes that occur in the necked region of the tension-test specimen. Voids form at the martensite/ferrite interface by decohesion or by cracking of the martensite particles during plastic straining. These voids then grow upon further plastic straining until the ductile ferrite ligaments joining the martensite particles fail by localized shear.[4] The total strain to fracture thus involves both the critical strain for void nucleation and the strain for tearing of the ferrite ligaments joining these voids.[21]

Increasing the percentage of martensite shortens the ferrite ligaments separating the martensite particles so that less plastic strain is required to rupture them. As a result, reduction of area decreases (Figure 13). Tempering decreases the hardness of the martensite phase, and thus increases the critical strain required for decohesion or cracking of the martensite particle. As a result, reduction of area increases after tempering for any given percent martensite (Figure 13). Even when the steels are compared at the same tensile-strength level, postuniform elongation and reduction of area increased after tempering (Figures 11 and 12). Because the tempered specimens would have to contain a higher volume fraction of martensite and shorter ferrite ligaments to achieve the same tensile strength as the as-quenched specimens, the increased strain required to cause decohesion or cracking of the softer martensite must more than compensate for the smaller strain required to tear the ferrite ligaments.

Conclusions

1) The tensile strength of ferrite-martensite (dual-phase) steels decreased upon tempering in a manner that can be predicted from the volume fraction and change in hardness of the martensite phase by use of a simple composite-strengthening model.

2) Because of relief of residual stresses, carbon segregation to dislocations in the ferrite, and the return of discontinuous yielding upon tempering, the change in yield strength was more complex, with the yield strength increasing when the specimens were tempered at 200°C, but decreasing upon tempering at 650°C.

3) After tempering at 200°C, continuous-yielding behavior typical of the as-quenched steels was still observed in those steels with a high volume fraction of martensite, whereas discontinuous-yielding was observed in those steels with a low volume fraction of martensite. After tempering at 650°C, discontinuous-yielding behavior was observed in all the steels.

STRAIN-AGING IN A DUAL-PHASE STEEL CONTAINING VANADIUM

L. Himmel, K. Goodman, and W.L. Haworth

Department of Metallurgical Engineering
Wayne State University
Detroit, Michigan 48202

The aging response of a dual-phase steel containing vanadium and prestrained in tension was investigated, including the kinetics of the yield-point development after low-temperature aging (50°-100°C) and the effects of aging time in this temperature range and at higher temperatures (150°-200°C) on the yield elongation. Two separate strain aging stages are observed. Low temperature aging causes the rapid development of a yield plateau, with an associated activation energy of about 80 kJ/ mole, and is characterized by the absence of sharp yield drops, relatively small yield elongations and the diffuseness of Lüders fronts during tensile elongation. Higher-temperature aging, with an associated activation energy of about 125 kJ/mole, is characterized by the appearance of welldefined yield drops during subsequent tensile straining together with the presence of well-defined Lüders fronts and large Lüders strains. A comparison of these results with previous studies reaffirms the importance of the mode of prestrain in determining strain-aging kinetics. Low-temperature aging is probably controlled by diffusion of carbon or nitrogen in ferrite, but it appears premature to identify the rate-controlling step in the case of high-temperature aging.

Plain-carbon steels and high-strength, low-alloy (HSLA) steels are subject to strain-aging, which reduces their ductility and increases their strength. Dual-phase steels, whose microstructure consists of a hard phase (mostly martensite) dispersed in a soft, ductile ferrite matrix have recently been developed to meet demands for high-strength steels with good formability, particularly in the automotive industry. Strain-aging has also been observed in dual-phase steels (1-5); however, the compositional and microstructural factors which influence or control the strain-aging behavior of steels in the dual-phase condition have not yet been established in detail. In this paper we present the results of a study of the aging response of a dual-phase steel containing vanadium and prestrained in tension. The kinetics of the yield-point development after low-temperature aging (50-100°C) and the effects of aging time in this temperature range and at higher temperatures (150-200°C) on the yield elongation (Lüders strain) were investigated. In addition, direct observations were made of the nucleation and propagation of Lüders bands in prestrained and aged specimens during tensile testing. Different activation energies were obtained for the low-temperature and high-temperature aging processes and corresponding differences in the initial yield behavior were also observed.

Procedure

The material investigated was a dual-phase steel supplied by the Jones and Laughlin Steel Corporation in the form of 1.9mm thick sheet. The chemical composition of this steel is given in Table I. The dual-phase microstructure was achieved by annealing in the intercritical (austenite plus ferrite) temperature range followed by forced air cooling. This produced a material containing about 15% martensite (and/or lower bainite) plus retained austenite by volume, and having a mean ferrite grain diameter (intercept length) of about 4.5 μm, equivalent to ASTM grain size 12.

Tensile specimens were prepared according to ASTM specification E-8 with the tensile axis parallel to the rolling direction of the sheet.

Table I Chemical Composition of Steel Used in This Investigation

Element		wt %
N		0.008
C		0.10
Mn		1.42
Si		0.58
P		0.014
S		0.005
Al		0.025
V		0.06
Mo		0.01
Cr		0.04
Ni		0.02
Cu		0.07
Nb	less than	0.005
Ti	less than	0.005
Zr	less than	0.005
Sn	less than	0.01

Tensile properties of the as-received material are given in Table II. The faces and edges of the specimens were lightly surface-ground prior to prestraining and aging. Several specimens were polished metallographically

(1 μm diamond finish) so that changes in surface topography could be monitored during tensile testing. All specimens were prestrained 2% in tension* at room temperature and stored briefly in an ice bath. They were then aged for various times at temperatures between 50°C and 200°C in a stirred, constant-temperature oil bath and subsequently quenched into water. Finally, the specimens were tested in tension at room temperature at a crosshead speed of about 6×10^{-4} mm-sec^{-1}. The initial portions of the stress-strain curves were recorded at a strain sensitivity of 1×10^{-4}.

Table II   Tensile Properties of the As-Received Material

Yield Strength (0.2 pct offset)	Tensile Strength	Uniform Elongation	Elongation to Fracture
56.0 Ksi	90.7 Ksi	23.2 pct	29.5 pct

Strain rate = 10^{-4} s^{-1}

Results

Representative load-extension curves are shown in Fig. 1 for prestrained specimens aged at low (50-100°C) and high (150-200°C) temperatures. The initial yield behavior is substantially different for these two groups of specimens, as seen by comparing the curves for specimens aged at 51°C and 149°C. In the low-temperature range, increasing aging times lead to the gradual development of a yield plateau, i.e., to a region of essentially zero strain-hardening rate, during initial yielding. No yield drops are observed, however, even after aging times (up to 1350h) sufficient to produce a yield elongation of more than 1%. After high-temperature aging, yield point elongations well in excess of 1% are observed after aging times as short as 1 min; furthermore, the yield plateau is now preceded by a distinct yield drop and terminated by a second yield drop immediately prior to the onset of rapid strain hardening.

Direct observation of the surfaces of polished specimens during the tensile test also revealed differences between low-temperature and high-temperature aging. The surface appearance of specimens aged for one hour at 81°C and 180°C, respectively, then pulled in tension into the region of yield-point elongation, is shown in Fig. 2. Sharply defined Lüders fronts are observed during tests on specimens aged between 150° and 200° (Fig. 2b) whereas on specimens aged between 50°C and 100°C the Lüders front is diffuse (Fig. 2a) and plastic deformation begins in several locations at the same time. After high-temperature aging, the initial yield drop typically coincides with the simultaneous nucleation of Lüders bands near the specimen shoulders at opposite ends of the gauge section. During the yield elongation at the lower yield stress, both Lüders fronts propagate at the same velocity toward the middle of the gauge section. When the fronts coalesce, a second distinct yield drop is observed (Fig. 1) and strain hardening begins. In specimens aged at low temperature, however, we were unable to detect the initiation or propagation of Lüders bands with the un-

---

*Preliminary experiments showed that the Lüders strain after aging was maximized by prestraining approximately 2% in tension. A similar effect has been observed by Tanaka et al. (4) in a 1.22% Mn - 0.5% Cr dual-phase steel.

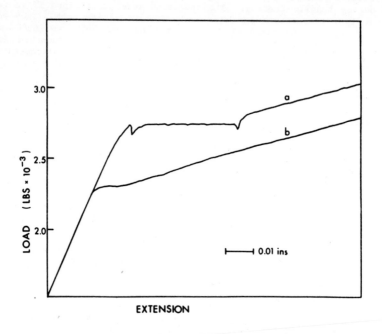

Fig. 1 Representative tensile behavior of prestrained steel after aging
   a) High temperature aging: 60 minutes at 149°C
   b) Low temperature aging: 108 minutes at 51°C

aided eye, despite the development of a yield elongation as high as 1¾% in a specimen aged 1200h at 51°C.

Activation energies for strain aging in the two temperature regimes were derived from the times required for the development of either the initial yield plateau or a given yield point elongation. Times for the development of the yield plateau as a function of aging temperature between 51°C and 100°C are shown in Fig. 3(a). We define the minimum aging time as that necessary to produce a clearly detectable region of zero strain hardening rate on the load-entension curve; the incremental strain corresponding to this region is about 5×10^{-4}. An Arrhenius plot of the aging times needed for the development of a detectable yield plateau after low-temperature aging is shown in Fig. 4. The activation energy obtained is 80 ±8 kJ/mole (19 ± 2 kcal/mole). Longer aging times resulting in a gradual lengthening of the yield plateau, as shown in Fig. 3(a). An Arrhenius plot of the aging times needed to develop a yield elongation of 0.2% after low temperature aging, also shown in Fig. 4, gives essentially the same activation energy, namely 75 ± 8 kJ/mole (18+2 kcal/mole).

After high temperature aging, the Lüders strain increases to 3% or more with increasing aging time as shown in Fig. 3(b). This behavior is similar to that observed by Rashid (6) in a conventionally processed HSLA steel containing vanadium. Insufficient data were obtained in the present

└──────┘ 1 cm

<u>Fig. 2</u> Surface appearance of polished specimens strained in tension
into the region of yield-point elongation. a) strained 0.1 pct
plastically after aging 1h at 81°C. b) strained 2¼ pct plastically after aging 1h at 181°C.

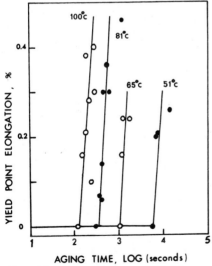
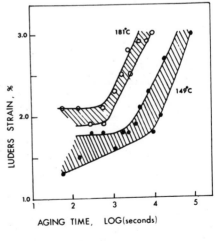

<u>Fig. 3(a)</u> Development of Yield
Plateau After Low Temperature Aging

<u>Fig. 3(b)</u> Development of Lüders
Strain After High Temperature Aging

309

study to derive an accurate activation energy for high temperature aging. Nevertheless, the activation energy appears to be substantially higher than that for low-temperature aging. An Arrhenius plot of the median times needed to develop a Lüders strain of 2.5% (Fig. 4) yields a value of about 125 kJ/mole (30 kcal/mole). This is significantly higher than the activation energy for diffusion of carbon or nitrogen in ferrite (7).

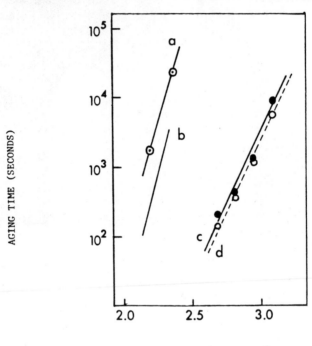

AGING TEMPERATURE (10^3/T°K)

Fig. 4 Kinetics of strain aging in dual phase steel

a) High temperature aging (this investigation): time to develop 2.5 pct yield elongation.
b) High temperature aging, from Davies (1).
c) Low temperature aging (this investigation): time to develop 0.2 pct yield elongation.
d) Low temperature aging (this investigation): time to develop yield point arrest.

Discussion

These observations demonstrate the existence of two separate strain aging stages having different kinetics and different activation energies. In specimens prestrained 2% in tension, low-temperature aging causes the rapid development of a yield plateau. Nevertheless, low temperature aging seems to result in comparatively weak or limited pinning of dislocations, as indicated by the absence of sharp yield drops on the load-extension curves, by the relatively small yield elongations or Lüders strains and by the diffuseness of the Lüders fronts. This low-temperature aging stage is

310

characterized by an activation energy essentially the same as that for the diffusion of interstitial carbon or nitrogen atoms in ferrite. The second aging stage, which is observed only at higher aging temperatures, is characterized by a significantly higher activation energy, and apparently gives rise to much stronger or more extensive dislocation pinning manifested by the well-defined yield drops which occur when Lüders bands are initiated (or when they coalesce), by the presence of sharp, clearly-visible Lüders fronts, and by the large Lüders strains.

Davies (1) investigated the kinetics of the development of a yield plateau in a similar dual-phase steel from the same supplier but obtained substantially different results. His experiments were carried out on both as-received material and on specimens that had been prestrained 2% by rolling. Although his stress-strain curves do not exhibit any yield drops, and are thus qualitatively similar to those we observed after low-temperature aging, Davies obtained an approximate activation energy of 138 kJ/mole (33 kcal/mole) for the development of a yield plateau after aging at temperatures between about 130° and 200°C. At equivalent temperatures, the aging times in Davies' experiments were at least a factor of 100 greater than those we observed. Moreover, Davies found that the yield plateau developed more slowly in the temper-rolled specimens than in the as-received material. This contrasts with the behavior we have observed after prestraining in tension, where the yield plateau develops much more rapidly than it does in the as-received material.

The differences in strain-aging kinetics of the steels tested by us and by Davies, respectively, appear to be due at least in part to the different modes of prestraining employed. Tanaka et al. (4) observed, for example, that rolling prestrain retards the development of Lüders strain in dual-phase steel after subsequent aging, whereas tensile prestrain accelerates the aging process. They showed that a relatively homogeneous microstructure is produced by tensile prestrain, and suggested that dislocations are then rapidly pinned during subsequent aging. The microstructure of cold-rolled material is less homogeneous, containing high local concentrations of dislocations which Tanaka et al. suggested are less amenable to rapid pinning during the aging step. More recently, similar differences in the aging behavior of dual phase steel prestrained by rolling or in tension have been observed by Krupitzer et al. (5).

The activation energies obtained previously for strain-aging or quench-aging in dual phase steels are presented in Table III together with the results of our present study. As seen from the table, activation energies ranging from about 75 to about 140 kJ/mole have been observed, depending on material, processing conditions and test conditions. These widely different activation energies are difficult to interpret, and various suggestions have been made as to the rate-controlling process responsible for a particular activation energy. For example, the rate controlling process may be diffusion of interstitial carbon or nitrogen, trapping of carbon and/or nitrogen by substitutional solute atoms or atom clusters, or microprecipitation at dislocations, to name just a few possibilities; the relaxation of residual stresses during aging may also influence the activation energy observed. The relative importance of such processes remains to be established in any particular case before the observed activation energy can be interpreted with confidence. Further investigation on carefully characterized material is needed before the aging behavior of dual-phase steels can be fully understood.

Table III Activation Energies Reported for Strain-Aging or Quench-Aging
in Dual-Phase Steels

Steel Type	Prestrain %	Prestrain Mode	Aging T °C	Aging Criterion	Reference	Q kJ/mole
Micro-alloyed	2	Tension	50-100	YP Return	This study	80
	2	Tension	50-100	0.2% Plateau	This study	75
	2	Tension	150-200	2.5% Plateau	This study	125
Micro-alloyed	2	Rolling	160-185	YP Return	Davies (1)	138
	0	-	130-200	YP Return	Davies (1)	138
C,N	0	-	65-170	$\Delta\sigma$	Nakaoka (3)	105-125
C,N	0	-	?	YP Return	Indacochea (8)	74-84

Conclusions

In a dual-phase steel containing vanadium and prestrained in tension,
two separate strain-aging stages are observed.

1) After low temperature aging (between 50 and 100°C), a yield plateau develops rapidly with an associated activation energy of
75-80 kJ/mole, corresponding to the diffusion of carbon or nitrogen in ferrite. No sharp yield drops are observed, the yield
elongation is small, and Lüders fronts during tensile elongation
are very diffuse.

2) During tensile elongation after high temperature aging (between
150 and 200°C) sharp yield drops are observed together with well-
defined Lüders bands and Lüders strains of up to 3%. The associated activation energy is approximately 125 kJ/mole.

3) A comparison of these results with previous work suggests that
the mode of prestrain is important in determining strain-aging
kinetics.

References

1. R.G. Davies, "Early Stages of Yielding and Strain-Aging of a Vanadium-
Containing Dual-Phase Steel," Met. Trans 10A, 1549 (1979).

2. T.E. Fine, R.V. Fostini, B.S. Levy, A.G. Preban and R. Stevenson,
"Evaluation of a New, Dual-Phase, Cold-Rolled Steel: Mechanical
Properties, Aging Responses, and Weldability," SAE Preprint 780136,
February 1978.

3. K. Nakaoka, K. Arabi and K. Kunihara, "Strength, Ductility and Aging
Properties of Continuously-Annealed Dual-Phase High-Strength Sheet
Steels," pp. 126-141 in Formable HSLA and Dual-Phase Steels, ed. A.T.
Davenport, TMS-AIME (1979).

4. T. Tanaka, M. Nishida, K. Hashiguchi and T. Kato, "Formation and Properties of Ferrite Plus Martensite Dual-Phase Structures," pp. 221-241

in Structure and Properties of Dual-Phase Steels, ed. R.A. Kot and J.W. Morris, TMS-AIME (1979).

5. R.P. Krupitzer, F. Reis, J.E. Franklin and R.E. Mintus, "The Effects of Strain-Aging, Temper Rolling and Galvanizing on the Properties of Dual-Phase Steels," Proc. 22nd Mechanical Working and Steel Processing Conference, ISS-AIME, November 1980 (in press).

6. M.S. Rashid, "Strain Aging Kinetics of Vanadium or Titanium-Strengthened High-Strength Low-Allow Steels," Met. Trans. 7A, 497 (1976).

7. E.O. Hall, Yield Point Phenomena in Metals, Plenum Press, New York, 1970.

8. Ernesto Indacochea, M.S. thesis, Colorado School of Mines, 1978: cited in D.K. Matlock, G. Krauss, L.F. Ramos and G.S. Huppi pp 62-90 in Structure and Properties of Dual-Phase Steels, ed. R.A. Kot and J.W. Morris, TMS-AIME (1979).

STRAIN AGING BEHAVIOR IN A

CONTINUOUSLY-ANNEALED, DUAL-PHASE STEEL

Ronald P. Krupitzer

Republic Steel Corporation, Research Center

Cleveland, Ohio

The strain aging of a commercially-produced dual-phase continuously-annealed steel has been studied as a function of temperature, time, and prestrain. Aging was conducted at temperatures from room temperature to 260°C (500°F) for materials prestrained from 0 to 10% by uniaxial tension or by cold rolling. As-annealed dual-phase steels were found to be essentially nonaging at room temperature and exhibited sluggish aging behavior at elevated temperatures. Deformation, either by cold rolling or by tensile prestrain, was found to markedly accelerate the increase in flow stress (ΔY) due to aging. Tensile prestrain accelerated the development of a yield plateau while rolling prestrain was found to retard Lüders formation. A wide range of activation energies was measured which suggests that no single diffusion process controls strain-aging over the range of times and temperatures studied. The effects of prestrain suggest that micro-residual stresses associated with the martensite strongly influence the aging behavior.

Introduction

The microstructural characteristics of dual-phase steels have been studied with great interest over recent years (1, 2). Generally, it is accepted that the matrix phase, ferrite, contributes in a significant way to the high ductility of these steels (3). The hard second phases, predominantly martensite and lesser amounts of retained austenite, impart both high strength to the microstructure and provide a source of mobile dislocations (arising from the normal volume expansion associated with the martensite transformation).

Dual-phase steels can be considered to provide a somewhat different environment for the diffusion of interstitial atoms than would be expected in ferrite-carbide structures. Dislocation density in dual-phase steels appears to be a maximum in the ferrite at or near the ferrite-martensite interfaces (4) while dislocations are more randomly distributed in ferrite-carbide structures.

Strain aging in plain carbon and in microalloyed (5, 6, 7) steels has been well studied and specific information about the roles of nitrogen and carbon in these materials has been obtained. For example, it has been shown that strain aging can proceed at different rates depending on the presence of nitrogen and/or carbon in solid solution (5), the degree of supersaturation, the presence of substitutional solutes (6, 7), and on the nature of the prestrain (8, 9). Prestrained samples age more rapidly than unprestrained steels when all applied strains are in the same direction. In this case the acceleration of aging due to prestrain is probably caused by the reduction in the mean diffusion distance of interstitial atoms to dislocations. However in cases when the prestrain and subsequent tensile test are not in the same direction (e.g., rolling prestrain - tensile test, transverse tensile prestrain - longitudinal tensile test), Lüders formation is retarded (8, 9). Experiments (8, 10, 11) have demonstrated that micro-residual stresses induced by certain types of prestrain account for the slow development of Lüders strain. Other parameters (changes in yield strength, tensile strength, etc.) are not so strongly affected by the nature of the prestrain.

The effect of prestrain on the aging of dual-phase steels becomes particularly interesting when one considers that local residual stresses are developed by the volume expansion (4) associated with the formation of martensite within the ferrite matrix. Other structural characteristics such as the presence of vanadium precipitates or clusters and the possibility of Snoek (7, 12, 13) rearrangements as an early stage of strain-aging have been discussed in the literature and need to be reviewed in the light of dual-phase steels.

Strain-aging behavior is important not only to enhance the fundamental understanding of dual-phase steel structure, but also to provide practical guidelines (14) for the use of dual-phase steels. In cases where strain-aging is not desirable (such as in-plant processing or coating operations) certain chemistry or processing modifications might be considered to retard aging. In cases where strain-aging is desirable (such as in a formed part) methods will be sought to enhance the strengthening obtained by aging.

The purpose of this study is to describe the strain-aging characteristics of a commercially-produced vanadium-treated dual-phase steel and to explain the behavior by applying classical theories of strain-aging to the dual-phase structure. It needs to be understood that the characteristics described here pertain only to the steel investigated since variations in chemistry or in the method of production would be expected to alter the aging behavior (14).

Materials and Procedure

The steel used in this investigation was a commercially-produced vanadium-containing dual-phase steel (MAXI PHASE 80*) with a chemical composition shown in Table I. The steel was continuously annealed in a protective
atmosphere of hydrogen and nitrogen and cooled by gas-jet cooling on a silicon electrical steel processing line. The average thickness of the samples
tested was 0.69 mm (0.027 in.). The mechanical properties of this steel in
the as-annealed (unaged) condition are shown in Table II. Although mechanical properties are shown for all three directions, it is clear that the
properties are relatively isotropic. For that reason only longitudinal samples were used in the strain-aging study.

Table I. Chemical Composition (Wt %)

C	Mn	Si	V	Al
0.10	1.40	0.49	0.075	0.06

Table II. Mechanical Properties in the As-Annealed Condition

Direction	Yield Strength MPa	ksi	Tensile Strength MPa	ksi	Uniform Elongation (%)	Total Elongation (%)	Lüders (%)	n	r
L	283	41.0	645	93.5	21.0	27.4	0	0.224	0.78
D	283	41.0	645	93.5	21.0	27.7	0	0.221	0.96
T	297	43.1	638	92.6	20.3	26.0	0	0.214	0.89

Samples were prestrained either in tension or by rolling prior to aging.
In all cases samples in the as-annealed condition were aged along with prestrained samples. One set of conventional 50.8 mm (2 in.) gage length tensile samples was prestrained at room temperature in uniaxial tension on an
Instron tensile testing machine to engineering strains of 0.01, 0.025, 0.05,
0.075, and 0.10. A second set of identical materials was prestrained by
cold rolling on a laboratory cold rolling mill (with 200 mm (8 in.) diameter
work rolls) to average strains of 0.02, 0.05, and 0.10. The rolling prestrain of 0.02 had also been chosen by Davies in a recent strain aging study
(13).

Aging was carried out at room temperature, 100°C (212°F) in boiling
water, and at 177°C (350°F) and 260°C (500°F) in either oil or salt. The
times and temperatures used are summarized in Table III.

Uniaxial tensile tests were used to evaluate the degree of strain-aging
after each treatment. In the case of unprestrained samples, the difference
between the 0.2% offset yield strength of the unaged specimens and that of
the aged specimens was used to determine ΔY (the increase in flow stress due
to aging). This is illustrated in Figure 1(a). For samples prestrained in
tension, ΔY was determined with single specimens by the difference between
the 0.2% offset yield strength after prestraining and aging and the flow

---

* MAXI PHASE 80 is a trademark of Republic Steel Corporation.

Table III. Temperatures and Times for Strain-Aging Study

Temperature	Time (Range)
Room Temperature	1 Month – 1 Year
100°C (212°F)	10 Seconds – 500,000 Seconds
177°C (350°F)	5 Seconds – 10,000 Seconds
260°C (500°F)	1.5 Seconds – 1,000 Seconds

stress after prestraining as shown in Figure 1(b). For samples prestrained by rolling, ΔY was determined by testing companion specimens and comparing the 0.2% offset yield strength immediately after rolling with that after aging as shown in Figure 1(c). Other aging parameters such % Lüders, ΔE (the loss in ductility due to aging), and ΔU (the increase in tensile strength due to aging) were also measured and are illustrated in Figure 1 (a to c).

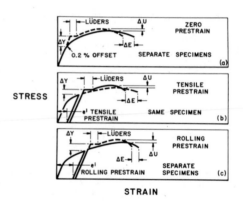

Figure 1. Methods of measuring aging parameters for (a) zero prestrain, (b) tensile prestrain, and (c) rolling prestrain.

Results and Discussion

As-Annealed Condition

The dual-phase steel selected for study was produced on a commercial continuous annealing line and as a result was subjected to very slight working by bending and unbending over large diameter processing rolls. These deformations were found to have no measurable effect on mechanical properties and it was therefore assumed that the aging behavior of steel produced on this line represents the aging characteristics of the unprestrained state.

The effects of aging between room temperature and 260°C (500°F) on ten-

sile properties are shown in Figure 2. At room temperature no significant
change in properties occurred for at least one year. Although this stability is not unusual for aluminum-killed batch-annealed mild steels, it is a
significant observation for a continuously annealed product, especially for
one not processed with an overaging thermal cycle. Rapid cooling after annealing can result in supersaturated carbon and nitrogen in the ferrite and
can cause appreciable room temperature strain aging (5, 15) in otherwise
"nonaging" steels. It is clear from Figure 2 that thermal activation is
required in order for strain aging to proceed in a vanadium-containing
rapidly air-cooled dual-phase steel.

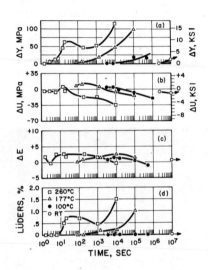

Figure 2.   Effect of time and temperature on the changes in (a)
            yield strength, (b) tensile strength, (c) total elon-
            gation, and (d) Lüders strain for continuously annealed
            (unprestrained) vanadium-contining dual-phase steel.

At elevated temperatures significant increases in yield strength and
Lüders strain are observed in Figure 2 and demonstrate a classical depen-
dency on temperature and time. Changes in yield phenomena are generally
associated with the classical Stage I of strain aging related to atmosphere
formation at dislocations. The yielding and Lüders phenomena, which are
characteristic of plastic flow in the ferrite, show the clearest evidence
of aging. It is evident from these results that prestrain is not required
for aging to occur at elevated temperatures in dual-phase steels.

Stage II aging behavior, classically observed as an increase in tensile
strength and a decrease in total elongation due to precipitation at dislocations (16), is not clearly observed in these steels. Some evidence of a
tendency to decrease both tensile strength and total elongation after prolonged time at temperature may be inferred from the figure, but these effects
are insignificant for the times and temperatures studied.

The aging of the unprestrained dual-phase steel of this investigation can be explained from information previously obtained about dual-phase structures. It has been shown that dislocation density is relatively low in the ferrite, except near the ferrite-martensite interface (4), where dislocations form on cooling because of the volume expansion associated with the martensite transformation. Rapid cooling is argued to allow insufficient time for pinning of dislocations by interstitials and therefore they remain mobile after annealing. Similarly, micro-residual stresses associated with the volume expansion of the martensite contribute to a low value of the as-annealed yield strength. The relatively high temperatures (over 100°C) required for pinning or return of Lüders suggest that carbon and not nitrogen is responsible for the aging, since nitrogen diffuses sufficiently fast below 100°C to cause aging when sufficient amounts are dissolved in the ferrite. It is interesting to note that at the highest aging temperature studied at which carbon is highly mobile, ΔY and Lüders first increase rapidly and then develop a plateau before increasing further. The explanation for this stabilization is not clear, but may reflect some tempering or softening of the martensite. Evidence of tempering can be seen in Figure 3 after 60 seconds at 260°C. The normally smooth surface of the martensite has been affected locally by tempering.

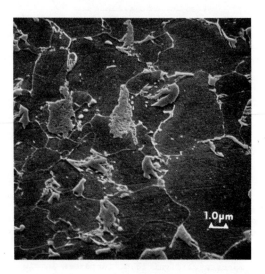

Figure 3.   The effect of aging at 260°C for 60 seconds on the microstructure of dual-phase steel.

Effects of Prestrain

Room Temperature Behavior.   The rapid work hardening characteristic of dual-phase steels is evident in Figure 4, which shows the effect of tensile and rolling prestrain on flow stress. The work hardening component of the flow curve, $\Delta\sigma_{cw}$, is also plotted in the figure and illustrates the large contribution to the flow stress by cold work. Approximately 5% prestrain produces an increase in flow stress of 250 MPa (35 ksi). Overall, rolling and tensile prestrain produced similar hardening over the 0 to 10% range of prestrain. More scatter was evident in the rolling data, probably because fewer samples were used for that portion of the study.

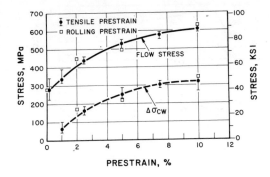

Figure 4. The effect of rolling and tensile prestrain on the flow
stress and on the amount of work hardening ($\Delta\sigma_{cw}$).

Although unprestrained dual-phase steels were found not to age at room
temperature, both tensile and rolling prestrain did produce significant
aging responses at room temperature. Figure 5 shows the effect of tensile
prestrain on ΔY, ΔU, and Lüders strain. Samples tested about 10 minutes
after prestrain showed a very slight increase in yield strength. This 10-
minute time period was used to avoid problems of anelastic stress relaxation
after prestraining. The development of a ΔY response over such a short
period of time cannot be explained by classical atmosphere formation, and
supports arguments by Nabarro (12) and Rashid (7) that an early stage of
aging exists in which small changes in mechanical properties can be observed
due to Snoek rearrangements. Negligible effects of prestrain on tensile
strength and Lüders strain were observed after the 10-minute delay. After
30 days, significant increases in yield strength of between 35 and 60 MPa
(5 and 8 ksi) were observed, and about half that increase was observed in
tensile strength. The shape of the ΔY versus prestrain plot is characteristic of that found in plain carbon (17) and microalloyed (6) steels; that
is, some prestrain can normally be found at which a maximum in ΔY will be
observed. Lüders strain in the form of a short yield plateau was found to
develop after tensile prestrain and increase with the level of prestrain.
The retarded aging behavior of the as-annealed dual-phase steel appears to
be appreciably altered by tensile prestrain and suggests that the role of
residual stresses may be particularly significant in understanding the
aging of dual-phase steels. This will be discussed in more detail in a
later section.

Rolling prestrain also affected room temperature aging of dual-phase
steel, but in a different way from tensile prestrain as shown in Figure 6.
A maximum in ΔY occurred at 5% rolling prestrain, as opposed to 1% for tensile prestrain. Essentially more rolling deformation than tensile is required to achieve a maximum aging increment. However, rolling reduction
appears to be more effective than tensile prestrain in suppressing aging at
higher (10%) rolling prestrain levels. It also can be observed from Figure
6 that tensile strength increased only slightly and that Lüders strain did
not appear after prestraining by rolling and aging at room temperature.
This is in contrast to the rapid Lüders formation which occurred at room
temperature after tensile prestrain.

321

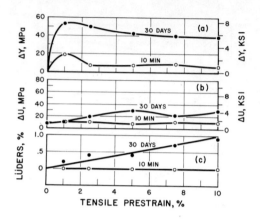

Figure 5. Influence of tensile prestrain on the increase in (a) flow stress, (b) tensile strength, and (c) Lüders strain at room temperature.

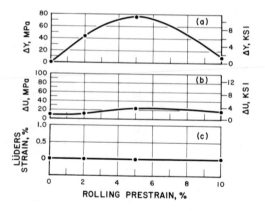

Figure 6. Influence of rolling prestrain on the increase in (a) flow stress, (b) tensile prestrain, and (c) Lüders strain at room temperature after 30 days.

Elevated Temperature Aging (100°C to 260°C). The effects of tensile prestrain on the increase in flow stress, ΔY, can be seen in Figure 7 for a range of times and temperatures. Clearly aging time significantly influences ΔY over the ranges studied, but does not appreciably alter the prestrain level at which a maximum in the curve occurs. The increase in yield strength due to aging is relatively insensitive to tensile prestrain (except for the initial rapid increase where the stress-strain curve changes from continuous to discontinuous yield) and demonstrates the insensitivity of ΔY to dislocation density. This type of dependence of ΔY on tensile prestrain is also found in mild steels (17).

322

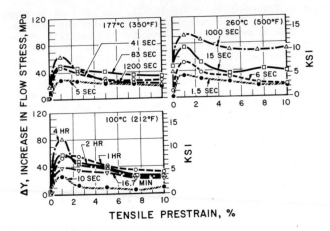

Figure 7. The effect of tensile prestrain on the increase in flow
stress after aging for various times and temperatures.

Rolling prestrain was also found to influence ΔY as shown in Figure 8.
The maximum in ΔY was found to occur at about 5% rolling prestrain. For
both rolling and tensile prestrains, ΔY was found to increase considerably
more rapidly than it did in the as-annealed dual-phase steels.

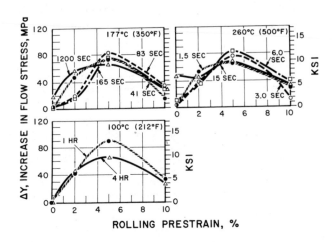

Figure 8. The effect of rolling prestrain on the increase in flow
stress after aging for various times and temperatures.

Strain Aging Kinetics

The effects of time and temperature on the changes in tensile properties are shown in Figures 9 to 11 for 1%, 5%, and 10% tensile prestrain, respectively. Clearly, the strain aging is thermally activated and occurs most rapidly at the highest aging temperature. Significant changes in ΔY and Lüders strain were found to occur at tensile prestrains of 1 to 10%. At 5% and 10% tensile prestrain (Figures 10 and 11), significant increases in tensile strength and reductions in ductility are also observed. These factors are commonly found in mild steels in the second stage of strain aging and suggest that strain-induced precipitation of carbides may be occurring at dislocations. The aging curves are not simple in shape and suggest that a number of mechanisms may be operating simultaneously.

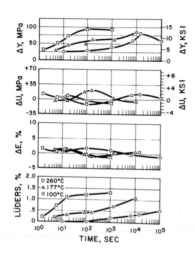 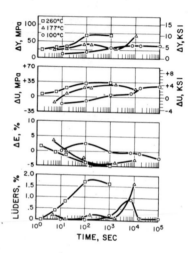

Figure 9. The effect of aging time and temperature on the tensile properties of dual-phase steel prestrained 1% in tension.

Figure 10. The effect of aging time and temperature on the tensile properties of dual-phase steel prestrained 5% in tension.

The effect of tensile prestrain on ΔY and Lüders strain after aging at 100°C is shown in Figure 12. The effect of increasing tensile prestrain is to significantly retard the increase in yield stress. Since increasing prestrain effectively raises the dislocation density and reduces the work hardening rate, more dislocations must be pinned at high prestrains to achieve the same level of ΔY which can be developed at low prestrains with fewer dislocations. Similarly, it appears that it takes somewhat longer for the materials with higher prestrain to develop Lüders or yield plateaus than it does for those with lower prestrains, although the relationship is not so clear as that for yield strength. At 177°C and 260°C, similar effects are observed and are shown in Figures 13 and 14, respectively. Aging is slowed down as the level of tensile prestrain is increased.

A comparison was made of the relative effects of rolling and tensile prestrain on the rate of aging. The lowest levels of prestrain, 2% rolling and 1% tensile, were selected for comparison with the unprestrained condition. At higher levels of prestrain, aging becomes retarded for both rolling and tensile prestrain as discussed in the previous paragraph, and the com-

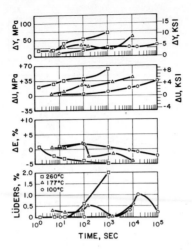

Figure 11. The effect of aging time and temperature on the tensile properties of dual-phase steel prestrained 10% in tension.

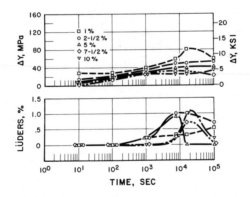

Figure 12. The effect of tensile prestrain and time on the increase in yield strength and on Lüders strain at 100°C.

parison between the three conditions should not be made. The results were approximately the same at all three temperatures and are illustrated for samples aged at 177°C in Figure 15. Samples which were prestrained either 1% in tension or 2% by rolling showed increases in yield strength before the as-annealed dual-phase steel. However, the onset of Lüders strain was found to be accelerated by tensile prestrain and retarded by rolling prestrain.

Since strain aging measurements were made over a wide range of times and temperatures, an attempt was made to determine whether an activation energy could be established for the process over the temperature range

325

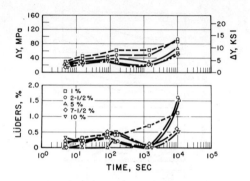

Figure 13. The effect of tensile prestrain and time on the increase in yield strength and on Lüders strain at 177°C.

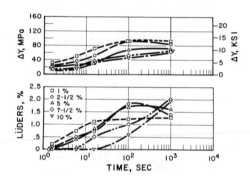

Figure 14. The effect of tensile prestrain and time on the increase in yield strength and on Lüders strain at 260°C.

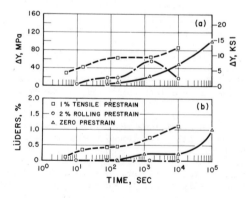

Figure 15. The influence of the type of prestrain on (a) the increase in flow stress and (b) Lüders strain at 177°C.

326

studied. An Arrhenius relationship was assumed in which

$$t = A \exp Q/RT \qquad\qquad [1]$$

and $\quad\quad \ln t = \ln A + Q/RT \qquad\qquad [2]$

where
- t = time to age to a specific ΔY or Lüders strain
- A = a constant
- Q = activation energy for the process
- T = absolute temperature
- R = gas constant

By plotting the natural logarithm of the time versus $1/T$, estimates of activation energies were made with the data available. Over the range of times and temperatures studied, activation energies varied between 16 and 33 kcal/mole (67 and 138 kJ/mole). Examples of activation energies calculated by this method for various aging conditions are shown in Figure 16. The activation energies for carbon and nitrogen diffusion in iron are about 18 to 20 kcal/mole (75 to 84 kJ/mole). Higher activation energies have been observed for aging in microalloyed (7) and dual-phase (13) steels in which activation energies of about 33 kcal/mole (138 kJ/mole) have been associated with carbon diffusion about vanadium clusters or precipitates. Evidently, no single process such as the diffusion of carbon in ferrite can completely explain the aging phenomena over the full range of temperatures, times, and prestrains studied.

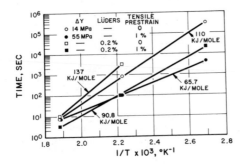

Figure 16. Examples of activation energies determined for different aging conditions.

Role of Residual Stresses

A close parallel can be drawn between the effects of residual stresses on the strain aging of plain carbon steels and the effects of prestrain on the aging of dual-phase steels. It has been well demonstrated in plain carbon steels that prestraining by rolling retards the return of the lower yield extension. Changes in other properties, such as yield stress, are not so noticeably affected by the type of prestrain (5). Also, the shape of the yield extension versus time curves often show similar features after tensile or rolling prestrain, although there is a considerable delay in the return of Lüders strain for the latter. Residual stresses developed by rolling prestrain have been used to explain these effects. Originally, the residual stress patterns were thought to be macro-stresses (through-the-thickness)

327

which were significant. Subsequent work by Tardif and Ball (8) has shown these effects to be primarily caused by micro-residual stress introduced by the rolling process.

In dual-phase steels, many of the characteristics of temper rolled steels exist immediately after annealing. Particularly significant is the fact that the return of Lüders strain is noticeably retarded in as-annealed dual-phase steels such that it does not occur at all at room temperature and very sluggishly at temperatures up to 260°C. On the other hand, a slight tensile prestrain radically accelerates the formation of Lüders strain so that slight yield plateaus even appear at room temperature. In a sense then, dual-phase steels behave like temper rolled (plain carbon) steels where it has been demonstrated that the retarding effects of rolling can be eliminated by tensile prestrain.

It is clear from the previous section that the diffusion of interstitials (carbon and perhaps nitrogen) are responsible for the aging which occurs in dual-phase steels. The process is not a simple one, however, and no single activation energy can be assigned to the entire process. The strain aging of the dual-phase steels of this investigation clearly suggest that in the as-annealed condition, micro-residual stresses are present, probably as a result of the volume expansion of martensite during transformation on cooling. Tensile prestrain either reduces or significantly alters this residual stress distribution and allows the easy formation of Lüders strain on aging. It appears that rolling prestrain of a dual-phase steel intensifies the residual stresses present after annealing and causes further retardation of Lüders formation. Also, ΔY, which increases after rolling or tensile prestrain, demonstrates that dislocation pinning (which controls ΔY in the early stages of aging) is similarly affected by both types of prestrain.

The increase in yield strength in dual-phase steels appears to be controlled by interstitial atmosphere formation. The wide range of activation energies observed indicate that carbon dominates the aging behavior of as-annealed dual-phase steels, but that carbon, nitrogen, and the relaxation of residual stresses may influence aging in tensile prestrained dual-phase steels. Also, Snoek relaxation processes and interactions of carbon with precipitates or clusters of vanadium may also influence the rate of aging. Lüders formation appears to be strongly influenced by the type of prestrain and is retarded by micro-residual stresses. These stresses occur naturally in continuously-annealed dual-phase steels and the effects of these stresses can be further enhanced by small rolling deformations.

Conclusions

1. A commercially-produced vanadium-containing dual-phase steel was found not to strain age in the as-annealed condition at room temperature for at least one year.

2. At elevated temperatures, the as-annealed dual-phase steel was found to increase in yield strength and develop yield plateaus, especially between 177 and 260°C. This study was conducted over a range of times and temperatures for which changes in tensile strength and total elongation were slight. Microstructural evidence of tempering of the martensite was found after 15 seconds at 260°C, however.

3. Tensile prestrain was found to accelerate strain aging for aging temperatures from room temperature to 260°C. Increases in yield strength and the development of Lüders strain were found over the full range of temperatures examined.

4. Rolling prestrain was found to retard the early formation of Lüders strain and accelerate the increase in yield strength compared to the rates of aging observed in the as-annealed condition.

5. The aging kinetics for the range of times and temperatures studied could not be associated with a single diffusion process. Estimates of activation energies were made and values ranged from 16 to 33 kcal/mole depending on the parameter used to measure aging (Lüders or ΔY) and the extent of aging.

6. The effects of rolling and tensile prestrain on Lüders return in dual-phase steels indicate that micro-residual stresses influence strain aging in dual-phase steels.

Acknowledgments

The author would like to acknowledge the work of William Darden and Dr. Joseph E. Franklin who assisted in the initial experiments and to John W. Zoha who completed the laboratory work. The microscopy of Dr. J. Michael Rigsbee and Patricia A. Labun was helpful in the interpretation of the results.

References

1. Formable HSLA and Dual Phase Steels, A. T. Davenport, ed., Proceedings of the Fall TMS-AIME Ferrous Metallurgy Committee Symposium, Chicago, October 26, 1977.

2. Structure and Properties of Dual-Phase Steels, R. A. Kot and J. W. Morris, eds., Proceedings of the TMS-AIME Heat Treatment Committee Symposium, New Orleans, February 19-21, 1979.

3. R. G. Davies, "Influence of Martensite Composition and Content on the Properties of Dual Phase Steels," Metallurgical Transactions A, (9A), 1978, pp. 671-679.

4. J. M. Rigsbee and P. J. Vander Arend, "Laboratory Studies of Microstructure and Structure-Property Relationships in Dual-Phase Steels," Formable HSLA and Dual-Phase Steels, loc. cit, pp. 58-88.

5. J. D. Baird, "Strain Aging of Steel — A Critical Review," Part I, Iron and Steel, 1963, pp. 186-192 and 326-334.

6. M. S. Rashid, "Strain Aging of Vanadium, Niobium, or Titanium-Strengthened High-Strength Low-Alloy Steels," Metallurgical Transactions A, (6A), 1975, pp. 1265-1268.

7. M. S. Rashid, "Strain Aging Kinetics of Vanadium or Titanium Strengthened High-Strength Low-Alloy Steels," Metallurgical Transactions A, (7A), 1976, pp. 497-503.

8. H. P. Tardif and S. C. Ball, "The Effect of Temper-Rolling on the Strain-Aging of Low Carbon Steel," JISI, 182, 1956, pp. 9-19.

9. B. B. Hundy, "Elimination of 'Stretcher Strains' in Mild-Steel Pressings," JISI, 178, 1954, pp. 127-137.

10. B. B. Hundy, "Inhomogeneous Deformation During the Temper-Rolling of Annealed Mild Steel," JISI, 181, 1955, pp. 313-315.

11. R. D. Butler and D. V. Wilson, "The Mechanical Behaviour of Temper Rolled Steel Sheets," JISI, 201, 1963, pp. 16-33.

12. F. R. N. Nabarro, "Report on Strength of Solids," Phys. Soc., London, 1948, p. 38.

13. R. G. Davies, "Early Stages of Yielding and Strain Aging of a Vanadium-Containing Dual-Phase Steel," Metallurgical Transactions A, (10A), 1979, pp. 1549-1555.

14. R. P. Krupitzer, et al., "Effects of Strain Aging, Temper Rolling, and Galvanizing on the Properties of Dual-Phase Steels," Proceedings of the 22nd Mechanical Working Conference, ISS-AIME, Toronto, October 29-30, 1980, pp. 1-21.

15. R. P. Krupitzer, et al., "Effects of Thermal History on the Properties of Hot Rolled, Low Carbon, Aluminum Killed Sheet," ISS-AIME, Mechanical Working and Steel Processing X, Chicago, 1971, p. 27.

16. D. V. Wilson and B. Russell, "The Contribution of Precipitation to Strain Aging in Low Carbon Steels," Acta Metallurgica, 8, 1960, pp. 468-479.

17. B. B. Hundy, "The Strain-Age Hardening of Mild Steels," Metallurgica, May 1956, pp. 203-211.

Section II-Mechanical Metallurgy
of Dual-Phase Steels

FORMABILITY OF DUAL-PHASE STEELS

J.D. Embury and J.L. Duncan
Faculty of Engineering
McMaster University
Hamilton, Ontario, L8S 4L7
Canada

The mechanical behaviour of dual-phase steels depends on a broad range of metallurgical phenomena. The initial hardening reflects the volume fraction of the martensitic phase via a back stress but this is also dependent on the local strength of the martensite. Formability is related not only to strain hardening and other continuum properties such as normal plastic anisotropy but may also be influenced by fracture phenomena associated with discrete microstructural events such as decohesion of the ferrite-martensite interface.

Various micromechanisms of duplex systems are discussed and the influence of the related mechanical behaviour in determining formability in different technological forming process is described. The low yield to tensile strength ratio of dual-phase steels is of particular benefit in stamping shallow components such as autobody skin panels; however, denting resistance requires a high yield strength emphasizing the importance of aging after forming.

Introduction

Dual-phase steels may be produced by a variety of processing routes. In essence the microstructure of these materials may be regarded as a fine scale aggregate of a hard martensitic phase dispersed in a soft ferrite. A number of recent studies have shown that the detailed nature and distribution of both the hard and soft phases can vary with changes in composition and processing route. Understanding the relationship between structure and properties in dual-phase steels is complicated by the fact that the hard and soft phases are "interactive" resulting in distribution of elastic stresses after processing or inequalities in plastic strain after deformation. Hence in modelling the mechanical behaviour, cognisance must be taken not only of the properties of the individual phases and the manner in which they contribute to some volume average but to local properties such as the strength of the martensite-ferrite interface or the local fracture behaviour of the dispersed hard phase.

The interaction between the phases and the variation in the detailed properties of the phases due to thermal treatments give rise to a broad range of metallurgical phneomena influencing the behaviour of the dual-phase steels. Given the complexity of the process involved it is not possible to produce a definitive review of formability based on current experimental evidence and our limited understanding of the micromechanisms of duplex systems. It is germane however, to sketch in broad terms the various forms of mechanical behaviour which relate to formability and to define those areas in which there are clear inadequacies both in experimental evidence and analytical perceptions. The mechanical properties which influence foromability are not necessarily those which are used to specify the material. For example, a dual-phase steel may be specified in terms of yield stress at particular small values of plastic strain and by total elongation; its formability will, however, be influenced by many other properties such as the work hardening rate, normal anisotropy and its ductility in a stretched sheared edge.

In order to provide a basis for discussing the forming characteristics of dual phase steels it is of value to consider first the tensile stress-strain relationships and then to consider more complex modes of deformation which pertain to the formability. In terms of mechanical behaviour, the dual-phase steels exhibit combinations of tensile strength and total elongation not attained by other high strength sheet steels. This is illustrated in Figure 1. The tensile test may also be used to specify the yield stress and rate of hardening at low plastic strains. Much of the effort on modelling the mechanical response has been devoted to the true stress-strain relation in simple tension and while this is important, it must be remembered that formability can rarely be predicted on the basis of tensile test data alone.

The paper by Becker, Hornbogen and Wendl (2) in this volume points out that where two microstructural components exist, a bulk property may depend on the constitutive law exhibited by each constituent and the fraction of the constituent; it may also reflect a unique local property such as the decohesion of the ferrite-martensite interface. Thus in modelling the stress-strain curve, the current level of flow stress is usually considered as the sum of a number of independent contributions in which plasticity laws for ferrite and martensite are employed additively. The compatibility between the hard and soft phases produces, however, a strain partition or the appearance of an unrelaxed plastic strain ε_p^* which results in a back stress σ_B of magnitude,

$$\sigma_B = Kf \; \epsilon_p^* \tag{1}$$

where K is a parameter related to the particle shape and contiguity and f is the volume fraction of hard phase. A number of authors have considered the role of elastic back stresses in enhancing the initial hardening rate. The

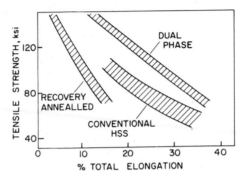

Fig. 1   The relation between tensile strength and total elongation for high strength steels, Ref. 1.

method of adding the flow stress components based on constitutive laws for the plasticity of the ferrite and the martensite has not been uniquely defined and this merits further critical discussion. Based on the model of Chang and Asaro (3) the total work hardening of the system σ_D can be expressed as the work hardening due to the matrix σ_{WHM} and the extra hardening due to the presence of the martensite σ_{WHP}. The initial flow stress σ_o, the back stress due to unrelaxed flow, σ_B and the total hardening can then be added to given the total flow stress σ_F of the form:

$$\sigma_F = \sigma_o + \sigma_B + (\sigma_{WHP}^2 + \sigma_{WHM}^2)^{1/2} \tag{2}$$

In this equation, σ_o may reflect a number of separate effects. Becker, Hornbogen and Wendl (2) show the effect of phosphorus in solid solution on the strength of the matrix. Other elements can also influence σ_o and often the strength of the ferrite matrix is not determined independently.

The stress at which plastic flow commences can be either greater or less than the fully annealed matrix strength depending on the influence of other microstructural features. Transformation of austenite might induce residual stresses lowering the yield stress as considered by a number of workers, e.g., Ref. 4. The experiments presented by Okamoto and Takahashi (5) using sub-zero treatment to eliminate retained austenite are extremely valuable.

The experimental determination of an initial yield stress, σ_o, in dual-phase steels is extremely difficult and Mathy, Gouzou and Greday (6) in this volume show that non-linearity occurs at very low elastic strain levels. It will also be very sensitive to aging and to any slight plastic strains induced in, for example, straightening operations.

After the onset of plastic flow the rate of hardening is determined by various events. Residual stress effects may be important initially but would die away rapidly. If the second phase does not deform it will influence hardening by virtue of the back stress, σ_B. This term in the monotonic law, equation (2), is only determined independently in reverse strain experiments as described earlier by Gerbase, Hobbs and Embury (4) and illustrated in Figure 2. The paper by Tseng and Vitovec (7) is of major interest because it confirms the concept proposed by Gerbase et al. (4) that

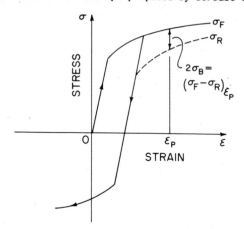

Fig. 2   Diagram illustrating the parameters in a reverse
         strain experiment, Ref. 4.

the   initial hardening reflects the volume fraction of the second phase via the back stress σ_B and also that the contribution of σ_B is fixed by the local strength of the martensite.   If the martensitic phase has a high carbon content it behaves as an elastic, rigid dispersion, whereas low carbon martensite is capable of flow at relatively low imposed strain. The partitioning of strain between the phases will clearly influence the rate of hardening and Tseng and Vitovec indicate that a wide range of strain partitioning is possible.

The initial work hardening rate is an important parameter and it would be useful to be able to summarize this in a single parameter.   Ramos, Matlock and Krauss (8) show that the conventional strain hardening component n is inappropriate.   Crussard (9) and Jaoul (10) have used various instantaneous rates for other stress strain curves.   Mathay, Goujou and Greday consider this problem in more detail (6).   It is possible that at present comparisons should be based on curves of normalized $(1/\sigma . d\sigma/d\varepsilon)$ or actual $(d\sigma/d\varepsilon)$ slope curves versus stress or strain.

In addition to understanding why the stress strain curve is rounded and the initial yield suppressed in these materials it is obvious that techniques of regaining a high yield stress after forming will be economically attractive.   The possibility of aging in paint bake cycles is described by Araki, et al. (11).   One area which has not received attention in dual-phase steels is the dimensional changes which may be associated with relaxation of back stress.   This has been studied by Shinoda et al. (12) for ferrite-pearlite steels and a result is illustrated in Figure 3.   In view of the importance of both dimensional stability and the role of relaxation in fatigue, similar studies for ferrite-martensite structures would be valuable.

At large strains the magnitude of the work hardening rate decreases and the strain rate sensitivity of the material may become of increasing importance.   Stevenson (13) has shown that uniform elongation is extended by increasing the rate sensitivity and that this effect is temperature dependent. Strain rate sensitivity data has been presented also by Waddington et al. (14), Figure 4, but rather more information on the structural and compositional origins of rate sensitivity is desirable, particularly in relation to the dependence of rate sensitivity on aging processes.

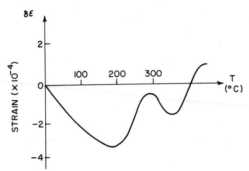

Fig. 3 Dimensional changes in the direction of extension versus annealing temperature for a pearlite steel pre-strained up to 6%, Ref. 12.

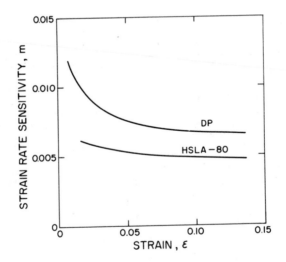

Fig. 4 Strain rate sensitivity versus strain for dual phase and HSLA steels, Ref. 14.

At large plastic strains, comparison between hardening rates in dual-phase and HSLA steels becomes important. Unfortunately the tensile test is unsuitable for gathering data in this range and other tests are discussed below. Plastic anisotropy can be determined from the tensile test and the paper presented by Okamoto and Takahashi (5) is important in indicating the relatively high r-values that can be obtained.

Large Strain Behaviour

The resistance to necking and tearing during forming and sheared edge ductility are both related to strain hardening, rate sensitivity and the accumulation of microstructural damage at strains above the maximum uniform strain level observed in the tensile test. The hardening behaviour at large strain can be derived from bulge tests (14) as illustrated in Figure 5 but

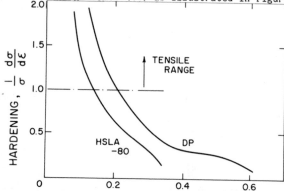

Fig. 5   Non-dimensional strain hardening versus effective strain for HSLA and dual phase steels derived from hydraulic bulge tests, Ref. 14.

generally little data of this kind are available. One might surmise that structures causing high hardening rates at low strain would be less effective than typical HSLA structures at high strain due to the breakdown of the mechanisms supporting the production of large back stresses and this may be a factor in lowering sheared edge formability in some dual-phase steels.   Other contributing factors may be decohesion of the ferrite-martensite interfaces and fracture of the harder phase due to high local strains but here again insufficient experimental work is available to give a definitive account.   The paper by Nishimoto et al. (15) is therefore extremely useful. These authors point out that tensile testing data do not provide a suitable index for ranking hole expansion tests and they relate failure to microcracks initiated during the punching operation.

Detailed metallography of sheared edges could provide some information about the relevant fracture processes. The angle of rotation θ in Figure 6 is an indication of the shear strain at failure and preliminary work by Waddington (16) indicates that this angle is rather less in a particular dual-phase steel than in an HSLA steel of comparable tensile strength.

Formability Requirements

Each metal forming process imposes its own formability requirements on the material.   Roll forming and folding operations are relatively insensitive to total elongation but require consistent levels of yield stress to avoid springback problems. Shallow stampings, on the other hand, require a high tensile/yield ratio but are more tolerant of small variations in yield strength (17). The mechanical properties which are desirable in a dual-phase steel will therefore depend on the way it is used and considera- tion is given here to various applications and to the type of forming processes employed.   In brackets and simple structural components, Yoshida (18) suggests that strength depends on $\sigma_y t^2$ where σ_y is the initial yield

338

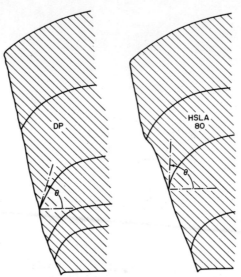

Fig. 6 Diagram showing the flow lines in sections cut from
sheet adjacent to a sheared edge, Ref. 16.

strength and t is the sheet thickness; in these parts most of the sheet is
undeformed and hence the strength of the part depends on the initial yield
strength and not the strength after a few percent strain. Forming considerations include minimum bend radius, springback and sheared edge
ductility. It is difficult to see real advantages for dual-phase steels
compared with HSLA steels with inclusion shape control for these
applications.

The interest, particularly amongst Japanese authors, in material in the
0.7 - 0.8 mm (0.028 - 0.31 in.) range suggests possible applications in body
skin panels. Material distributions derived from Yoshida's work (18)
indicate large usage in this range in compact and small cars, Figure 7. The
formability requirements for such parts require some explanation. It is
often stated that the total elongation for the material must be high, in the
40-50% range, and yet a study of strain distribution shows that within the
panel, strains rarely exceed a few percent, Figure 8. A closer examination
of the mechanics of forming body panels (17) shows that the basic requirement in the material is the tensile/yield ratio and not total elongation.
This comes about because of the requirements of shape control. In order to
form a satisfactory panel, the whole sheet must be plastically deformed.
This is achieved by holding the edge of the sheet and stretching it as much
as possible over the punch, Figure 9. The maximum stretching occurs in the
side wall and if the tensile strength is exceeded in the wall it will split.
Due to the action of friction on the punch radius, this tension is not fully
transmitted around the punch and if the tensile/yield ratio of the material
is insufficient, the side wall may split before the panel is sufficiently
stretched to set the shape. It may be shown that the required tensile/yield
ratio is governed by the limit,

$$(\text{tensile/yield}) > \exp(\mu\theta) \qquad (3)$$

where μ is the friction coefficient and θ the angle at which the sheet is
pulled over the die as shown in Figure 10.
Formability in skin panels is therefore related to the tensile/yield
ratio and to friction between the sheet and the tooling. The tensile test.

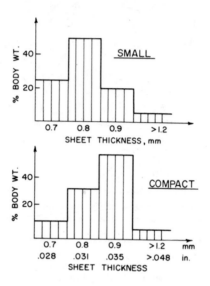

Fig. 7   Distribution of sheet metal by weight for different
         gauges in Japanese autobodies, Ref. 18.

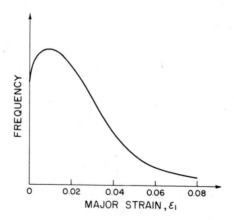

Fig. 8   Distribution of strain in small units of area in sheet
         formed into autobody parts such as door and roof
         panels (redrawn from Ref. 19).

will indicate the former but not the latter. For this reason a simulative
test was developed (20) which is illustrated in Figure 11.   The sheet is
rigidly gripped at each end and stretched over a punch until failure occurs.

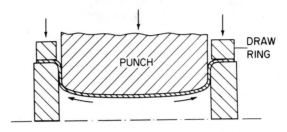

Fig. 9 Section view of punch and draw ring used in a typical
shallow stamping.

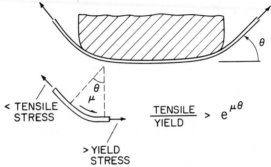

$$\frac{\text{TENSILE}}{\text{YIELD}} > e^{\mu\theta}$$

Fig. 10 Schematic view of stretching sheet over a punch illustrating
the effect of friction and wrap angle.

in the side-wall region A. Gauge length measurements in the region B give a
maximum strain after friction ε_B and this is taken as the formability
indicator. Analytical modelling indicates that this parameter should
indicate the potential for achieving good shape control in the material in
forming smoothly contoured parts.

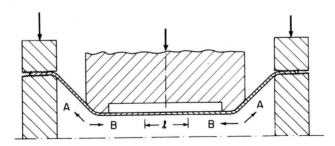

Fig. 11 Diagram of simulative formability test for stamping
a shallow panel, Ref. 20.

If the sheet is plastically deformed the springback is small and
independent of the yield stress of the material is illustrated in Figure 12
from Ref. 17. This is important because maintaining a constant yield from

341

lot to lot of sheet is not possible. Although the springback is small it is not zero and it may be shown that it is proportional to,

$$(d\sigma/d\epsilon)/E$$

as shown in Fig. 12, where $d\sigma/d\epsilon$ is the slope of the stress strain curve at the small level of plastic strain existing in the part and E is the Young's Modulus. In dual-phase steels, the slope will be very much greater than other materials and special care will have to be taken in calculating springback allowances.

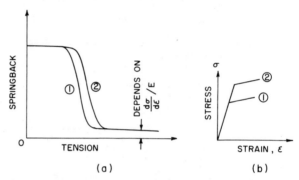

Fig. 12(a)   Diagram showing variation of springback with tension for two materials stretched over a smoothly contoured punch.

Fig. 12(b)   Assumed stress strain curves for materials having different yield strengths but identical plastic work hardening rates, $d\sigma/d\epsilon$.

The importance of the r-value in these parts is far from obvious. The mechanism whereby high r-value aids deep drawing is well understood; however, shallow stampings such as autobody panels are formed by stretching and not by deep drawing. The r-value is probably related to control of shape and the elimination of wrinkling in shallow parts as suggested by Yoshida (18) but here again more study of both a theoretical and practical nature is indicated.

After forming, the important characteristic of the part is its strength and particularly its resistance to denting. Yoshida (18) suggests that dent resistance depends on,

$$\sigma_p t^2$$

where σ_p is the flow stress for the strain level in the formed part, and not the initial yield strength as in the case above for bending. To achieve significant gauge reduction, large increases in flow stress are required. It is not practical to increase σ_p by employing large strains and hence large amounts of work hardening during the forming operation and hence there is considerable interest in increasing the yield stress by aging after forming. The effect of the stress strain curve is illustrated in Figure 13. The tensile/yield ratio of the aged material is similar to that for an HSLA steel and this would be most unsuitable for forming skin panels. The characteristics of dual-phase in having a low yield stress is highly, desirable for forming, but poor for dent resistance. Aging in the paint

342

baking process clearly obtains a desirable combination of forming and service characteristics. (The historical preference for unstabilized rimmed steel for body panels may have similar origins. In this material the yield is supporessed by roller levelling before forming and subsequently restored by natural aging.) Distortion during aging is a problem which, as indicated above, should be investigated.

Concluding Remarks

Although the modelling of the tensile properties based on concepts such as a law of mixtures and the back stresses which arise due to compatibility have been reasonably successful, a number of important problems remain to be elucidated.

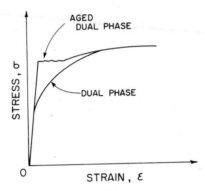

Fig. 13 Diagram showing the effect of aging on the stress strain curve.

From a variety of papers presented in this symposium it is clear that the tempering and strain aging processes are very complex. Studies of the directionality of properties both after some initial deformation path and after strain aging may be of value in clarifying the role of macroscopic stress distributions on mechanical behaviour of the dual-phase steels and in separating the influence of initial residual stresses due to phase transformation from those built up during a given strain path. Also, the possibility of using some form of aging or strain aging process to augment the strength level of formed parts is of considerable practical significance.

When we consider more complex mechanical responses such as sheared edge cracking it is clear that limitations to the interface strength, the mechanisms of strain hardening at high strain and crack propagation all influence the sheared edge sensitivity and fracture characteristics. Further research in this area will help to decide whether dual-phase or other HSLA steels are more appropriate for specific applications.

The characteristics of dual-phase steels seem well suited for stamping smooth panels. The application could save weight in finished components provided the yield strength can be increased after forming. The important aspects appear to be the development of large tensile/yield ratio for good formability and shape control, obtaining surface conditions which reduce friction and promote consistent sliding, repeatable strain hardening rates at small strains for springback control, increasing the r-value for

343

wrinkling control and above all being able to age after forming to obtain dent resistance.

The papers presented in this volume deal with a broad variety of these topics and should help to stimulate the intelligent and selective application of dual phase steels in the future.

Acknowledgements

The authors are grateful to their colleagues at McMaster University for their input on this topic and to the Natural Sciences and Engineering Research Council of Canada for their support of research.

References

1. R.G. Davies and C.L. Magee, "Physical Metallurgy of Automotive High Strength Steels", pp. 1-19 in Structure and Properties of Dual-Phase Steels, R.A. Kot and J.W. Morris, ed.; AIME, Warrendale, Pa, 1979.

2. J. Becker, E. Hornbogen and F. Wendl, this book.

3. Y.W. Chang and R.J. Asaro, "Bauschinger Effects and Work-hardening in Spheroidized Steels", Metal Science J., 12 (1978) p. 277.

4. J. Gerbase, J.D. Embury and R.M. Hobbs, "The Mechanical Behaviour of Some Dual Phase Steels - With Emphasis on the Initial Work Hardening Rate", in Structure and Properties of Dual-Phase Steels, R.A. Kot and J.W. Morris, ed.; AIME, Warrendale, Pa., 1979.

5. A. Okamoto and M. Takahashi, this book.

6. H. Mathy, J. Gouzou and T. Freday, this book.

7. D. Tseng and F.H. Vitovec, this book.

8. L.F. Ramos, D.K. Matlock and G. Krauss, "Communication on the Deformation Behaviour of Dual-Phase Steels", Met. Trans. A, 10 (1979), p. 259.

9. C. Crussard, "Rapport entre la forme exacte des courbes de traction des métaux et les modifications concomitantes de leur structure", Rev. Met., 10 (1953) pp. 697-710.

10. B. Jaoul, "Etude de la forme de courbes de déformation plastique", J. Mech. Phys. Solids, 5 (1957), pp. 95-114.

11. K. Araki, S. Fukunaka and K. Uchida, "Development of Continuously Annealed High Strength Cold Rolled Sheet Steels", Trans. ISIJ, 17 (1977) pp. 701-709.

12. T. Shinoda, A. Sawada and T. Mori, "The Back Stress Decrease on Annealing and Hydrogen Embrittlement of Deformed Eutectoid Carbon Steel", Mat. Sci. and Engineering, 41 (1979), pp. 103-111.

13. R. Stevenson, "Improving the Formability of Low Carbon Sheet Steel by Control of Interstitial Carbon Content and Temperature", Met. Trans. A, 11 (1980) p. 1909.

14. E. Waddington, R.M. Hobbs and J.L. Duncan, "Comparison of a Dual-Phase Steel with Other Formable Grades", J. Appl. Metalworking, 1 (2) (1980), pp. 35-47.

15. A. Nishimoto, Y. Hosoya and K. Nakaoka, this book.

16. E. Waddington, McMaster University Report, March, 1980.

17. J.L. Duncan and J.E. Bird, "Die Forming Appoximations for Aluminum Sheet", Sheet Metal Ind., 15 (1978), pp. 1015-1025.

18. K. Yoshida, "Some Recent Trends in Steel Sheets and Forming Techniques of Autobody", Trans. ISIJ, 19 (1979), pp. 257-267.

19. H. Ishigaki, "Deformation Analysis of Large Sized Panels in the Press Shop", p. 315 in Mechanics of Sheet Metal Forming, D.P. Koistinen and N-M. Wang, ed.; Plenum Press, New York, N.Y., 1978.

20. E. Waddington and J.L. Duncan, "Formability Test for Shallow Stampings", Proc. 8th Canadian Conf. Applied Mechanics, Moncton, N.B., June, 1981.

THE EFFECT OF MICROSTRUCTURE ON THE DEFORMATION BEHAVIOR

AND MECHANICAL PROPERTIES OF A DUAL-PHASE STEEL

R.D. Lawson, D.K. Matlock, and G. Krauss
Department of Metallurgical Engineering
Colorado School of Mines
Golden, Colorado 80401

ABSTRACT

A plain carbon Mn-Si steel has been intercritically annealed at 760 and 810C and cooled at various rates after annealing. Quantitative metallography of specimens etched to differentiate epitaxial and retained ferrite was used to construct microstructure maps as a function of cooling rate. Correlation of the microstructure with the results of tensile testing showed that significant amounts of epitaxial ferrite and small amounts of martensite in a matrix of retained ferrite are associated with good combinations of strength and ductility. A review of analyses of strain hardening from stress-strain curves shows that the Jaoul-Crussard and Swift analyses best delineate the early stages of strain hardening in the C-Mn-Si dual-phase steel. The slope of the second stage of strain hardening, according to the Jaoul-Crussard analysis, is related to the uniform true plastic strain and thus the ductility of dual-phase steels. This correlation shows that the lower the slope of the second stage of work hardening the greater the ductility. Continuum models of dual-phase steel deformation underestimate uniform true plastic strain of dual-phase steels with low volume fractions of martensite because of their inability to reproduce the strain hardening and inhomogeneity of plastic deformation at low plastic strains.

Introduction

The relationship of microstructure to the work hardening behavior of dual-phase steels continues to be of considerable interest. Previous studies have experimentally evaluated the strain dependence of the work hardening behavior of dual-phase steels, and several theoretical treatments have been developed to account for the deformation behavior and mechanical properties of these multi-phase alloys (1-4). Some of the first models applied the rule of mixtures to distribute stress in a microstructure of ferrite and martensite assuming an equal distribution of strain between the two phases (3,4,5). Other models partitioned strain between the ferrite and martensite by means of a continuum mechanics approach (6,7,8). Yet another approach has consisted of the use of micromechanics to model the accumulation of dislocations in two-phase mixtures with increasing strain (9,10,11). A recent brief review (12) of the various approaches has indicated that combinations of a continuum mechanics model and a micromechanical model to account for dislocation substructure development with strain may best represent dual-phase deformation behavior.

The above models all assume dual-phase steel microstructures consisting only of ferrite and martensite. However, microstructural studies show that a dual-phase steel may contain not only ferrite and martensite, but also mixtures of ferrite and cementite and fine precipitates in the ferrite retained during intercritical annealing (1,13). In fact, two types of ferrite, the retained ferrite containing precipitates, and precipitate-free ferrite formed by epitaxial growth into the austenite on cooling, may be present in dual-phase steel (1,13,14,15). Accompanying the microstructural diversity, the work hardening behavior also appears to be complex, as reflected in the various stages of work hardening detected by Jaoul-Crussard (1) or Swift (2) analyses of dual-phase steel stress-strain curves. Neither the microstructural nor the work hardening complexities have as yet been incorporated into models of dual-phase steel deformation behavior.

The purpose of the present paper is to explore further the relationship between microstructure and work hardening behavior in a dual-phase steel. Microstructures as a function of cooling rate after intercritical annealing have been characterized, measured, and recorded on microstructure maps. Stress-strain curves have been analyzed, and correlations between mechanical properties, microstructure, and two

348

continuum mechanics deformation models are presented and discussed. A major purpose of this investigation was to evaluate the significance of the various stages of work hardening that have been observed in dual-phase steels.

Experimental Procedure

The composition in weight percent of the steel, designated as CSM 2, examined in this investigation is: 0.063 pct. C, 1.29 pct Mn, 0.24 pct. Si, 0.009 pct. P, and 0.019 pct. S. The steel was received in the as-hot rolled condition in the form of 2.79 mm (0.110 in) thick sheet. Sheared coupons were surface ground to a final thickness of 1.91 mm (0.075 in) and machined into flat longitudinal tensile samples with gage dimensions of 31.8 mm (1.25 in) x 6.4 mm (0.25 in) according to ASTM procedures for subsized tensile specimens (16).

The heat treatment consisted of intercritical annealing at 810 or 760C in neutral salt baths for 10 minutes followed by quenching to room temperature in water, oil, or in forced air in a cooling tower with controlled air flow. Cooling rates were determined by embedding a 0.508 mm (0.020 in) Inconel-sheathed, chromel-alumel thermocouple in a coupon of the same width and thickness as the tensile samples welded to a calibration fixture (17). The cooling rates represent an average between 700 and 300C.

The tensile samples were tested on a closed-loop servo-hydraulic MTS testing system at a strain rate of 0.04 min^{-1} within 10 minutes after heat treatment. Signals provided by a 25.4 mm (1 in) x 50% strain gage extensometer and a 20,000 lb capacity load cell were used as input for a data processing system which stored the data as load-elongation on a magnetic disk. Strain hardening rates were calculated with a computer program which calculated the average of a forward and backward three point approximation about each stress-strain value.

Metallographic specimens were cut from the grip section of the tensile samples and prepared by standard metallographic procedures. An etching technique (14) applied to each specimen to differentiate retained and epitaxial ferrite consisted of etching in 2% Nital and 4% Picral followed by immersion in a boiling alkaline chromate solution.

Point counting methods were used on photomicrographs to determine the volume fractions of the different phases (18).

349

The ferrite grain size of CSM 2 was determined (19) to be equivalent to ASTM No. 10 (average grain diameter 10.7 μm). The grain size was not affected by intercritical annealing treatments.

Results

Stress-Strain Curves and Mechanical Properties

Figure 1 shows a family of true-stress, true-strain curves of specimens cooled at different rates after intercritical annealing at 810C. The decrease in ultimate strength, the increase in true uniform plastic strain, and the transition from continuous to discontinuous yielding with decreasing cooling rate are all apparent. Specimens intercritically annealed at 760C showed a similar set of stress-strain curves.

Figures 2 and 3 show the variation in mechanical properties as a function of cooling rate for specimens intercritcally annealed at 820C and 760C, respectively. In both figures e_{TOT} is the total engineering strain, e_U is the uniform engineering strain, S_{UTS} is the ultimate tensile strength, $S_{0.002}$ is the yield stress evaluated at 0.2% engineering strain, and e_{yp} is the Luders strain. Specimens cooled at a high rates show no

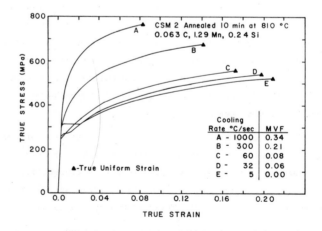

Figure 1: Summary of true stress, true strain curves for steel CSM 2 cooled at different rates from 810C.

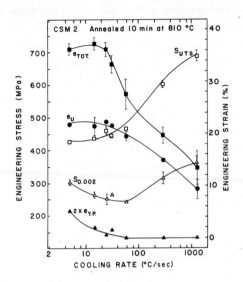

Figure 2: Summary of tensile data as a function of cooling rate for specimens annealed 10 minutes at 810C.

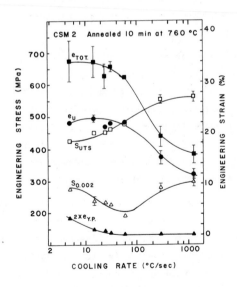

Figure 3: Summary of tensile data as a function of cooling rate for specimens annealed 10 minutes at 760C.

351

yield point elongation. With decreasing cooling rates, yielding becomes progressively more discontinuous and passes through a stage of inhomogeneous deformation band propagation superimposed on continuous uniform deformation (1). The latter phenomenon does not produce a horizontal or constant stress segment on the stress-strain curve but does produce inflection points as shown, for example on Curve D of Figure 1. The extent of the regions of detectable slope change of the initial portions of the stress-strain curve have been plotted as yield point elongation, in Figures 2 and 3. Also of interest in Figures 2 and 3 is the minimum in yield stress, developed at intermediate rates of cooling. The divergence of yield and ultimate strengths produced by intermediate cooling rates has been associated with the good combinations of strength and ductility achievable in dual-phase steels.

Microstructural Characterization

Figure 4 shows the differences in microstructure developed by cooling at various rates from 810C. The dark-etching phase is martensite, the white-etching phase is epitaxial ferrite, and the gray phase is ferrite that was retained during intercritical annealing. The boundaries between the white and gray etching ferrites define the extent of the austenite pool formed during intercritical annealing. Figure 4 shows that as the cooling rate decreases, more of the austenite transforms to epitaxial ferrite and less to martensite.

Figure 5 shows a similar series of microstructures produced by cooling from 760C at various rates. Much less of the austenite transforms to epitaxial ferrite with decreasing cooling rate, an indication that the lower intercritical annealing temperature produces higher hardenability than does the higher annealing temperature, a result also noted in a Nb-microalloyed dual-phase steel (13,14). The increased hardenability of austenite formed by low-temperature intercritical annealing is attributed to the higher carbon concentration of that austenite.

Figures 6 and 7 show microstructure maps determined from quantitative metallographic analysis of microstructures produced by intercritical annealing at 810C and 760C, respectively. The fact that the etching differences between retained and epitaxial ferrite define the extent of the austenite pool is confirmed by the approximately constant total volume fraction of transformed phases measured for all cooling rates. After 810C annealing the volume fraction of austenite (or transformed phases) is 0.40

352

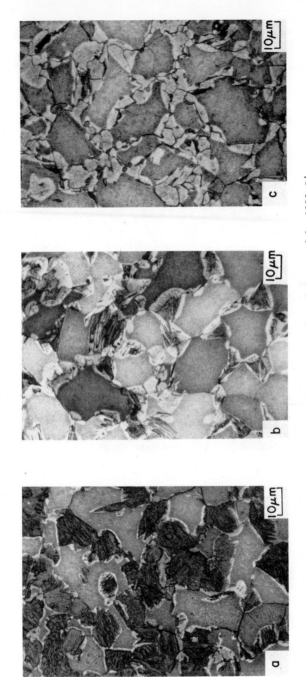

Figure 4: Light micrographs of steel CSM 2 cooled at: (a) 1000 C/sec, (b) 300 C/sec, and (c) 60 C/sec from 810C. Nital-Picral- Alkaline Chromate etch.

Figure 5: Light micrographs of steel CSM 2 cooled at: (a) 1000 c/sec, (b) 300 C/sec, and (c) 60 C/sec from 760C. Nital-Picral-Alkaline Chromate etch.

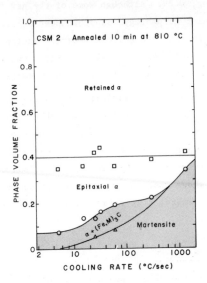

Figure 6: Quantitative microstructure map showing the effect of cooling rate on the microstructure of steel CSM 2 annealed 10 minutes at 810C.

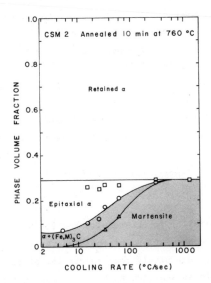

Figure 7: Quantitative microstructure map showing the effect of cooling rate on the microstructure of steel CSM 2 annealed 10 minutes at 760C.

and after 760C annealing, 0.29. Although some of the new ferrite formed on cooling has an acicular morphology, as shown best in Figure 4(b), it was difficult to differentiate the acicular ferrite during quantitative metallographic analysis, and therefore all ferrites formed by transformation of the austenite are classified as epitaxial ferrite.

The microstructure maps show quantitatively the effect of cooling rate on microstructure. As noted previously for other dual-phase steels (1,15), good strength-ductility combinations are associated with microstructures containing significant amounts of epitaxial ferrite and martensite and some ferrite-carbide mixtures. Comparison of Figures 2 and 3 with Figures 6 and 7 shows that the cooling conditions that produce the good combinations of mechanical properties for the CSM 2 steel are in the range of 60 to 120C/sec.

Effect of Martensite on Strength

Figures 8 and 9 show the relationship of the ultimate tensile strength and yield strength to the volume fraction of martensite in specimens annealed at 810 and 760C, respectively. In each annealing series, the ultimate tensile strengths are directly proportional to the volume fraction of martensite. Other investigators have reported similar results (2,3,20). The slopes of the ultimate tensile strength versus volume fraction martensite curves, however, appear to be quite dependent on the temperature of intercritical annealing. The ultimate tensile strengths of specimens annealed at 810C increase much more rapidly with increasing martensite volume fraction than do those of specimens annealed at 760C. The following equations, obtained by linear regression analysis of the data in Figures 8 and 9, show the greater effect of martensite volume precent, V_m, on the ultimate strength of specimens annealed at 810C.

$$810C: \quad S_{UTS} \text{ (MPa)} = 8.49 \ V_m + 412 \qquad (1)$$
$$760C: \quad S_{UTS} \text{ (MPa)} = 3.87 \ V_m + 432 \qquad (2)$$

Cribb and Rigsbee (2), similar to our results for the 810C specimens, also found a strong dependence of ultimate strength on the amount of second phase (largely martensite but significant quantities of acicular ferrite/bainite in specimens annealed at higher temperatures) in V-microalloyed dual-phase steels annealed between 775 and 900C.

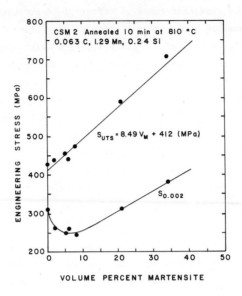

Figure 8: Ultimate tensile strength and 0.2% offset yield stress as a
function of volume percent martensite for steel CSM 2 annealed
10 minutes at 810C.

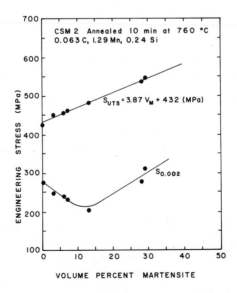

Figure 9: Ultimate tensile strength and 0.2% offset yield stress as a
function of volume percent martensite for steel CSM 2 annealed
10 minutes at 760C.

Figures 8 and 9 show that the 0.2% offset yield strengths of the intercritically annealed steel are a minimum at low volume fractions of martensite. The high values of the yield strength at zero martensite volume fraction represent the lower yield stress associated with discontinuous yielding and the minimum develops as increasing amounts of martensite produce the transition to more continuous yielding behavior as shown in the stress-strain curves of Figure 1. Increasing amounts of martensite beyond that which produces the minimum, increase the work hardening rates and therefore, the 0.2% offset yield strength (1). The minimum in yield strength versus volume percent martensite has also been reported (21), but some investigators (22) show only a linear relationship between yield strength and percent martensite, perhaps because specimens with very low amounts of martensite were not tested.

An important new consideration with respect to comparison of the data shown in Figures 8 and 9 is the fact that the balance of epitaxial and retained ferrite changes as a result of intercritical annealing temperature. The microstructure maps, Figure 6 and 7, show graphically that for a given martensite content, the 810C specimens contain much more epitaxial ferrite than do the 760C specimens. The implications of variations in the amounts of the two types of ferrite on the work hardening behavior of dual-phase steels will be discussed in a later section.

Discussion

The engineering properties presented in the previous section, e.g. ultimate tensile strength, yield strength, and total ductility, actually reflect complex interactions of the yielding and strain hardening behavior of the microstructural components of dual-phase steel. This section first considers various systematic empirical approaches to analyzing the work hardening behavior of dual-phase steels, and then correlates the results presented above with models developed for deformation of multi-phase materials.

Application of Constitutive Equations to Describe the Strain Hardening Behavior of Dual-Phase Steels

In past investigations (1,2) distinct stages in the deformation behavior of dual-phase steel were revealed through analysis of stress-strain curves with various empirical equations. The stress-strain curves of CSM 2 in Figure 1 were analyzed with four common constitutive equations: the Hollomon (23), the Voce (24), the Ludwik (25), and the

358

modified Swift (26) equations. In Table I the four constitutive equations and the analytical equations derived from each one are presented. Figure 10 shows the Hollomon analysis for the stress-strain curves of Figure 1. Similar to previous observations (1) the non-linear log true stress-log true strain plots show that the Hollomon equation does not adequately describe the stress-strain behavior of any of the specimens over the entire range considered.

The Voce equation in Table I assumes that the stress approaches an asymptotic value, σ_0', exponentially with increasing strain unlike the other three equations which are parabolic functions. In contrast to the parabolic equations, the Voce equation has been found to provide the best fit for $2\,1/4$ Cr-1 Mo steel (30), deformed Cu (31), and 5182-0 Al (32). Use of the Voce equation is further supported by the criticism (33) that the other equations in Table I will be dependent on any prestrain or prestress which may exist in the sample prior to testing. This criticism does not apply to the Voce analysis since the adjustable parameters σ_0', and ε_0 are functionally related in the Voce equation. Analysis of the stress-strain curves in Figure 1 with the Voce equation is shown in Figure 11 as the $\log \frac{d\sigma}{d\varepsilon_p}$ versus ε_p, where ε_p is true plastic strain. As in the case of the Hollomon analysis the non-linear nature of the curves in Figure 11 shows that the Voce equation also does not provide a useful description of the stress-strain behavior for any of the four specimens. Because the Voce analysis does not amplify the early strain region the curves for specimens subjected to different cooling rates from the intercritical annealing temperature tend to superimpose.

Analysis of the CSM 2 dual-phase steel stress-strain curves by the Jaoul-Crussard method (27) which results in the analytical expression for the Ludwik equation in Table I is shown in Figure 12. Three distinct stages in the deformation behavior are revealed by this analysis and become more pronounced as the cooling rate from the intercritical annealing temperature is decreased. Matlock et. al. (1) observed similar families of curves in Jaoul-Crussard plots for increasing grain size, increasing aging time and increasing annealing temperature as well as decreasing cooling rate. The onset of the second stage of strain hardening behavior, which corresponds to the first change in slope with increasing strain and the minima in strain hardening as shown in curve D have been associated with a gradual shift from homogeneous to inhomogeneous deformation short of producing prominent Lüders strain (1). Stage 2 deformation is observed to occur in the approximate range of

359

Table I. Constitutive and Analytical Equations Used to
Analyze the Strain Hardening Behavior of Dual-Phase Steels

Equation, Name, and Reference	Constitutive Equation	Analytical Equation
Hollomon(23)	$\sigma = k\,\varepsilon_p^n$	$\log \sigma = \log k + n \log \varepsilon_p$
Voce(24)	$\sigma = \sigma_o' - \exp[-k''(\varepsilon_p - \varepsilon_o)]$	$\log \dfrac{d\sigma}{d\varepsilon_p} = (\log k'' + 0.43k''\varepsilon_o) - 0.43k''\,\varepsilon_p$
Ludwik(25)	$\sigma = \sigma_o + k'\,\varepsilon_p^m$	$\log \dfrac{d\sigma}{d\varepsilon_p} = \log (k'm) + (m - 1) \log \varepsilon_p$
Modified Swift(26)	$\varepsilon_p = \varepsilon_{po} + c\sigma^{m'}$	$\log \dfrac{d\sigma}{d\varepsilon_p} = (1 - m') \log \sigma - \log (cm')$

σ and ε_p are the true-stress and true plastic strain, respectively.

k, n, σ_o', k'', ε_o, m, k', σ_o, c, m', and ε_{po} are all constants.

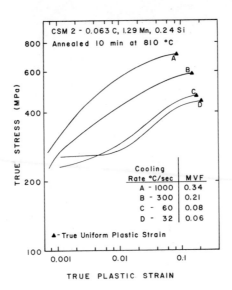

Figure 10: Analysis of the stress–strain curves of Figure 1 for steel CSM 2 according to the Hollomon equation in Table I.

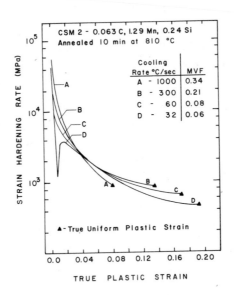

Figure 11: Analysis of the stress–strain curves of Figure 1 for steel CSM 2 according to the Voce equation in Table I.

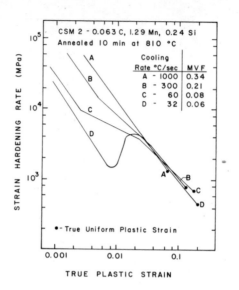

Figure 12: Jaoul-Crussard analysis of the stress-strain curves of Figure 1 for steel CSM 2 according to the Ludwik equation in Table I.

0.5 to 4% strain. Above 4% true plastic strain a third stage in the deformation behavior occurs. The slopes in stage 3 for each of the four processing conditions are approximately equal, indicating similar deformation mechanisms at high strains.

Analysis of the same stress-strain curves by the modified Swift equation is shown in Figure 13 as $\log \dfrac{d\sigma}{d\varepsilon_p}$ versus $\log \sigma$. Stages in the deformation behavior are also shown by the Swift analysis and at approximately the same strain hardening intervals as in the Jaoul-Crussard analysis. Cribb and Rigsbee (2) showed similar changes in the slope of stage 2 in both Swift and Jaoul-Crussard plots for two vanadium dual-phase steels treated as a function of annealing temperature and cooling rate. In contrast to the Jaoul-Crussard analysis, Figure 11, the Swift analysis in Figure 13 shows the variation in strength produced by the different cooling rates. The true stress at maximum load, which is the last value of stress on each of the Swift plots, agrees closely with the condition for instability, i.e. $\dfrac{d\sigma}{d\varepsilon_p} = \sigma$, also plotted on Figure 13.

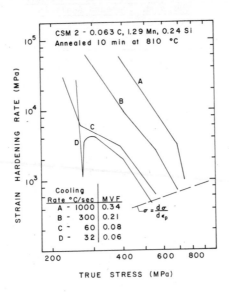

Figure 13: Analysis of the stress-strain curves of Figure 1 for steel CSM 2 according to the Modified Swift equation in Table I.

Correlation of Ductility with Stages of Strain Hardening

The preceding discussion of the various constitutive equations for plastic deformation shows that no single set of parameters could describe the deformation behavior of dual-phase steels over the entire plastic strain range. Further, each constitutive analysis emphasizes different aspects of the strain hardening behavior, the Swift and Jaoul-Crussard analyses more precisely defining the various stages of deformation. This section uses a unique set of parameters for each stage to construct the stress-strain curves of dual-phase steels with a wide variety of microstructures. In particular a correlation which shows the importance of low strain strain hardening behavior to total ductility is developed. Discussion of the relationship of the empirically defined stages to the interacting deformation mechanisms that are influenced by the various microstructural components of dual-phase steels is deferred to the final section.

Table II shows the values of m and k for the three stages of deformation defined by the Jaoul-Crussard analysis for the stress-strain

363

Table II. Work Hardening Parameters for
the Jaoul-Crussard Analysis

Cooling Rate C/sec	Stage No.	m	k' (MPA)	σ_o (MPA)	True Plastic Strain Interval (%)
1000	1	-0.110	- 671	1212	0 - 0.28
	2	-0.110	- 671	1212	0.28 - 3.96
	3	-0.110	- 671	1212	3.96 - 6.61
300	1	-0.23	- 91	692	0.05 - 0.50
	2	0.21	1033	27	0.62 - 3.75
	3	-0.025	-3939	4820	3.75 - 13.23
60	1	-0.22	- 30	366	0.02 - 0.39
	2	0.57	1296	202	0.39 - 3.27
	3	0.17	967	-152	3.27 - 18.58
32	1	NA	NA	NA	NA
	2	NA	NA	NA	NA
	3	-0.08	-1006	1682	3.28 - 20.00

curves shown in Figure 12. Included also in Table II are the strain
intervals for each stage. Figure 14 shows the variation in tensile
strength with the work hardening parameters m_{II} and m_{III}. For stage 2
deformation, the decrease in m_{II} with increasing tensile strength (or
correspondingly martensite volume fraction) is related to higher strain
hardening rates and more homogeneous deformation at small strains. Figure
14 also shows that for stage 3 deformation m_{III} decreases slightly with
increasing tensile strength.

The effect of early strain hardening on ductility is indicated in
Figure 15 which correlates the slope in stage 2 of the Jaoul-Crussard
plot, $(m_{II}-1)$, with true uniform plastic strain. This correlation shows
that the uniform ductility increases with a decrease in the absolute value
of the slope in region II.

To demonstrate the importance of the early strain hardening behavior,
as characterized by m_{II}, on ductility, a simplified model based on
correlations suggested by Figure 12 is now presented. This analysis is
based on the criteria which leads to satisfying the condition for
instability, i.e. necking, given by:

$$\frac{d\sigma}{d\varepsilon} = \sigma_{up} \tag{3}$$

364

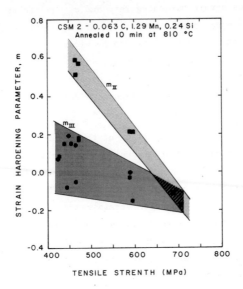

Figure 14: Variation of the strain hardening parameter, m, in regions II and III of the Jaoul-Crussard plot with ultimate tensile strength for steel CSM 2.

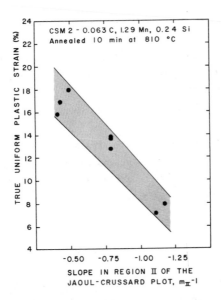

Figure 15: True uniform plastic strain as a function of the slope in region II of the Jaoul-Crussard plot for steel CSM 2, showing the increase in ductility with smaller slopes in region II.

where $\frac{d\sigma}{d\varepsilon}$ is the true-strain hardening rate and σ_{up} is the true-stress at instability which occurs in the third stage of deformation as observed in Figure 12. The stress at instability can be viewed as resulting from a sum of the initial flow stress plus the increments of stress due to strain hardening in each of three distinct stages of deformation. The stress increments and corresponding strain ranges are defined in Figures 16 and 17.

Figure 16 is a schematic comparison of the stress-strain curves used to model the deformation behavior of dual-phase steels for three different processing conditions. The stress-strain curve for specimen A in Figure 16 was drawn to represent a parabola which is described by a single Ludwik equation over the entire true plastic strain range. The stress-strain curves for specimens B and C were drawn as composites of three parabolas, one each for stages 1,2, and 3, with different values for the parameters σ_o, m and k for each stage. The parabolas representing stage 3 for curves A,B, and C are parallel with identical values of m_{III} and k_{III}. The extension of this parabola for curve C is shown by the dotted line. The tangent points of the parabolas define the three stages of the stress-strain curves. The corresponding Jaoul-Crussard plots are also shown in Figure 16.

The following assumptions about the deformation behavior are implied in Figure 16:

1) The slope in region III for all of the specimens is the same. Therefore, the parameters m_{III} and k_{III} remain constant as the slope in region II changes.

2) ε_{II}, the strain which defines the transition from stage 2 to stage 3, remains constant.

3) The stress at zero true plastic strain σ_o, remains constant despite changes in the flow stress resulting from processing.

As shown in Figure 17, which is the stress-strain curve of material C in Figure 16, the true stress as a function of strain in region III can be described as:

$$\sigma = \sigma_o + \Delta\sigma_I + \Delta\sigma_{II} + \Delta\sigma_{III} \tag{4}$$

where σ_o is the initial flow stress and $\Delta\sigma_I$, $\Delta\sigma_{II}$, and $\Delta\sigma_{III}$, are the increments of stress resulting from strain hardening in each of the three deformation regions. The increments of stress can be calculated for each strain region with the corresponding Ludwik equation, Table I. In Figure

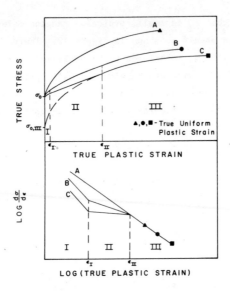

Figure 16: A schematic comparison of the true stress-strain curves of dual-phase steel subjected to different processing histories with the corresponding Jaoul-Crussard plots exhibiting the relationship between true uniform strain and the change in slope of the Jaoul-Crussard plot in region II.

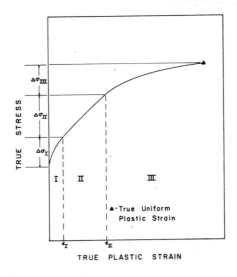

Figure 17: An illustration of a composite stress-strain curve which can be described by different Ludwik equations in three distinct regions of the true plastic strain range.

16, it can be seen that ε_I is small, typically ranging between 0.002 to 0.004 for dual-phase steels as shown in Figure 12. Therefore to simplify the following analysis $\Delta\sigma_I$ is assumed to be negligible. Note however that $\Delta\sigma_I$ can be included without changing the basic procedure to be presented here. Combining Equations 3 and 4 and with the above assumptions, the conditions for instability can be written as:

$$\frac{d\sigma}{d\varepsilon} = \sigma_o + \Delta\sigma_{II} + \Delta\sigma_{III} \tag{5}$$

As shown in Figure 17 with $\Delta\sigma_I$ and ε_I equal to zero, the Ludwik equations in stages 2 and 3 reduce to:

$$\Delta\sigma_{II} = k_{II}(\varepsilon_{II})^{m_{II}} \tag{6}$$

and

$$\Delta\sigma_{III} = k_{III}(\varepsilon_{up})^{m_{III}} - k_{III}(\varepsilon_{II})^{m_{III}} \tag{7}$$

where ε_{up} is the true uniform plastic strain. For strains greater than ε_{II} the stress-strain curve in region III is described by:

$$(\sigma_{III})_{\varepsilon \rangle \varepsilon_{II}} = \sigma_{o,III} + k_{III}(\varepsilon_p)^{m_{III}} \tag{8}$$

Therefore the strain hardening rate when ε_p is equal to ε_{up} is:

$$\frac{d\sigma}{d\varepsilon_p})_{\varepsilon_{up}} = m_{III} \, k_{III} \, (\varepsilon_{up})^{(m_{III}-1)} \tag{9}$$

Combining Equations 6,7, and 9 with Equation 5 results in the following expression for the true uniform strain in terms of the Ludwik parameters in each stage:

$$\sigma_o - k_{III}(\varepsilon_{II})^{m_{III}} + k_{II}(\varepsilon_{II})^{m_{II}} = k_{III}[m_{III}(\varepsilon_{up})^{(m_{III}-1)} - \varepsilon_{up}^{m_{III}}] \tag{10}$$

Applying the boundary condition that at ε_{II} the strain hardening rates in regions II and III are equal gives k_{II} as a function of m_{II}:

$$k_{II} = \frac{k_{III} \, m_{III} \, (\varepsilon_{II})^{m_{III}}}{m_{II} \, (\varepsilon_{II})^{m_{II}}} \tag{11}$$

Substituting Equation 11 into 10 produces the following expression:

368

$$m_{II} = \frac{k_{III}\, m_{III}\, (\varepsilon_{II})^{m_{III}}}{k_{III}[m_{III}(\varepsilon_{up})^{(m_{III}-1)} - (\varepsilon_{up})^{m_{III}}] - [\sigma_o - k_{III}(\varepsilon_{II})^{m_{III}}]} \qquad (12)$$

If the assumptions listed previously apply, i.e. k_{III}, m_{III}, ε_{II} and σ_o are all constants, then Equation 12 describes the change in uniform plastic strain with m_{II} where the slope of the Jaoul-Crussard plot in region II is $(m_{II}-1)$. Although m_{II} is not a simple function of ε_{up} it can be shown that for all possible values of m_{III} (i.e. $1 < m_{III} < 0$ or $0 < m_{III}$) as $(m_{II}-1)$ decreases ε_{up} increases as observed in Figure 15.

To evaluate the correlation between m_{II} and ε_{up} described in Equation 10, representative values of the Ludwik parameters m_{III} and k_{III}, were obtained from the data listed in Table II for the sample cooled at 300 C/s, which exhibits stress-strain behavior similar to curve B in Figure 16. With these parameters and by using the 0.2% offset yield stress for this material, 312 MPa, as σ_o, the correlation of ε_{up} with $m_{II}-1$ shown in Figure 18 was obtained. The effects of varying σ_o on this correlation are presented by the dotted lines in Figure 18. Also shown is a comparison of the measured values from Figure 15 with the behavior predicted by this

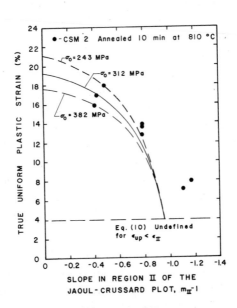

Figure 18: Correlation of the true uniform plastic strain and the low strain strain-hardening behavior (as described by the slope in region II of the Jaoul-Crussard plot) from Equation 12.

analysis. The correlation of ε_{up} with the slope in region II of the Jaoul-Crussard plot predicted from Equation 12 is shown in Figure 18 to describe well the behavior in samples with true uniform plastic strains greater than approximately 11%.

The assumption that σ_o will remain constant as the cooling rate is changed, is not realistic as observed by the increase in the 0.2% offset yield strength with martensite volume fraction in Figure 8. The effect of increasing σ_o is to lower the uniform strain for a given slope in region II as shown by the family of curves in Figure 18. The effect of σ_o on the predictions of Equation 12 is most significant at high uniform strains. However, at low uniform strains the differences between the curves of Equation 12 and the measured values result mainly from the assumption that the slope in region III is constant.

Variations in the slope in region III may result from the effects of processing on microstructure. If the strain hardening at high strains were controlled by the deformation of the retained ferrite this would imply a constant slope in region III. However, the slight decrease in m_{III} with increasing tensile strength indicates the deformation behavior of the martensite on the composite strain hardening becomes more important with an increase in martensite volume fraction. Increasing the martensite volume fraction also decreases the amount of epitaxial ferrite which may have a lower strength and higher ductility than the retained ferrite (15). Therefore, the combined effects of variations in the volume fractions of both martensite and epitaxial ferrite apparently lead to a more rapid decrease in the strain hardening rate with strain in region III. For example, as the martensite volume fraction increases, a more negative slope in region III of the Jaoul-Crussard plot results. The effects of the relative strain hardening behaviors of the martensite and ferrite will be discussed in the next section in terms of continuum mechanics models of two phase materials which have been applied to describe mechanistically the early strain hardening behavior of dual-phase steels.

Evaluation of Models for Deformation of Dual-Phase Steels

Figure 19 shows experimentally determined and theoretically calculated curves of true uniform plastic strain versus volume percent martensite. The experimental curve, the continuous line drawn through the

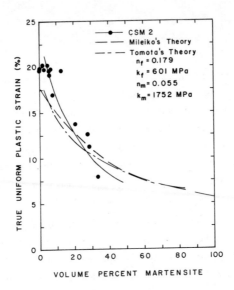

Figure 19: Variation of true uniform plastic strain with volume fraction martensite measured for a C-Mn-Si dual-phase steel and as predicted by the models of Mileiko(28) and Tomota et al(29) with the Hollomon parameters listed.

closed circles, was obtained from the tensile tests and metallographic examination of the CSM 2 alloy subjected to the heat treatments described earlier. The other curves were obtained by calculation with Mileiko's and Tomota's theories of two-phase deformation.

Mileiko's model (28) uses a plastic instability approach to describe the stress-strain curve of continuous fiber composites from the stress-strain behavior of the individual components. The stresses are partitioned by the rule of mixtures and the strain in the fiber and the matrix are assumed to be equal. Necking of the composite requires necking in both components at the same strain. The Hollomon equation, Table I, is assumed to describe the stress-strain curve for the fiber, matrix and composite. From this analysis the volume fraction of fiber, V_F, is related to the uniform strain by the following expression:

$$V_F = \frac{1}{1 + \beta \left[\dfrac{\varepsilon_C - \varepsilon_F}{\varepsilon_M - \varepsilon_C}\right] \varepsilon_C^{(\varepsilon_F - \varepsilon_M)}} \tag{13}$$

where

$$\beta = \frac{\sigma_F \, \varepsilon_M^{\varepsilon_M} \, \exp \varepsilon_F}{\sigma_M \, \varepsilon_F^{\varepsilon_F} \, \exp \varepsilon_M}$$ (14)

the true uniform strain in the fiber, matrix and composite are given by ε_F, ε_M, and ε_C, respectively, and σ_F and σ_M are the true tensile strengths in the fiber and matrix, respectively. If it is assumed that the martensite in a ferrite/martensite dual-phase steel acts like fibers (with a grain aspect ratio of approximately 1) aligned in the tensile direction then Mileiko's model can be applied. In this case ε_F in Equations 13 and 14 is replaced by ε_m, the strain in the martensite, and ε_M is replaced by ε_f, the strain in the ferrite.

The continuum mechanics model developed by Tomota et al (29) divides the deformation of a two-phase alloy into three stages. In the first stage both phases deform elastically and are assumed to have equal elastic constants. The yield stress of the composite, Y(i), is assumed to be equal to the yield stress of the soft matrix, σ_y^I. Plastic flow is assumed to occur only in the matrix and the applied stress, σ_{33}^A, determined by the flow stress of the matrix and image stresses resulting from unrelaxed plastic strain discontinuity between the matrix and the hard second phase, is given by:

$$\sigma_{33}^A = \sigma^I + fA\varepsilon_p^I \quad ; \quad \sigma_{33}^A \geq \sigma_Y^I$$ (15)

with

$$A = \frac{E(7 - 5\upsilon)}{10 - (1 - \upsilon^2)}$$ (16)

E = Young's modulus
υ = Poisson's ratio
σ^I = flow stress of the matrix
f = volume fraction of second phase
ε_p^I = true plastic strain in the matrix

The strains are partitioned according to the rule of mixtures:

$$\varepsilon_p = (1 - f) \, \varepsilon_p^I + f\varepsilon_p^{II}$$ (17)

where ε_p^{II} is the plastic strain in the hard phase and is zero in stage 2. When the composite reaches a certain stress, Y(ii), the yield stress

372

of the second phase, σ_y^{II}, is reached. Y(ii) can be determined by solving the following simultaneous equations:

$$Y(ii) = \sigma_y^{II} - (1 - f) \, A\varepsilon_p^I(ii) \qquad (18)$$

and

$$Y(ii) = \sigma^I + fA\varepsilon_p^I(ii) \qquad (19)$$

where $\varepsilon_p^I(ii)$ is the critical plastic strain in the matrix at the start of the plastic deformation in the harder phase. In stage 3 both phases are plastically deforming and the internal stress causes further plastic deformation in the second phase. Therefore the applied stress in stage 2 can be determined by solving the following two simultaneous equations relating stress to strain increment, $\delta\varepsilon$:

$$\sigma^I = \sigma_{33}^A - fA[\varepsilon_p^I(ii) + (\delta\varepsilon_p^I - \delta\varepsilon_p^{II})] \qquad (20)$$

$$\sigma^{II} = \sigma_{33}^A + (1-f)A[\varepsilon_p^I(ii) + (\delta\varepsilon_p^I - \delta\varepsilon_p^{II})] \qquad (21)$$

Thus by choosing empirical equations, such as the Hollomon equation, to represent σ_f^I and σ_f^{II} and by choosing strain increments for $\delta\varepsilon_p^I$ the stress-strain curves for the composite can be constructed. This theory also allows for the calculation of the strain in each phase by applying Equation 17.

The following values of the Hollomon parameters were used to calculate the Mileiko and Tomota et al curves in Figure 19

$$\begin{aligned}
n_m &= 0.055 \\
K_m &= 1752 \text{ MPa} \\
n_f &= 0.179 \\
K_f &= 601 \text{ MPa}
\end{aligned} \qquad (22)$$

where the subscripts m and f refer to the martensite and ferrite, respectively. The values were taken from investigations (6) of 0.2C martensite and 1.5Mn ferrite with a 9μm grain size. Figure 19 shows that the Mileiko and the Tomota et al models predict similar effects of martensite content on the uniform elongation of dual-phase steels. Although Tomota´s model distributes strain between the ferrite and martensite in a more realistic manner than Mileiko´s model, both predict very similar behaviors for the uniform strain as a function of martensite volume fraction. This similarity may be explained by the fact that with the particular Hollomon parameters chosen to describe the ferrite and martensite deformation in Figure 19, Tomota´s model shows that, although

373

plastic strain accumulates only in the ferrite in the very early stages of deformation, after the martensite does begin to deform, the strain accumulation in both phases is very similar. Therefore, Mileiko´s model which is based on equal strain accumulation in the martensite and ferrite predicts very similar results. However, if the work hardening characteristics of the two-phases exhibit a larger difference, Tomota´s model would show a more significant deviation from the equal strain model as suggested by Bhadeshia and Edmonds (4) to be the case for a high-silicon dual-phase steel.

Although Figure 19 shows that there is reasonably good agreement between theory and experiment, the theoretical models significantly underestimate the true uniform plastic strains in deformed dual-phase steels containing low martensitic volume fractions. For example, in a dual-phase steel with 8 volume percent martensite, a microstructure that yields good combinations of strength and ductility as noted earlier, the actual true uniform strain is about 20% while that predicted by the models is only 16%.

Figure 20, similar to Figure 19, shows another comparison of predicted and experimentally observed true plastic strains versus

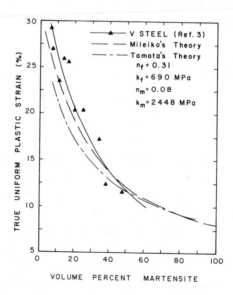

Figure 20: Variation of true uniform plastic strain with volume fraction martensite measured for a V-microalloyed dual-phase steel(3) and as predicted by the models of Mileiko(28) and Tomota et al(29) with Hollomon parameters listed.

parameters (indicated in the Figure) determined by Davies were used to calculate the Mileiko and Tomota curves. Again, the true uniform plastic strains at low volume fractions of martensite are underestimated by the models.

In order to account for the differences between actual and predicted deformation behavior and to show the importance of the early stages of deformation on uniform plastic strain, Jaoul-Crussard analyses of an experimentally determined and calculated stress-strain curve of a dual-phase steel annealed at 810C and cooled at 60C/sec were performed. Figure 21 shows the measured stress-strain curve and the stress-strain curve calculated from the Hollomon parameters used for Figure 19 and the 0.08 martensite volume fraction measured for this specimen. The total uniform plastic strain for the theoretical curve was obtained by calculating n for the ferrite-martensite composite curve determined according to the Tomota model and equating that n to the total uniform plastic strain. Figure 22 shows the Jaoul-Crussard analyses of the stress-strain curves of Figure 21. The experimental curve shows three well defined stages of strain hardening as discussed earlier. In particular a second stage with a low slope or low rate of change of strain hardening with strain is clearly present. The theoretical curve however, shows no region of work hardening

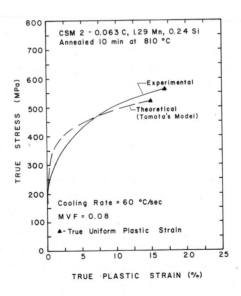

Figure 21:   The measured true stress-strain curve for a C-Mn-Si dual-phase steel with a martensite volume fraction of 0.08 and the true stress-strain curve predicted by the model of Tomota et al(29).

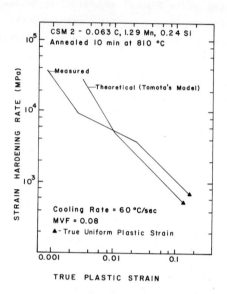

Figure 22: The Jaoul–Crussard analyses of the true stress–strain curves measured for a C–Mn–Si dual-phase steel with a martensite volume fraction of 0.08 and predicted by the model of Tomota et al(29).

with a slope as low as that determined experimentally. Our analysis, presented as Equation 12, shows that the slope of region II in a Jaoul–Crussard plot vitally affects the amount of uniform strain that can be developed in a dual-phase steel. The inability of the Tomata model to predict a second stage with low slope, therefore, results in the underestimation of strain at low martensite volume fractions.

Previous work (1) has shown that the second stage of work hardening is associated with the degree of heterogeneity of strain hardening in dual-phase structure, be it produced by variations in annealing parameters, aging or changes in distribution and amount of martensite. The Tomota continuum model assumes uniform deformation throughout each of the phases in the composite and that the strain hardening is due to an average image stress, neglecting the contribution due to fluctuating local stresses. Thus although an attempt is made to partition stress, the heterogeneity of deformation that is so important in dual-phase steel has not yet been modeled.

376

A further consideration that must be made in modeling dual-phase deformation behavior is the effect of several types of ferrite that may co-exist with martensite under certain conditions. The strain hardening behavior of epitaxial and retained ferrite might be quite different and may significantly affect the stage 2 strain hardening region of the stress-strain curve as delineated by the Jaoul-Crussard analysis, and therefore, the uniform and total elongations of a given specimen. Huppi et al (15) have shown the importance of epitaxial ferrite on the balance of strength and ductility of a V-microalloyed dual-phase steel. Figure 23 which correlates both the ultimate tensile strength and total strain with volume fraction of epitaxial ferrite was obtained from specimens with approximately the same volume fraction of martensite (0.12 to 0.15), and therefore, the differences in strain hardening behavior of the two types of ferrite are largely responsible for the observed variations in the mechanical properties.

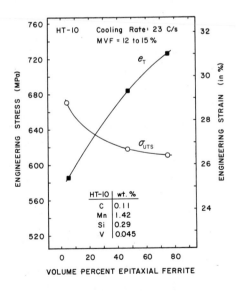

Figure 23: Variation in ultimate tensile strength and total elongation with volume percent epitaxial ferrite for a V-microalloyed steel with a constant volume fraction martensite(15).

377

Conclusions

Systematic study of the deformation behavior of a C-Mn-Si dual-phase steel has been performed. The following conclusions were developed from this investigation.

1. Strain-dependent stages in strain hardening were developed in a dual-phase steel as a function of intercritical annealing temperature and cooling rate.

2. A comparison of the cooling rate dependent microstructures, as presented on quantitative microstructure maps, with the corresponding engineering properties, showed that good combinations of strength and ductility are associated with significant amounts of epitaxial ferrite in conjunction with martensite and retained ferrite.

3. The ultimate tensile strength varied linearly with martensite volume fraction for two intercritical annealing temperatures. The increase in ultimate stress with increasing martensite volume fraction was greater for specimens annealed at the higher temperature. The latter specimens contained greater amounts of epitaxial ferrite on cooling.

4. A comparative analysis of techniques for analyzing stress strain curves showed that the Jaoul-Crussard and Swift analyses best delineated strain hardening stages at low plastic strains.

5. The strain hardening rates at low strains were directly related to uniform plastic strain by means of the slope of the second stage of strain hardening as delineated by the Jaoul-Crussard analysis.

6. Continuum models of dual-phase deformation behavior do not accurately account for the strain accumulation in the second stage of strain hardening and therefore underestimate the true uniform plastic strain at the low volume fractions of martensite associated with good combinations of strength and ductility in dual-phase steels.

7. Refinements in modeling of dual-phase steel deformation behavior must take into account the inhomogeneous deformation at low strains and the effect of the relative amounts of epitaxial and retained ferrite on strain hardening.

Acknowledgements

This work was supported by AISI and the steel was supplied by Dr. Arnold Marder of the Bethlehem Steel Corporation. The assistance of Mark Geib and David Korzekwa of the Colorado School of Mines is greatly appreciated.

References

1. D.K. Matlock, G. Krauss, L.F. Ramos and G.S. Huppi, "A Correlation of Processing Variables With Deformation Behavior of Dual-Phase Steels," pp. 62-90 in Structure and Properties of Dual-Phase Steels, (R.A. Kot and J.W. Morris, Eds.), TMS-AIME, New York (1979).

2. W.R. Cribb and J.M. Rigsbee, "Work-Hardening Behavior and It's Relationship to the Microstructure and Mechanical Properties of Dual-Phase Steels," pp. 91-117 in Structure and Properties of Dual-Phase Steels, (R.A. Kot and J.W. Morris, Eds.), TMS-AIME, New York (1979).

3. R.G. Davies, "The Deformation Behavior of a Vanadium-Strengthened Dual Phase Steel," Metallurgical Transactions, 9A (1978) pp. 41-52.

4. H.K.D.H. Bhadeshia and D.V. Edmonds, "Analysis of Mechanical Properties and Microstructure of High-Silicon Dual-Phase Steel," Metal Science, February 1980, pp. 41-49.

5. R.G. Davies, "The Mechanical Properties of Zero-Carbon Ferrite-Plus Martensite Structures," Metallurgical Transactions, 9A (1978) pp. 451-455.

6. K. Araki, Y. Takada, and K. Nakoka, "Work Hardening of Continuously Annealed Dual-Phase Steels," Trans ISIJ, 17 (1977) pp. 710-717.

7. I. Tamura, Y. Tomota and M. Ozawa, "Strength and Ductility of Fe-Ni-C Alloys Composed of Austenite and Martensite with Various Strengths," pp. 611-615 in Proceedings 3rd Int. Conf. on the Strength of Metals and Alloys, Cambridge, 1973, Vol. 1.

8. G.R. Speich and R.L. Miller, "Mechanical Properties of Ferrite - Martensite Steels", pp. 145-182 in Structure and Properties of Dual-Phase Steels, (R.A. Kot and J.W. Morris, Eds.), TMS-AIME, New York (1979).

9. L.F. Ramos, D.K. Matlock and G. Krauss, "A Comparison of the Effects of Ferrite Grain Size on the Deformation Behavior of Dual-Phase Steels and Quenched Armco Iron", Accepted for publication, Metallurgical Transactions, 1981.

10. L.M. Brown and W.M. Stobbs, "The Work-Hardening of Copper Silica", Phil. Mag., 73 (1971) pp. 1185-1232.

11. M.F. Ashby, "The Deformation of Plastically Non-Homogeneous Alloys," pp. 137-192 in Strengthening Methods in Crystals, (A. Kelly and R.B. Nicholson, Eds.), Elsevier Publishing Co., 1971.

12. D.A. Korzekwa, R.D. Lawson, D.K. Matlock and G. Krauss, "A Consideration fo Models Describing the Strength and Ductility of Dual-Phase Steels," Scripta Metallurgica, 14 (1980) pp. 1023-1028.

13. M.D. Geib, D.K. Matlock and G. Krauss, "The Effect of Intercritical Annealing Temperature on the Structure of Niobium Microalloyed Dual-Phase Steel," Metallurgical Transactions, 11A (1980) pp. 1683-1689.

14. R.D. Lawson, D.K. Matlock and G. Krauss, "An Etching Technique for Microalloyed Dual-Phase Steels," Metallography, 13 (1980) pp. 71-87.

379

15. G.S. Huppi, D.K. Matlock and G. Krauss, "An Evaluation of the Importance of Epitaxial Ferrite in Dual-Phase Steel Microstructures," Scripta Metallugica, 14 (1980) pp. 1239-1243.

16. ANSI/ASTM E8-79a, 1980 Annual Book of ASTM Standards, Part 10, ASTM, Philadelphia, 1980, pp. 197-216.

17. G.S. Huppi, "Dual-Phase Microalloyed Steels: Temperature and Cooling Rate Effects," M.S. Thesis No. T2124, Colorado School of Mines (1979).

18. E.E. Underwood, ASM Metals Handbook, Vol. 8, (T. Lyman and H.E. Boyers, Eds.), Ohio (1973), pp. 37-47.

19. W.R. Cribb, "Quantitative Metallography of Poly-Phase Microstructures," Scripta Metallurgica, 12 (1978) pp. 893-898.

20. L.F. Ramos, D.K. Matlock and G. Krauss, "On the Deformation Behavior of Dual-Phase Steels," Metallurgical Transactions, 10A (1979) pp. 259-261.

21. J.M. Rigsbee and P.J. VanderArend, "Laboratory Studies of Microstructures and Structure-Property Relationships in ´Dual-Phase´ HSLA Steels", pp. 58-86 in Formable HSLA and Dual-Phase Steels, (A.T. Davenport, Ed.), TMS-AIME, New York (1979).

22. R.G. Davies, "On the Ductility of Dual-Phase Steels", pp. 25-39 in Formable HSLA and Dual-Phase Steels, (A.T. Davenport, Ed.), TMS-AIME, New York (1979).

23. J.H. Hollomon, "Tensile Deformation," Transactions TMS-AIME, 62 (1945) pp. 268.

24. E. Voce, "The Relationship Between Stress and Strain for Homogeneous Deformation", Journal of the Institute of Metals, 74 (1948) pp. 537.

25. P. Ludwik, Element der Tehnologischen Mechanic, Julius Springer, Berlin, 1909, pp. 268.

26. H.W. Swift, Journal of the Mechanics and Physics of Solids 1 (1952) pp. 1-18.

27. B. Jaoul, "Etude de la Forme des Courbes de Deformation Platique," Journal of the Mechanics and Physics of Solids, 5 (1957) pp. 95-114.

28. S.T. Mileiko, "The Tensile Strength and Ductility of Continuous Fibre Composites," Journal of Material Science, (1969) pp. 974-977.

29. Y. Tomota, K. Kuroki, T. Mori and I. Tamura, "Tensile Deformation of Two-Ductile-Phase Alloys: Flow Curves of Ferrite-Austenite Fe-Cr-Ni Alloys," Materials Science and Engineering, 24 (1976) pp. 85-94.

30. R.L. Klueh and T.L. Hebble, "A Mathematical Description of the Stress-Strain Behavior of Annealed 2 1/4 Cr - 1 Mo Steel," Transactions of ASME Journal of Pressure Vessel Technology, May 1976, p. 118.

31. H.J. Kleemola and M.A. Nieminen, "On the Strain Hardening Parameters of Metals," Metallurgical Transactions, 5 (1974) p. 1863.

32. R. Stevenson, "Inferring Microscopic Deformation Behavior from the
     Form of Constitutive Equations for Low Carbon Steel and 5182-0
     Aluminum," presented at ASTM Symposium on Mechanical Testing for
     Deformation Model Development, Bal Harbor, Florida, November 1980.

33. R.E. Reed-Hill, W.R. Cribb and S.N. Monteiro, "Concerning the
     Analysis of Tensile Stress-Strain Data Using Log $d\sigma/d\varepsilon$p Versus Log σ
     Diagrams," Metallurgical Transactions, 4 (1973) pp. 2665-2667.

ANALYSIS OF MECHANICAL PROPERTIES OF A LOW-ALLOYED

MO-STEEL WITH DIFFERENT PHOSPHORUS ADDITIONS

J. Becker, E. Hornbogen and F. Wendl
Institut für Werkstoffe
Ruhr-Universität Bochum
D-4630 Bochum, West-Germany

The effect of phosphorus additions on the tensile properties of low alloyed steels was investigated. Dual-phase structures with about 25% martensite were obtained by annealing at different temperatures. Additions of phosphorus between 0 and 1 wt% P changed the properties of the phases as well as of the interfaces. Up to about 0.3 wt% P uniform elongation and tensile strength of the bulk alloy are improved as they are predominantly determined by the properties of the ferrite components of the microstructure. At high P-contents the embrittling effect especially on interfaces becomes dominant. An attempt has been made to explain the results in general terms using the concept of partial properties. This implies that equations have to be found by which the bulk properties can be obtained from the properties of the microstructural components.

Introduction

In recent years a new group of structural steels has been developed which consists of 70-90 % ferrite and the rest martensite (1-4). A way to characterize this microstructure quantitatively has been proposed in earlier papers (4,5). Accordingly a complete description requires the densities of grain boundaries $\alpha\alpha$ and $\alpha_M\alpha_M$ and of interface boundaries $\alpha\alpha_M$ in addition to volume fractions $f_\alpha + f_{\alpha_M} = 1$, and grain diameters. A dual-phase structure is characterized as follows (Fig. 1):

1. As in a duplex structure the number of α- and α_M- grains per unit volume is equal.

2. As in a dispersion the second phase α_M forms isolated zones surrounded by α. There exists a maximum volume fraction $f^*_{\alpha_M} \leq 0.3$ above which α_M starts to form $\alpha_M\alpha_M$-boundaries and at $f_{\alpha_M} = f_\alpha = 0,5$ changes to an ideal duplex structure.

3. A network structure and a dual-phase structure have in common that α_M- zones are all connected by α-grain boundaries.

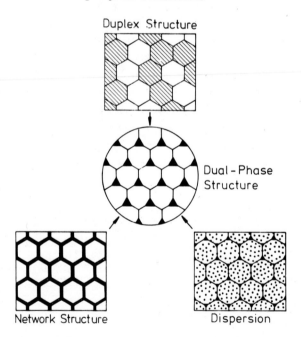

Fig. 1 —   Schematic representation of basic two-phase microstructure species.

The types of boundaries are not only used as topological features. Their properties in addition to those of the crystalline phases will have to be used to explain the bulk properties of dual-phase steels. These properties have attracted attention because of the good formability, together with a high work hardening coefficient which provides high strength after forming (6). Several attempts have been made to explain mechanical properties of coarse two-phase structures from the properties of its microstructural components (7-9). It is the purpose of this paper to investigate the effect of phosphorus on strength and ductility of a dual-phase steel. Phosphorus is the most potent substitutional solid solution hardener of α-iron. In addition, it is attempted to apply and extend the principle of partial properties for an improved general understanding of coarse two-phase alloys. Partial properties are the particular properties of the microstructural components. In the case of phosphorus as the alloying element, the conflicting effects on improvement of yield stress and work hardening ability, and induction of inter- and transcrystalline embrittlement will have to be considered.

For all alloys an equal volume fraction of martensite ($f_{\alpha_M} \approx 0,25$) was developed by suitable heat treatments. As phosphorus partitions more into the α-, than into the γ-phase a strong effect of this element on the partial properties of the ferrite is anticipated (10). The heat treatment was designed to yield a rather coarse grained structure in order to allow observations of micromechanical properties by optical microscopy and microhardness. These experimental observations together with earlier micromechanical models (11-15) will be used for an attempt to aquire an improved understanding of the bulk mechanical properties of dual-phase steels.

Materials and Experimental Conditions

Four vacuum melted and subsequently hot-rolled alloys were used for this work. The chemical composition is given in Table I.

Table I. Composition Wt%

Steel No.	C	Mn	Si	Mo	P	S	N	Al
1	0,11	1,54	0,49	0,15	0,004	0,006	0,005	0,05
2	0,12	1,52	0,50	0,15	0,10	0,006	0,005	0,03
3	0,11	1,55	0,52	0,15	0,24	0,006	0,005	0,03
4	0,11	1,61	0,58	0,15	0,98	0,006	0,006	0,05

For the determination of the transformation temperatures the alloys were
heated in the dilatometer to 1100°C at a rate of 3°C/min. The solution
treatment for alloys 1 to 3 was done under Argon for 1 hr at 1230°C. To
obtain the desired dualphase structure the alloys were cooled to a temperature
in the $(\alpha+\gamma)$-field and held for one hour at the temperature at which the
correct volume fraction of ferrite f_{α} = 0,75 formed and then subsequently
water quenched (Fig. 2). The high-phosphorus steel was subjected to a corresponding heat treatment. However no homogeneous γ-solid solution is obtained
at 1230°C.

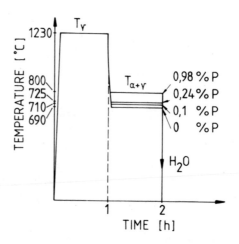

Fig. 2 - Heat treatments required to produce a microstructure containing
25 % γ (α_M) phase

The volume fractions of the phases were determined by light microscopy.
Direct evidence of the partial mechanical properties could be obtained by
measurements of microhardness (DPH: 0.25 N and 300 N for macrohardness).

The tensile tests were conducted with cylindrical specimens and strain
rates of $\dot{\varepsilon}$ = 1.33 x 10^4 sec^{-1} for all experiments. All fracture surfaces
were investigated by SEM, as well as in cross section by light microscopy
after nickel plating the surface.

Experimental Results

The transformation temperatures of all the steels as determined by
dilatometry are shown in Fig. 3. The $\gamma \rightarrow \alpha$ transformation temperature is
shifted to higher values as the P-content is increased. For the 0.98% P-alloy

only the end of the γ → α-reaction could be determined (F_E = 730°C) because
the dilatometer allowed heating only to 1100°C.

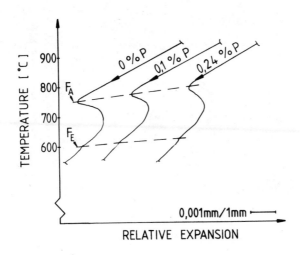

Fig. 3 - Dilatometric investigation of transformation behavior.
(Cooling rate: 3°C/min; F_A-start of ferrite formation,
F_E-end of ferrite formation).

The microstructures as obtained by the different intercritical annealing
treatments are shown in Fig. 4. The influence of phosphorus on the two microstructural components are demonstrated in Fig. 5. The microhardness of the
ferrite (α) increases with increasing P-content, while that of martensite
(α_M) is much higher initially, but decreases with increasing P-content. The
microhardness closely resembles that of the ferrite.

The engineering stress-strain curves give good evidence for the favorable effect of even rather high phosphorus additions on uniform elongation
and tensile strength (Fig.6a). Yield stress $R_{p0.2}$, tensile strength R_m and
true uniform elongation all increase up to 0.24% P, however note that the
maximum reduction in area is observed at 0.1% P (Fig. 6b-d).

The microscopic investigations of the fractured specimens yielded the
following results (Fig. 7). A dimple rupture was found for both the 0 and
0.1% P-alloys, however less dimples appear in the P-containing alloy. The
light micrographs provide evidence for crack paths exclusively through the
highly deformed α-phase (Fig. 8). Interfaces ($\alpha\alpha_M$) do not crack, and a film

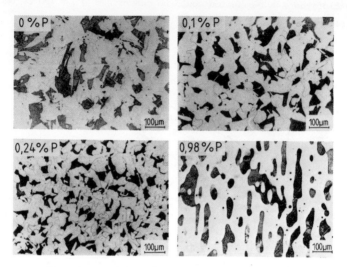

Fig. 4 -   Microstructures of the heat treated alloys.

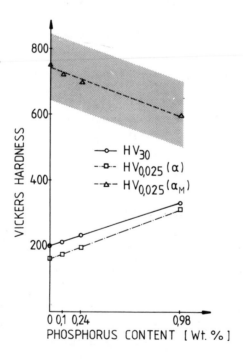

Fig. 5 -   Effect of phosphorus on microhardness of the phases
                α and α_M and on macrohardness of the bulk alloy.

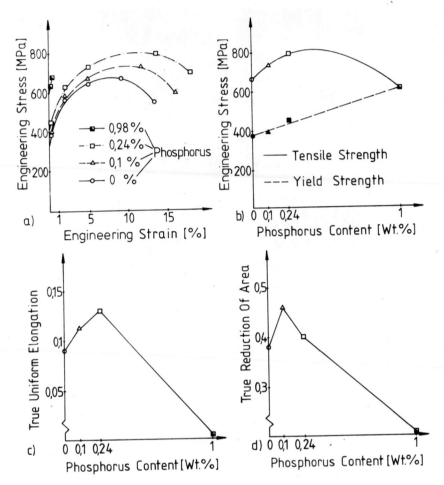

a) — Engineering Stress [MPa] vs Engineering Strain [%]

    0,98 %
    0,24 % } Phosphorus
    0,1 %
    0 %

b) — Engineering Stress [MPa] vs Phosphorus Content [Wt.%]

    —— Tensile Strength
    --- Yield Strength

c) — True Uniform Elongation vs Phosphorus Content [Wt.%]

d) — True Reduction Of Area vs Phosphorus Content [Wt.%]

Fig. 6 - Effect of phosphorus content on tensile properties for the steels
with dual-phase structure.

of α is always connected with the interface. For 0.24 wt% P and above the
behavior changes (Fig. 8c,d). The portion of cracked $\alpha\alpha_M$-interfaces increases,
and in addition, transcrystalline cleavage of α-phase occurs in the high
phosphorus alloy (Fig. 7c,d). Light microscopy confirms that the martensite
is neither deformed extensively nor cracked in all alloys.

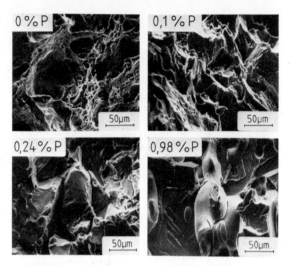

Fig. 7 - Fracture surfaces (SEM) of the dual-phase steels as a function of phosphorus content (for interpretation see Fig. 14).

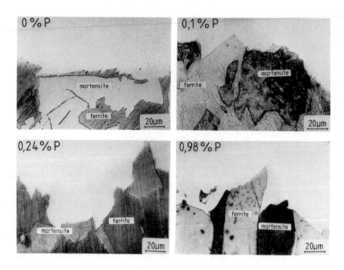

Fig. 8 - Cross section of the fracture surfaces of the dual-phase steels as a function of phosphorus content (light microscopy).

390

Discussion of the Results

A method is required which allows an explanation of the deformation and fracture behavior of dual-phase structures from the properties of its microstructural components. Special attention will be paid to uniform elongation and tensile strength, because they are of the greatest practical importance. The concept of partial properties has been applied for a long time to explain properties of fiber composites and lamellates. Recently it has been applied to interpret critical and subcritical crack propagation in metals and in polymers which consist of two microstructural components with different mechanical properties. It could be shown that a bulk property P can depend on volume fractions f_i and partial properties $\bar{p}_i$ according to the following principal relations (13-15):

$$P = \bar{p}_\alpha f_\alpha + \bar{p}_\beta f_\beta \tag{1}$$

the rule of mixtures,

$$P = \bar{p}_\alpha f_\alpha \tag{2}$$

the bulk property depends on one partial property only and the volume fraction of α,

$$P = \bar{p}_\alpha \tag{3}$$

the bulk property depends on one partial property independent of the volume fraction of α.

The partial properties are those of the phases α and α_M, where α_M is the martensite which originated during cooling from the intercritical annealing temperature. A relation for the elongation at fracture ε_f can be derived using the following prerequisites. For a lamellate of α and β loaded perpendicular to the plane of the plates the microstructural elements are arranged in a sequence and consequently under constant stress and different strain (Fig. 9).

$$\bar{\sigma}_\alpha = \bar{\sigma}_\beta \tag{4a}$$

$$\bar{\varepsilon}_\alpha \neq \bar{\varepsilon}_\beta \tag{4b}$$

$$\varepsilon = \bar{\varepsilon}_\alpha \cdot f_\alpha + \bar{\varepsilon}_\beta \cdot f_\beta \tag{4c}$$

391

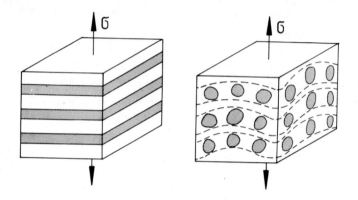

Fig. 9 - Dispersion of α_M in α as approximate micromechanical analogy
of a lamellate loaded perpendicular to the plane of the plates.

The rule of the mixtures (Equ. 1) describes the total strain ε. If only
plastic deformation is considered, and if the yield stress of β is not sur-
passed before α ruptures, equation 4c can be reduced to:

$$\varepsilon \approx \overline{\varepsilon}_\alpha \cdot f_\alpha \qquad (4d)$$

These conditions are representative for the dual-phase steels (Fig. 5),
if it is assumed in addition that a coarse dispersion can be approximately
treated as a lamellate (Fig. 10). A corresponding model can be applied to a
structure in which the softer phase forms a continuous network along former
grain boundaries. It does not apply to a duplex structure which can be regarded as an interwoven skeleton of α and β, for which the conditions of
constant stress (Equ. 4a) does no longer apply (Fig. 1). The fracture strain
of a dual-phase steel follows from equation 4c and 4d (Fig. 10):

$$\varepsilon_f = \overline{\varepsilon}_{f\alpha} \cdot f_\alpha + \overline{\varepsilon}_{\alpha_M} \cdot f_{\alpha_M} \qquad (5a)$$

Plastic deformation of martensite can be neglected in this connection.
$\overline{\varepsilon}_{\alpha_M} \approx 0$. It is, however, of importance, that this phase is not completely
undeformable in order to explain why rupture of $\alpha\alpha_M$-interfaces does not
occur in low-phosphorus alloys even at very large amounts of plastic deformation in the ferrite:

392

$$\varepsilon_f \approx \bar{\varepsilon}_{f\alpha} \cdot f_\alpha \qquad (5b)$$

This relation is valid up to a volume fraction f_{α_M} above which different martensite zones touch each other and start to form a skeleton. For the topological conditions of a dual-phase structure the critical volume fraction of martensite $f^*_{\alpha_M} \approx 0.3$ is known as the point of percolation (16). From Equation 5b information can be obtained on the partial property $\bar{\varepsilon}_{f\alpha}$, i.e. the ferrite as microstructural component. It is known from earlier investigations that phosphorus shows the highest specific solution hardening effect of all elements substituted into the b.c.c. lattice of α-iron (17) (Fig. 11). Figure 6a points out that the bulk properties can be explained as being determined by this effect of phosphorus dissolved in the ferrite component of the microstructure. This is combined with the effect which the nondeforming martensite exerts on deformation of ferrite. By impediment of slip in the interfaces a larger number of slip systems have to be activated and a higher rate of work hardening must occur in the ferrite crystallites, as compared to a monophase polycrystal or a single crystal (Fig. 12). Therefore the deformation behavior is a function of the phosphorus content x_P, the volume fraction of martensite f_{α_M}, and in

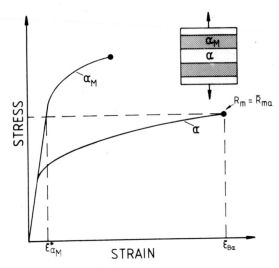

Fig. 10 - Schematic stress-strain curves representing the partial tensile properties of the two microstructure components α and α_M of a dual-phase structure, for the condition $R_{p\alpha_M} > R_{m\alpha}$.

393

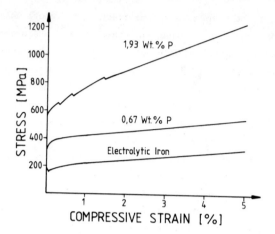

Fig. 11 - Stress plastic strain curves of iron and iron-phosphorus
alloys (compression test) (17).

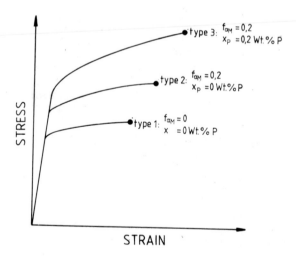

Fig. 12 - Schematic comparison of the bulk stress-strain curve- of a
homogeneous ferritic microstructure (type 1), a dual-phase
microstructure without phosphorus (type 2), and a dual-phase
microstructure with phosphorus (type 3).

addition the grain size which was not subject of this investigation.

There exist critical limits for the two variables. The first is the amount of martensite at the point of percolation $f^*_{\alpha_M} \approx 0.3$. The second is determined by embrittling effects of phosphorus especially on interfaces and α-crystals. It had been assumed that the material ruptures at a tensile strength $\bar{R}_{m\alpha}$ which is smaller than that of the martensite $\bar{R}_{m\alpha_M}$ (Fig. 10).

$$R_m < \bar{R}_{m\alpha_m} \qquad (6a)$$

$$R_m = \bar{R}_{m\alpha} \qquad (6b)$$

At very high phosphorus contents cracking is observed along the $\alpha\alpha_M$-interfaces (Fig. 8d). The tensile strength then becomes codetermined by the properties $\bar{R}_{m\alpha\alpha_M}$ of interfaces which are embrittled by segregation of phosphorus (Fig. 13).

$$R_m = \bar{R}_{m\alpha\alpha_M} < \bar{R}_{m\alpha} < \bar{R}_{m\alpha_M} \qquad (6c)$$

This provides an example for an equation of the general type which has been introduced as Equation 3 (Equ. 6b and c). If the discussion of the general Equations 1 to 3 is resumed, it can be concluded that properties of grain boundaries and interfaces have to be included for a complete description of bulk properties. There are now five partial properties which will have to be related

$$P = F\ (\bar{P}_\alpha,\ \bar{P}_\beta,\ \bar{P}_{\alpha\alpha},\ \bar{P}_{\beta\beta},\ \bar{P}_{\alpha\beta}) \qquad (7a)$$

For example the rule of mixtures will have to be expanded in a way which includes the portions of boundaries ($f_{ij} = \frac{\Sigma A_{ij} \lambda}{V}$, where ΣA_{ij} is the sum of the boundary areas, λ their thickness, V the total volume.

$$P = \bar{P}_\alpha f_\alpha + \bar{P}_\beta f_\beta + \bar{P}_{\alpha\alpha} f_{\alpha\alpha} + \bar{P}_{\beta\beta} f_{\beta\beta} + \bar{P}_{\alpha\beta} f_{\alpha\beta} \qquad (7b)$$

Equation 6c represents a special case of equation 7b, namely intercrystalline fracture along $\alpha\alpha_M$-interfaces.

$$P = \bar{P}_{\alpha\beta} \qquad (7c)$$

A special feature of the dual-phase structure is that these intercrystalline cracks find no path which continues through the total cross section (Fig. 14). As long as the ferrite matrix stays ductile the bulk properties are ductile in spite of local interfaces cracking. A schematic representation of the principal possibilities of crack path through dual-phase structures

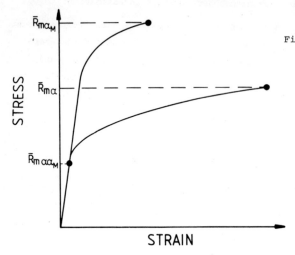

Fig. 13 - Schematic stress-strain curves representing the partial tensile properties of the phases α and α_M. In addition The tensile strength of the embrittled $\alpha\alpha_M$-interface is shown.

Phosphorus Content	Fracture in a P-alloyed Dual-Phase Steel	Partial Properties of Microstructural Components			
		ferrite	martensite	grainboundary	interphase
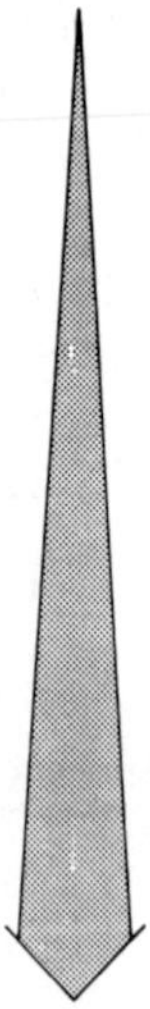	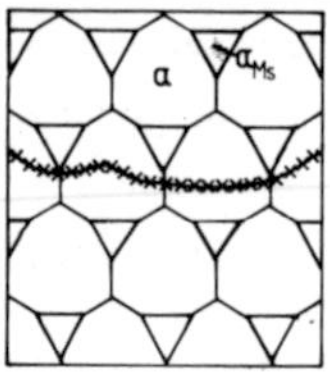 α α_{Ms}	low strength	high strength	high strength	high strength
		ductile	–	–	–
	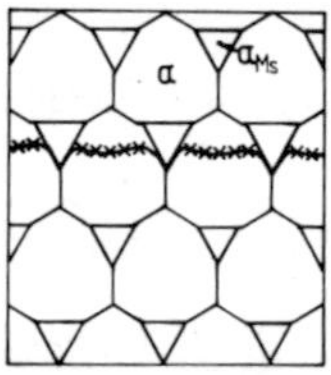 α α_{Ms}	low strength	high strength	–	low strength
		ductile	–	–	embrittled
	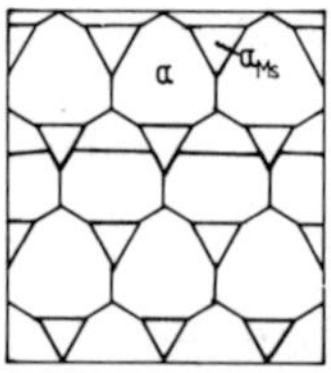 α α_{Ms}	low strength	high strength	–	low strength
		brittle	–	–	embrittled

•••••• ductile rupture   —— brittle fracture

Fig. 14 - Schematic representation of the effect of phosphorus content on fracture mechanism and partial properties of microstructural components (compare Figs. 7 and 8).

are shown in Figure 14.

Increasing additions of phosphorus improve uniform elongation by causing a higher work hardening-ability of the α-phase, until embrittlement of $\alpha\alpha_M$-boundaries overrides this effect. The same is true for an increasing amount of martensite which leads to a lowering of ductility as soon as compatibility of the microstructure requires that hard martensite shares the high amounts of deformation in ferrite. Along these lines the optimum composition and micro-structure of dual-phase steel can be determined.

Acknowledgement

We wish to thank Climax Molybdenum Company of Michigan for providing the steels used in these investigation.

References

1) S. Hayani and T. Furukawa, "A Family of High Strength Cold Rolled Steels", pp. 311-321 in Proc. Microalloying '75 on High Strength Low-Alloy Steels, Washington D.C., October 1975.

2) Proc. AIME Symposium, "Formable HSLA and Dual-Phase Steels", Chicago, Illinois, October 1977.

3) Proc. Seminar on Vanadium Cold Pressing and Dual-Phase Steels, Vanadium International Technical Comitee, October 1978.

4) Proc. AIME Symposium, "Structure and Properties of Dual-Phase Steels", New Orleans, Lousiana, February 1979.

5) J. Becker, E. Hornbogen and P. Stratmann, "Dualphasen-Gefüge", Z. Metall-kde., 71 (1980) pp. 27-31.

6) M.S. Rashid, "GM 980X-A Unique High Strength Sheet Steel with Superior Formability", SAE Reprint 760206, February 1976.

7) H. Fischmeister and B. Karlsson, "Plastizitätseigenschaften grob-zweipha-siger Werkstoffe", Z. Metallkde., 68 (1977) pp. 311-326.

8) J. Gurland, "A Structural Approach to the Yield Strength of Two-phase Alloys with Coarse Microstructures", Mater. Sci. Eng., 40 (1979) pp.59-71.

9) J.Y. Koo, M.J. Young, and G. Thomas, "On the Law of Mixtures in Dual-Phase Steels", Met. Trans.A., 11A (1980) pp. 852-854.

10) R. Vogel "Über das System Eisen-Phosphor-Kohlenstoff", Arch. Eisenhüttenw., 93 (1929) pp. 369-381.

11) P. Stratmann and E. Hornbogen, "Mechanical Properties of Microduplex Structures in Iron Alloys", pp. 607-623 in Proc. ICSMA 4, Nancy (1977).

12) P. Stratmann and E. Hornbogen, "Mechanische Eigenschaften zweiphasiger Duplex- und Dispersionsgefüge in Nickelstählen", Stahl und Eisen, 99 (1979) pp. 643-648.

13) E. Hornbogen und K. Friedrich, "Die Anwendung partieller Eigenschaften zur Beschreibung des bruchmechanischen Verhaltens grob-zweiphasiger metallischer und hochpolymerer Werkstoffe", Radex-Rundschau, H. 1/2 (1980) pp. 98-107.

14) E. Hornbogen and K. Friedrich, "The Use of Partial Properties to Interpret Bulk Fracture Mechanical Properties of Semicrystalline Polymers", pp. 1422-1432 in Proceedings of the International Conference on Pure and Applied Chemistry, IUPAC, Mainz (1979) Vol. 3.

15) E. Hornbogen and K. Friedrich, "On the Use of Partial Properties to Interpret the Bulk Propagation Behaviour of Coarse Two-Phase Materials", J. Mater. Sci., 15 (1980) pp. 2175-2182.

16) D. Stauffer, "Perkolation - der einfachste Phasenübergang?" Phys. Bl., 36 (1980) pp. 124-126.

17) E. Hornbogen, "Observations of Yielding in an Iron - 3,17 At% Phosphorus Solid Solution", Trans. of the ASM, 56 (1963) pp. 16-24.

THE BAUSCHINGER EFFECT AND WORKHARDENING OF DUAL-PHASE STEELS

D. Tseng and F.H. Vitovec
Department of Mineral Engineering
University of Alberta
Edmonton, Alberta, Canada

The back stress hardening of a dual-phase steel (DPS) has been investigated using the Bauschinger effect technique. The DPS was prepared by intercritically annealing a Nb-V HSLA steel. Permanent softening, commonly observed in particle hardened alloys, is not found in this DPS. It workhardens continuously on reverse straining until the forward and the reverse flow stress are equal. Evaluation of the data shows that the Mroz-Sowerby model for combined kinematic and isotropic hardening does describe the properties of this DPS over the entire range of volume fractions of martensite (fm). The contribution to work hardening by isotropic hardening increases with decreasing fm, thus at low fm particle hardening dominates while at large fm long range internal stresses characteristic of the iso-strain model are controlling factors.

Introduction

Two basic approaches have been used for the modeling of "Dual Phase Steels" (DPS); one is based on continuum mechanics, the other on micro-mechanical concepts (1). The continuum mechanics models assume that the composite material consists of two phases, each having its own stress-strain characteristics (2). The stronger phase (martensite) may be deformable (2, 3) or it may be hard and purely elastic. Accordingly, the assumptions regarding the partitioning of plastic strain between the two phases also varies (4) from assuming that both, ferrite and martensite undergo identical deformation (isostrain model) (2, 3) to the case where all plastic deformation occurs in the ferrite only and martensite is distributed as nondeforming particles (5).

Recent studies have indicated that each model may apply to a specific regime of plastic strain and volume fraction of martensite (1). Similar models have been used in the past to explain the mechanical behavior of materials under reverse strain and to describe the associated Bauschinger effect (BE). Therefore, it seems possible to shed some light on the mechanism of deformation of DPS by studying their behavior under reversed strain.

Models for the Bauschinger Effect

When work hardenable materials are deformed by forward and then reverse loading they usually exhibit a decreased reversed yield stress. This decrease of the yield stress is identified as BE. Since the early observation by Bauschinger this effect has been found with many pure metals and alloys and in both, single crystals and polycrystals. It is generally believed that this effect is caused by internal stresses that develop as a result of inhomogeneous deformation.

For the quantitative assessment of the BE it is convenient to plot both, forward and reverse stresses in the positive direction as a function of cumulative strain. Figure 1 identifies significant stresses and strains which are used in evaluations of the BE. The curve shows the typical reduction in yield stress and the well rounded nature of the initial plastic portion of the reverse flow curve. In some instances the reverse curve is displaced to lower stress and eventually becomes parallel with the forward curve. The stress increment of parallel displacement is termed "permanent softening". Orowan suggested that the rounding off of the initial portion of the reverse curve was caused by weak, permeable obstacles to dislocations, and that the permanent softening results from back stresses (6). He also suggested that on reverse straining the back stresses should vanish at about 1 to 3% strain. This has been verified by Wilson (7), who measured x-ray line shifts and found that at the critical reverse strain, ε_{Rp}, where the mean internal stress decayed to zero, the stress difference $\Delta\sigma_p$ between the forward and reverse curves was approximately twice the mean internal stress which existed prior to reverse straining. These observations have provided the basis for many recent BE studies.

Ibrahim and Embury (8) showed that if a material has a basic flow stress σ_o, the forward flow stress σ_F at a given strain and the reverse flow stress σ_R, at the same strain are given by

$$\sigma_F = \sigma_0 + \sigma_f + \sigma_B$$

$$\sigma_R = \sigma_0 + \sigma_f - \sigma_B \tag{1}$$

The stress σ_f has been identified as resulting from various dislocation interactions which contribute to hardening (forest hardening), and σ_B as a back stress or residual stress. Combination of the expressions for forward and reverse stress gives

$$\frac{\sigma_F - \sigma_R}{\sigma_F - \sigma_0} = \frac{2\sigma_B}{\sigma_F - \sigma_0} \tag{2}$$

which is identifed as Bauschinger effect parameter (BEP) provided that $\sigma_R = \sigma_{RM}$, as defined by permanent softening (Figure 1). The BEP is used as a measure of the fraction of the total strain hardening due to elastic back stresses.

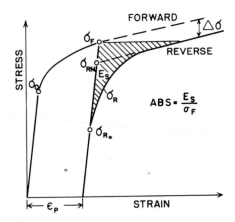

Figure 1 – Schematic illustration of the flow curve for forward and reverse straining. E_s is the area between the elastic unloading curve and the reverse curve up to a stress equal to the forward stress. ABS is the average Bauschinger strain, $\Delta\sigma$ is permanent softening, σ_0 is the initial yield stress, σ_{Ro} is the yield stress on reverse flow, σ_{Rm} is the reverse flow stress extrapolated from the parallel section of the forward and reverse flow curves.

While the above concepts were primarily developed to identify dislocation strain hardening mechanisms continuums mechanics models initially disregard microstructural details. The early model used by Masing (9) to interpret the BE viewed a solid as being composed of a set of n elastic – plastic elements each having a different yield point but all being strained equally (isostrain hypothesis) (10). Strain hardening concepts were later added.

In Masings model the properties of each element remain unchanged during forward and reversed strain and kinematic hardening is caused by the development of long range residual stresses. However, isotropic hardening processes may also occur which are retained on load reversal. This case has been dealt with my Mroz (11) and discussed in detail by Sowerby et al. (12, 13). Figure 2 illustrate the Mroz's model for kinematic hardening.

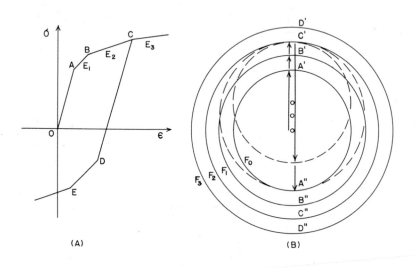

(A)   (B)

Figure 2 - Mroz-Sowerby model: (A) piecewise linear stress-strain curve, (B) yield loci in stress space and translations during forward loading and reverse hardening.

The stress-strain curve is approximated by linear segments OA, AB, BC, CD of constant tangent moduli. In stress space this approximation can be represented by introducing surfaces, F_0, F_1, F_2 defining regions of constant work hardening or tangent moduli, where F_0 denotes the yield surface and F_1, F_2, ... are surfaces defining regions of constant work hardening moduli. For simplicity it is assumed that the material is isotropic. In this case the stress surfaces are represented by a number of concentric circles as shown in Figure 2. The elastic segment OA of the stress-strain curve is represented in stress space by a point moving from O in a vertical direction to meet F_0 at A' (Figure 2b). A further increase of the load moves the circle F_0 to meet circle F_1 at B'. This translation describes the segment AB with modulus E_1 of the stress-strain curve. Similarly region BC in Figure 2a is represented by F_0 and F_1 translating together to meet F_2 at C'. On unloading the stress point moves back across the diameter of F_0 to meet A''. This translation defines the domain CD in the stress-strain curve. Similar translations continue when the reverse load is increased. This model encompasses all characteristics of Masings model.

Isotropic hardening can be described by Mroz's model as expansion of the yield surfaces. Sowerby et al. introduced a combination parameter, m, which partitions the stress increment $\Delta\sigma$ into isotropic and kinematic hardening components. It is assumed that during translation by one stress increment isotropic hardening of magnitude m.$\Delta\sigma$ would occur leaving (1 - m).$\Delta\sigma$ as kinematic hardening. This is essentially equivalent to an expansion of the circles F_i in stress space. The combination factor may be the same or different for forward and reverse loading. It may be interpreted by microscopic models which take both, kinematic and isotropic hardening into consideration.

Experimental

The steel used in this study was a niobium-vanadium containing HSLA steel. Its composition in weight percent was 0.10 C, 1.55 Mn, 0.014 P, 0.006 S, 0.27 Si, 0.040 Nb, 0.035 V and 0.034 Al. Specimen blanks were cut parallel to the rolling direction of the 18.3 mm thick plate. To establish a reference microstructure all blanks were normalized at 900°C for 30 min. and air cooled. Cylindrical test specimens of 6 mm diameter and 12 mm gage length were machined from the blanks. This aspect ratio was adequate for compressive deformation up to 6 percent strain without buckling. To produce dual phase structures with different volume fraction of martensite batches of specimens were annealed in an argon atmosphere at 730°C, 760°C, 820°C, and 865°C respectively for 30 min., and water quenched. The mechanical tests were performed in an MTS system with Wood's metal grips. Care was taken during mounting by cooling the specimens to avoid tempering of the martensite.

The volume fraction of martensite (fm) and the ferrite grainsize in each individual specimen was measured with an image analyzer. For this purpose specimens were etched in a modified picral solution (14). The volume fraction of martensite ranged from 20 percent to nearly 100 percent depending on the intercritical annealing temperature. Examples of microstructures are shown in Figure 3.

Test Results and Discussion

Tensile tests as well as load reversal tests with different tensile prestrain were performed on a total of 28 specimens. Figure 4 shows flow stresses at specified strains as a function of volume fraction of martensite. The flow stresses increase linearly with increasing martensite content similarly as observed in previous studies by R.G. Davies (15, 16).

Figure 5 shows tension-compression stress-strain curves for the steel annealed at the four intercritical temperatures (ICAT). To show the BE the compression part of the curve is also plotted by inversion with respect to the point at which the unloading elastic line meets the strain axis. None of the curves shows a permanent softening. The absence of permanent softening has been observed previously with different dual-phase alloy systems (17, 18). Not shown here are the curves for the steel in normalized condition which exhibited permanent softening.

Analysis of the data is now pursued in terms of continuum mechanics models. A check of the applicability of the kinematic model is possible on the basis of the yield stress for forward and reverse deformation which should follow the relation

$$\sigma_{Ro} = 2\sigma_o - \sigma_F \tag{3}$$

The experimental data showed that the conventionally defined yield stress

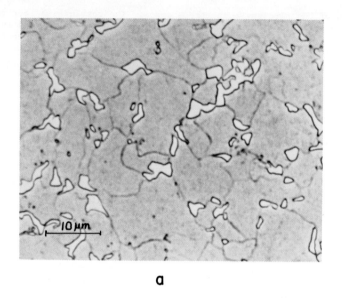

a

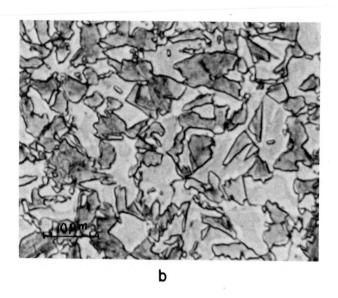

b

Figure 3 - Microstructures of the steel: (a) intercritical annealing temperature 730°C, (b) intercritical annealing temperature 820°C.

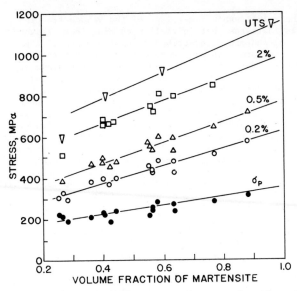

Figure 4 - Flow stresses at different strains as a function of the volume fraction of martensite.

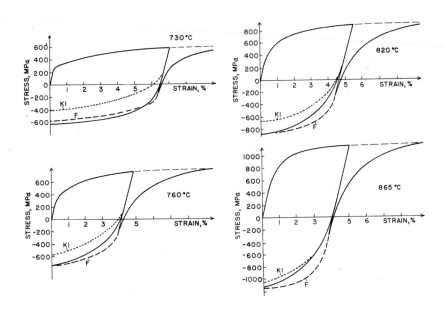

Figure 5 - Tension - compression stress strain curves for different inter-critical annealing temperatures.

values follow the relationship regardless of the volume fraction of
martensite. This indicates that initiation of plasticity is governed by
kinematic hardening and can be described by a simple parallel element
model. As plastic deformation progresses deviation from the kinematic
hardening may occur. This is illustrated by the curves drawn in the lower
half of the stress-strain space of Figure 5. In each case the ideal
kinematic curve, (KI) and the initial forward curve (F) have been drawn
for comparison with the experimental curve. The kinematic curves were
computed from the forward ones, $\sigma = f(\varepsilon)$, by writing $\frac{1}{2}\sigma = f(\frac{1}{2}\varepsilon)$. One
observes a systematic deviation of the experimental curve from the
theoretical ones with different ICAT's. At high ICAT where the microstructure almost entirely consists of low carbon martensite the experimental curve follows closely the kinematic one. As the volume fraction of
martensite decreases, i.e. at lower ICAT the experimental curve starts out
at kinematic hardening, but shows a more rapid strain hardening with
increasing reversed strain and more or less rapidly approaches and finally
intersects the initial forward curve.

To illustrate this behavior the reverse plastic strain to the point
of intersection with the forward curve, ε_{RC}, has been plotted as a function
of prior forward strain in Figure 6.

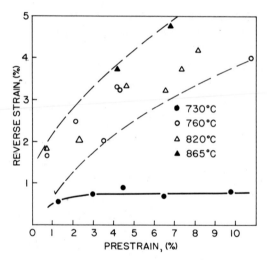

Figure 6 - Reverse plastic strain to point of intersection with initial
forward curve as a function of tensile prestrain.

This strain may be used here as an indicator of the magnitude of reversed
strain at which the residual stresses become relaxed. At low ICAT, i.e.
at low fm, ε_{RC}, is small, thus indicating rapid reverse hardening. The
data for the other ICAT's fall all together into one scatter band which
has a trend of increasing critical strain with increasing prestrain. It
is interesting to note that the trend expressed in Figure 6 is also
exhibited by the average Bauschinger Strain (ABS). Values of the ABS as
defined in Figure 1 are plotted as a function of forward prestrain in
Figure 7. Here too the data for the higher martensite fractions fall into
a single scatter band while the data for low martensite content (ICAT 730°C)
fall outside and below the band. The increase of Bauschinger strain with
increasing prestrain correlates with the variation of the magnitude of back

stress with prestrain as observed with materials which exhibit permanent softening (18).

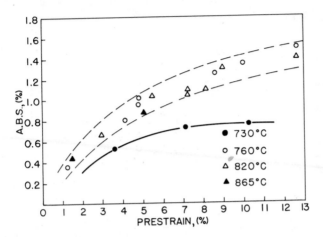

Figure 7 - Average Bauschinger strain as a function of tensile prestrain.

The reversed strain curves were further evaluated using the Mroz-Sowerby combined kinematic-isotropic hardening model. For this purpose the combination parameter m was computed from the experimental data over a range of reversed strain. Curves for different ICAT's and corresponding volume fraction of martensite are shown in Figure 8.

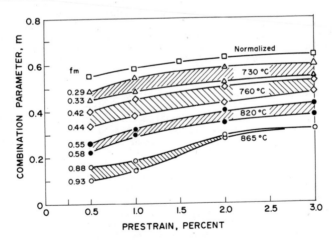

Figure 8 - Combination parameter m as a function of reverse strain for different volume fractions of martensite.

In general, the m-values increase rapidly initally with increasing prestrain. Beyond 0.5 percent prestrain the combination parameters rise only slowly with further strain. Furthermore, as the volume fraction of martensite decreases, m increases reaching maximum values for the steel in normalized condition. Figure 9 slows a plot of the m-values as a function of volume fraction of martensite.

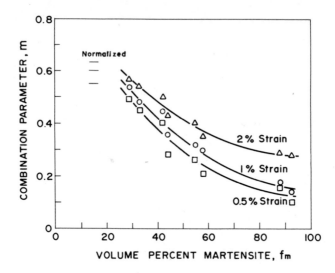

Figure 9 - Combination parameter for isotropic-kinematic hardening for different reverse strains as a function of the volume fraction of martensite.

At a large fm, m is low indicating dominance of kinematic hardening. With decreasing martensite content m increases, reaching a value of 0.5, at about 25 volume percent martensite. The photomicrograph of Figure 3 shows that at this low fm martensite occurs in small islands, causing essentially particle hardening. This type of hardening mechanism has been analysed in detail by several authors (19, 20, 21).

Figures 10 and 11 show photomicrographs of the structure in highly deformed regions of test specimens intercritically annealed at 730°C and and 820°C. Figure 10 shows that the martensite islands at low volume fractions do not participate in deformation while the low carbon martensite at high volume fractions, Figure 11, deforms plastically as extensively as the ferrite grains. This trend has also been verified by quantitative microscopy.

The data indicate regimes of volume fractions of martensite for two different models. At martensite contents larger than about 40 percent the isostrain model applies. At low volume fractions martensite contributes to hardening by acting as hard nonpermeable elastic dispersion. Dislocation mechanisms for both hardening processes have been extensively studied by several investigators (22, 23, 24). Mroz-Sowerby's combined kinematic-isotropic hardening model does describe the properties of DPS over the entire range of fm. It compares in approach with the model proposed by Tamura et al. (4) for DPS.

Summary and Conclusions

The steel investigated showed upon intercritical annealing to dual-

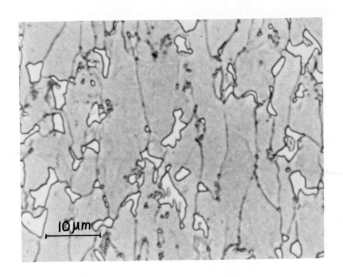

Figure 10 - Microstructure of highly deformed region of the steel annealed
at 730°C (fm = 0.26).

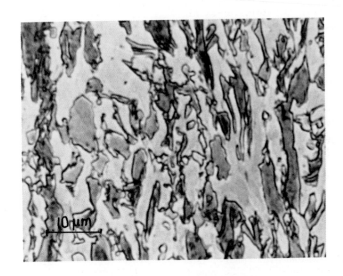

Figure 11 - Microstructure of highly deformed region of the steel annealed
at 820°C (fm = 0.56).

phase structure a pronounced Bauschinger effect but no permanent softening. This same steel exhibited permanent softening in normalized condition.

At high intercritical annealing temperatures and the resulting high volume fraction of relatively low carbon martensite the reverse stress strain curve follows closely the theoretical one for kinematic hardening, as described by the isostrain model.

At low intercritical annealing temperatures and resulting low volume fraction of high carbon martensite, strain hardening on reversed loading is very rapid. The behavior is characterized by hard particle dispersion strengthening.

The mechanical behavior of dual-phase steels on reverse straining can be well described by Mroz-Sowerby model for combined kinematic and isotropic hardening. The contribution of isotropic hardening increases with decreasing intercritical annealing temperature and corresponding decreasing volume of fraction of martensite.

Acknowledgement

The financial support of this work by the National Science and Engineering Council of Canada is gratefully acknowledged.

References

1. D.A. Korzekwa, R.D. Lawson, D.K. Matlock, and G. Kraus, "A Consideration of Models Describing the Strength and Ductility of Dual-Phase Steels," Scripta Metallurgica 14 (1980) pp.1023 - 1028.

2. S.T. Mileiko, "The Tensile Strength and Ductility of Continuous Fibre Composites," J.Materials Science 4 (1969) pp.974 - 977.

3. Y. Tomota, K. Kuroki, T. Mori, and I. Tamura, "Tensile Deformation of Two-Ductile-Phase Alloys: Flow Curves of α-γ Fe-Cr-Ni Alloys." Mats. Sci. and Eng. 24 (1976) pp.85-94.

4. I. Tamura, Y. Tomota, and M. Ozawa, "Strength and Ductility of Fe-Ni-C Alloys Composed of Austenite and Martensite with Various Strength." pp.611 - 615 in Proc.3rd Int.Conf. on the Strength of Metals and Alloys. Vol.1, Cambridge, 1973.

5. L.M. Brown, W.M. Stobbs, "The Workhardening of Copper-Silica II. The Role of Plastic Relaxation." Phil.Mag. 23 (1971) pp.1201 - 1232.

6. E. Orowan, "Causes and Effects of Internal Stresses", pp.59 - 80 in Internal Stresses and Fatigue in Metals, G.M. Rassweiler and W.L. Grube, ed.; Elsevier, New York, N.Y., 1959.

7. D.V. Wilson, "Reversible Work Hardening in Alloys of Cubic Metals," Acta Met. 13 (7) (1965) pp.807 - 814.

8. N. Ibrahim and J.D. Embury, "The Bauschinger Effect in Single Phase b.c.c.Materials," Mats.Sci.and Eng., 19 (1975) pp.147 - 149.

9. G. Masing, Lehrbuch der Allgemeinen Metallkunde, p.405; Springer Verlag, Berlin-Goettingen-Heidelberg 1950. See Also: G. Masing and W. Mauksch, Wiss. Veroeff. Siemens 4 (1) (1925) p.74.

10. R.J. Asaro, "Elastic-Plastic Memory and Kinematic - Type Hardening," Acta Met. 23 (10) (1975) pp.1255-1265.

11. Z. Mroz, "An Attempt to Describe the Behavior of Metals under Cyclic Load Using a More General Workhardening Model", Acta Mechanica 7 (2/3) (1969) pp.199 - 212.

12. R. Sowerby and Y. Tomita, "On the Bauschinger Effect and its Influence on U.O.E. Pipe Making Process," Int.J.mech.Sci, 19 (1977) pp.351 - 359.

13. R. Sowerby and D.K. Uko, "A Review of Certain Aspects of the Bauschinger Effect in Metals", Mats.Sci.Eng. 41 (1979) pp.43 - 58.

14. F.S. Le Pera, "Improved Etching Technique to Emphasize Martensite and Bainite in High Strength Dual Phase Steels" J.Metals 32 (3) (1980) pp.38 - 39.

15. R.G. Davies, "Influence of Martensite Composition and Content on the Properties of Dual Phase Steels," Metallurgical Trans.9A (5) (1978) pp.671 - 679.

16. R.G. Davies, "The Deformation Behavior of a Vanadium Strengthened Dual Phase Steel," Met.Transactions 9A (1) (1978) pp.41 - 52.

17. G.D. Moan and J.D. Embury, "A Study of the Bauschinger Effect in Al-Cu Alloys", Acta Met. 27 (5) (1979) pp.903 - 914.

18. S. Saleh and H. Margolin, "Bauschinger Effect During Cyclic Straining of Two Ductile Phase Alloys", Acta Met. 27 (4) (1979) pp.535 - 544.

19. D.V. Wilson and A. Konnan, "Work Hardening in a Steel Containing a Coarse Dispersion of Cementite Particles", Acta Met. 12 (5) (1964) pp.617 - 628.

20. G. Moan, C.M. Sargent and J.D. Embury, "On the Reversible Component of Plastic Flow in Two-Phase Materials", pp.41 - 44 in the Microstructure and Design of Alloys. Vol.1 The Institute of Metals, London, 1973.

21. Y.W. Chang and R.J. Asaro, "Bauschinger Effects and Work-Hardening in Spheroidized Steels," Metal Science 12 (6) (1978) pp.277 - 283.

22. D. Kuhlmann-Wilsdorf and C. Laird, "Dislocation Behavior in Fatique-II Friction Stress and Back Stress as Inferred from an Analysis of Hysteresis Loops," Mats.Sci.Eng. 37 (1979) pp.111 - 120.

23. H. Fischmeister and B. Karlsson, "Plasticitatseigenschaften grobzweiphasiger Werkstoffe," Z.Metallkunde 68 (1977) pp.311 - 327.

24. D.J. Lloyd, "The Bauschinger Effect in Polycrystalline Aluminum Containing Coarse Particles," Acta Met. 25 (4) (1977) pp.459 - 466.

411

A STUDY OF THE EARLY STAGES OF PLASTIC DEFORMATION OF DUAL-PHASE

STEELS, INCLUDING MICROPLASTICITY (*).

Henri Mathy   Jacques Gouzou   Tony Gréday

Centre de Recherches Métallurgiques

11, rue E.Solvay - B-4000 Liège (Belgium).

Dual-phase steels are characterized by a high strain hardening coefficient as compared to that of classical steels. The study of the behavior of dual-phase steels in the first stages of plastic deformation will help to better define and improve the microstructure to be achieved. Neither the Hollomon, nor the Ludwik laws are appropriate to describe the tensile curve. Microplasticity tests were performed on dual-phase steels with various morphologies of the second phase : they show that parabolic microplasticity occurs in the deformation process of the dual-phase steels earlier than in classical steels. The parabolic laws are related to different types of dislocation sources, the activities of which are enhanced when the ferrite is substructured ; such ferritic grains are promoted by intercritical rolling. Finally, microstructural examinations have revealed the ability of ferrite to accommodate the strains in the macroscopic deformation range.

---

*   This work was supported by IRSIA - Belgian Institute for Scientific Research in Industry and Agriculture.

413

Introduction

The continuous yielding shape of the tensile curve which characterizes dual-phase steels (1) implies a specific behavior of these steels in the first stages of plastic microdeformation. In order to study such behavior, microplasticity tests have been performed on three different kinds of dual-phase microstructures.

A Cr-Mo steel, the chemical composition of which has been suggested earlier (2), has been used. Plates of 12mm thick have been hot-rolled, the hot-rolling of most of them has been finished in the intercritical Ar_3-Ar_1 range of temperatures (3). Extended microstructural examinations have been carried out in both initial and deformed states in order to relate the mechanical properties of the steel to its microstructure.

Materials and Experimental Procedures.

The steel used in this study was prepared by air-induction melting of 36 kg (83 lbs) heats. The chemical composition of this steel is, in weight percent :

C	Mn	Si	Cr	Mo	Al	N
.07	1.15	.80	.50	.40	.040	.007

The square base ingots (115x115mm - 4.5x4.5 in) were forged down to 60mm thick (2.35 in) and then hot-rolled down to 12mm (.47 in) thick plates according to three different processing schedules as indicated in Table I.

In order to develop three different microstructures, the plates of types A, B or C have been submitted to the thermomechanical treatments which will be described hereafter. It should be noted that the plates of group A will be our reference steel and that the plates of groups B and C are as hot-rolled dual-phase steels.

1. Group A. After intercritical rolling and accelerated cooling, the plates were heat-treated at a temperature just below that of the Ar_1 point (700°C-1292°F), in order to induce tempering of the martensite islets dispersed in the ferrite microstructure. As indicated in Table I, there are in this series two different conditions (A1 and A2) according to the different cooling rates used for the final cooling of the tempered specimens (respectively .03°C/sec and 1.5°C/sec - .054°F/sec and 2.7°F/sec).

2. Group B. This group of plates was hot-rolled in the austenite region (finishing temperature of rolling : 850°C - 1562°F) and the final cooling rates were : for type B1, 1,5°C/sec between 800° and 500°C (-2.7°F/sec between 1472° and 932°F) and for type B2, 30°C/sec (54°F/sec), between 850°C (1562°F) and the temperature of the Ar_1 point, and then as for B1.

3. Group C. This group of plates was hot-rolled in the intercritical range of temperatures A_3-A_1, the last pass (reduction 25 percent) being performed at 750°C (1382°F). The final cooling rates used respectively for types C1 and C2 correspond to those already mentioned for the types B1 and B2. The type C3 plates received a rolling reduction of 35 percent in the last pass and were cooled at a rate of 30°C/sec (54°F/sec).

414

Table I - Hot-Rolling and Heat Treating Processing Schedule.

Plate	Last Rolling Pass			Cooling Rate [*] after Rolling.		Heat Treatment
	Temp. °C	°F	Reduction %	°C/sec	°F/sec	
A1	750	1382	25	30	54	HT 1 [**]
A2						HT 2 [***]
B1	850	1562	25	1.5	2.7	none
B2				30→700°C 1.5→R.T.	54→1292°F 2.7→R.T.	none
C1	750	1382	25	1.5	2.7	none
C2			25	30→650°C 1.5→R.T.	54→1202°F 2.7→R.T.	none
C3			35	30	54	none

[*] Cooling rate between 800° and 500°C (1472°-932°F) if not otherwise stated.

[**] HT 1 : 700°C - 1/2h / 0.03°C/sec → 500°C - 1.5°C/sec → R.T.
    (1292°F)   (0.054°F/sec) → 932°F - 2.7°F/sec → R.T.

[***] HT 2 : 700°C - 1/2h / 1.5°C/sec
     (1292°F)   (2.7°F/sec).

Microplasticity tests have been performed using a uniaxial dead-weight tensile machine which was described previously (4). It was specially built for measuring microstrains and its extensometer was designed very carefully. It allows a linearly increasing stress to be applied to the specimen, with a very good stability of temperature, and the deformation of the specimen is measured throughout the test. The round tensile specimen is 5.5mm (.21 in.) in diameter, with a gage length of 100mm (3.94 in.).

Standard mechanical properties have also been measured using triplicate round tensile specimens, 4mm (.16 in.) in diameter and a gage length of 25mm (1 in.). Moreover, curves of impact strength vs. temperature were obtained using Charpy V specimens (10x10mm) in the rolling direction (notch perpendicular to the rolling plane).

Microstructural examinations were carried out on optical, scanning and transmission electron microscopes. Moreover a quantitative characterization of the sizes and amounts of the microstructural phases has been obtained using an image analyser.

Results

General Description of the Microstructures.

The mean ferritic grain size is of the order of 5-8 microns (10.5-12.5 ASTM) for all microstructures examined.

The plates of group A are characterized by a dispersion of tempered martensite islets in a soft ferrite microstructure (Fig. 1.a). The dispersed martensite islets are elongated in the direction of rolling, A1 and A2 plates microstructures differing only by the size of the carbides in the tempered martensite.

a. Plate A2 (20% second phase)   b. Plate B2 (17% second phase)

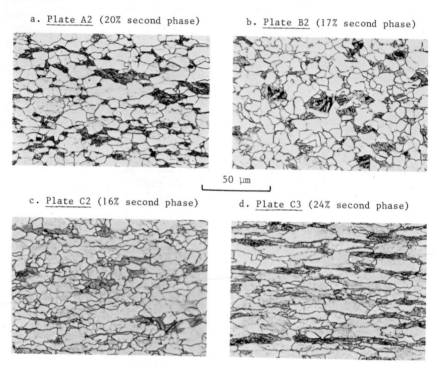

50 μm

c. Plate C2 (16% second phase)   d. Plate C3 (24% second phase)

Fig. 1 - Examples of microstructures.

A polygonal ferrite in which martensite islets are dispersed is the typical microstructure of plates of the group B (Fig. 1b). The microstructure of the plates of group C consists of elongated ferrite grains and martensite + bainite islets (Fig. 1c and d). A more detailed description of some relevant features of those microstructures will be given later.

Standard Mechanical Properties.

The three groups of plates differ in their mechanical properties (Table II). Group A does not exhibit continuous yielding, and the YS/$_{UTS}$ ratio is between 0.8 and 0.9. The more slowly cooled steel (A1) has a yield strength (L.Y.S.) lower than steel A2, while remaining higher than most of the yield strengths of the dual-phase steels (groups B and C). Ultimate tensile strengths of group B are higher and their YS/$_{UTS}$ ratio is very low (0.46). It must be noted that an increase in ductility and a decrease of the impact strength of these plates has been measured. Tensile strengths of group C are still higher (especially for condition C3) and the YS/$_{UTS}$ ratio is in the range 0.55-0.6 ; the elongation is reduced (C3 is characterized by a total elongation of 20 percent) ; in this last case, the transition temperature is very low.

416

Table II - Mechanical Properties

Plate	Y.S. MPa	UTS MPa	EL_T %	R.A. %	$\rho_{20°C}$ J/cm^2	T_{tr} °C 50 J/cm^2
A1	450*	560	29.5	75	260	- 70
A2	520*	595	28	77.5	260	- 80
B1	290	635	34	54.5	100	- 20
B2	285	620	31.5	60	140	- 60
C1	375	670	25.5	50.5	110	- 35
C2	360	655	27	49.5	140	- 60
C3	460	755	20.5	48	125	- 80

* Lower yield strength - the other ones are conventional 0.2% yield strength.

It is very common to represent tentatively the stress-strain behavior of steels using the Ludwik law (5) : $\sigma = \sigma_0 + k\varepsilon^m$, or its simplified version according to Hollomon (6) : $\sigma = k'\varepsilon^n$. It can be easily shown that this is totally inadequate to describe the deformation process of dual-phase steels. From the application of the Hollomon law to plates of types B1 and C3 (Fig. 2), it must be concluded that a straight line cannot be fitted on the experimental data. This result justifies the use of more sophisticated deformation techniques and interpretation of the data.

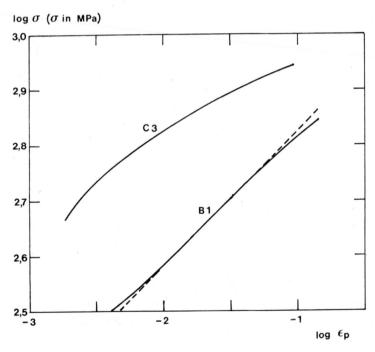

Fig. 2 - Analysis of stress-strain curves according to Hollomon law.

417

Microplasticity Tests.

Figure 3 gives an example of a microplasticity curve, obtained with steel Bl. This curve is seen to exhibit three distinct regions :

- in region OA, the strain increases linearly, while remaining slightly above pure elastic strain : the microstrain which corresponds to this first period is due to small movements of the existing dislocations ; for clarity both elastic strain and linear microstrain were subtracted from the experimental record, so that region OA corresponds to a part of the x-axis in Figure 3 ;

- starting from a critical stress σ_1 (point A), microstrains are observed, which increase according to a second-order parabolic law ; this stage of the test corresponds to the progressive generation of dislocations and obeys a very simple relationship :

$$\varepsilon_p = K_1 \ (\sigma - \sigma_1)^2$$

in which K_1 represents the ability of the steel to generate dislocations above the stress σ_1.

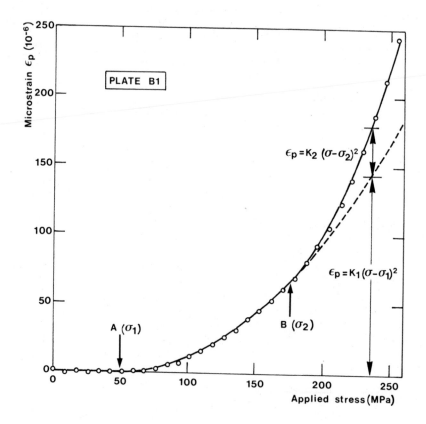

Fig. 3 - Microplasticity curve for plate Bl.

- starting from point B (stress σ_2) a third period of the microstrain curve consists in another parabolic law ; this third period also corresponds to the progressive generation of dislocations, and obeys the same type of relationship :

$$\varepsilon_p = K_2 \ (\sigma - \sigma_2)^2$$

This second parabolic law did not exist in all specimens : Figure 4, which relates to steel A1, gives an example of a microplasticity curve exhibiting one parabolic law only.

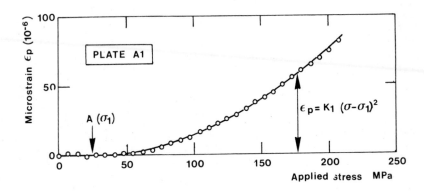

Fig. 4 - Microplasticity curve for plate A1.

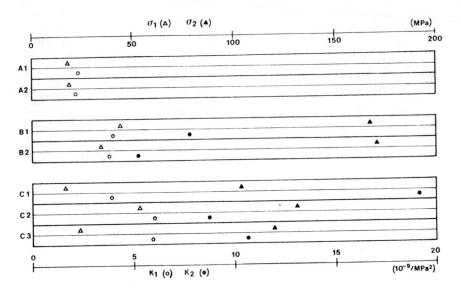

Fig. 5 - Variation with the cooling process of the microplasticity parameters, for the three groups A, B and C.

419

Figure 5 shows the variation with the cooling process, of σ_1 and K_1, σ_2 and K_2, presented by open or solid symbols respectively, for the three groups A, B and C. It is readily observed that :

- both states from group A present only one parabolic region, as opposed to the plates from groups B and C, which all show two distinct parabolic regions ;

- the existence of two such parabolic regions, corresponding to different stress levels, suggests that two types of dislocation sources come successively into operation in groups B and C, the first continuing to operate after the sources of the second type have come into action ; it seems likely that the sources of the first type operate in ferrite and that those of the second type act at the interfaces between ferrite and the second phase ;

- both states in group A present a rather low value of σ_1, which means that dislocation sources are readily formed in these steels ; the value of K_1 is also low, which implies that the density of the sources is not high and that each of them generates only a small number of dislocations;

- plates from group B present a distinctly higher value of σ_1 : dislocation sources in ferrite do not form so easily as in group A, which is readily explained by the ferrite-martensite microstructure observed in this group B ; these plates also present a rather high value of K_1 : dislocation sources in ferrite are more numerous than they were in group A, because of the existence of high internal stresses at the ferrite-martensite interfaces (Fig. 6a) ;

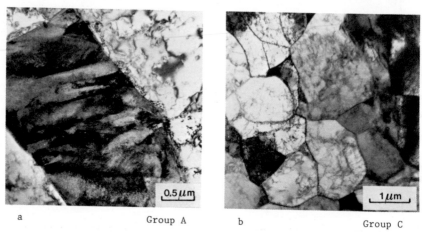

a   Group A   b   Group C

Fig. 6 - Possible sites for dislocation sources.
       a. Interface between the second phase and the ferrite.
       b. Grain and subgrain boundaries.

- The plates from group C present a still higher value of K_1 than in group B : this is due presumably to the fact that the substructured ferrite (Fig. 6b) includes a larger number of favourable sites where dislocation sources may generate ; the high value of σ_1, for plate C2, is also noticeable whereas it is rather low for the other plates in group C ;

- the value of σ_2 in group C is clearly lower than in group B : this is presumably due to the fact that most of the interfaces between ferrite and the
second phase are disturbed to a large extent by intercritical rolling applied to the plates of group C ;
- the value of coefficient K_2 is higher in group C than in group B : there is
a greater number of favourable sites for dislocation sources in group C,
owing to the already mentioned disturbed interfaces. Moreover, the amount
of interface is larger in this group C as a result of the elongated shape
of the second phase islets.

Behavior of the Phases during the Deformation Process.

Clearly revealing by any metallographic technique the microscopic deformation of the phases in the microdeformation domain is very difficult. For
this reason, we examined such a behavior in the range of large strains, i.e.
during uniform strain and necking. This is exemplified in Figure 7 for the
plates of groups B and C. It must first be emphasized that the coherency at
the ferrite-martensite interfaces is maintained throughout the tensile test.
Microcracks never generate there, contrary to what occurs at interfaces between non-metallic inclusions and the metallic phase (Fig. 7-b showing a MnS
inclusion).

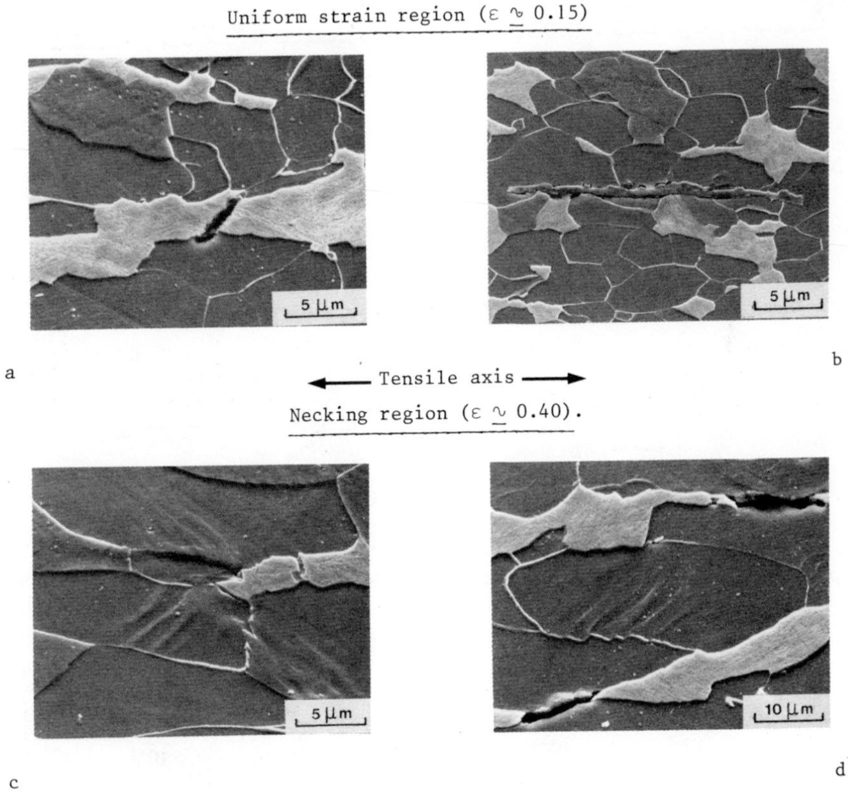

Uniform strain region ($\varepsilon \simeq 0.15$)

a

◄—— Tensile axis ——►

b

Necking region ($\varepsilon \simeq 0.40$).

c

d

Fig. 7. Behavior of the ferritic and martensitic phases in the domain of
large strains.

421

When the martensitic islets are locally unable to support the stresses in-duced during the deformation, they tend to be sheared (see on Fig. 7-a the mi-crocrack at 45° to the tensile axis). The surrounding ferrite can accommodate very large local strains : Figure 7c and d show deformation bands, associated with a rupture in the second phase islets, which spread out (again at 45° to the tensile axis) in the ferrite grains. They tend either to restore the con-tinuity which was compromised by the cracks (Fig. 7-d), or to relax the stres-ses at the tip of a martensitic islet (Fig. 7-c). Broken segments of ferritic grain boundaries are shifted, building up a stair-like configuration (Fig. 7-c and 7-d).

Analysis of Microplasticity and Standard Tensile Test Results.

Besides the mechanical tests, the results of which are given in Table II, other standard tensile tests were also carried out on flat specimens (100 x 15x2mm -3.93x.59x.08 in.) on which a grid (10x10mm ap. 0.4x0.4 in.) was prin-ted in order to measure the plastic strain. These data are useful to discuss somewhat the strain-hardening behavior of the steels, by the application of the Jaoul-Crussard analysis (7,8). Therefore, log $(d\sigma/d\epsilon_p)$ is plotted versus log ϵ_p (Fig. 8).

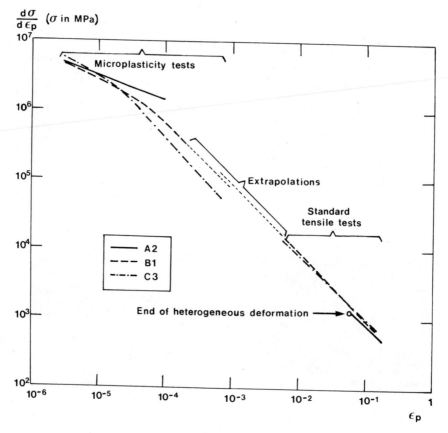

Fig. 8 - Compared results of microplasticity and standard tensile tests.

As already mentioned, the plates of the group A exhibit discontinuous yielding. For this reason the first reported data point of the plates of group A relates to the end of heterogeneous deformation. However, it can be seen that afterwards plates A roughly behave as the other groups. This common behavior corresponds to a relationship of the type : log $(d\sigma/d\varepsilon_p)$ = K - log ε_p. Strain hardening of the plates of group A has been measured to be inferior to groups B and C. This is confirmed by the fact that the straight line fitting the data points of plates A is located below the corresponding lines of the two other groups.

The results obtained in the field of microplasticity and reported also on Figure 8 clearly reveal the previously described complex set of behavior of the conditions studied. In the limits of the experimental conditions, the specimens pertaining to group A can be described by a single straight line. On the contrary, the other groups must be depicted by two linear segments, with a larger negative slope ; this means that the plates of groups B and C offer a gradually smaller resistance to deformation than the ones belonging to group A do.

Discussion.

In a preceding section the microplasticity behavior of the dual phase steels studied has been fully described. In this section we shall compare (Table III) these results to those obtained with three more conventional steels (9) : mild steel (.1% C - .7% Mn - .1% Si), experimental low-carbon vanadium steel (.01% C - 1.1% Mn - .3% Si - .15% V) and a structural vanadium steel (.2% C - 1.4% Mn - .3% Si - .15% V). These three steels were Al-killed and their properties measured after a normalizing heat treatment.

Table III - Comparison of Microplasticity Critical Parameters.

Type of Steels	Y.S. (MPa)	UTS (MPa)	σ_1 (MPa)	$K_1 \times 10^9$ (MPa^{-2})	σ_2 (MPa)	$K_2 \times 10^9$ (MPa^{-2})
Reference steels(9)						
- Mild steel	290	410	60	8.7	-	-
- Exp. low carbon vanadium steel	300	380	150	6.2	-	-
- Structural vanadium steel	460	640	140	1.9	-	-
Present work (*)						
- Group A	485	580	19	2.3	-	-
- Group B	390	630	40	4	169	6.5
- $\overline{C1 + C2}$	370	660	35	5.5	117	13.9
- C3	460	755	24	6	120	10.5

(*) Mean values of properties are presented here.

First of all, it must be emphasized that a second parabolic region in the curve of microplastic deformations only occurs for the two groups (B and C) of steels which exhibit dual-phase properties.

This second law of deformation is associated with dislocation sources located at the ferrite-second phase interfaces, which become active later in the deformation process and promote the deformability of those interfaces where microcracks have not been observed even at very large strains. Moreover, another important feature consists in the important difference of the σ_1 values at which the first parabolic deformation law is operating. This first parabolic law is likely to be associated with dislocation sources located in the grain boundaries of the ferrite phase. It can then be concluded that this type of source is more effective in generating dislocations in dual-phase steels than in microalloyed vanadium steels. This conclusion is enhanced when one takes into consideration the substructure existing in the ferrite grains of plate C3, the tempering treatment applied to plates of group A and even the quasi entirely ferritic microstructure of mild steel : all these steels exhibit a σ_1 value of the same order of magnitude as it is likely that the mean free path for dislocation movement in the ferrite grains is also of the same order of magnitude ; on the contrary the microalloyed steels contain finely dispersed microprecipitates operating as obstacles to the dislocation movement.

The excellent behavior of the ferrite phase of the dual-phase steels during the first stages of the deformation process is still occuring at high strain levels, as it is confirmed by the appearance of a stair-like configuration at grain boundaries and of deformation bands (Fig. 7).

Let us detail now the effects of the second type of dislocation sources. These sources become active later in the deformation process for the plates of group B than for the plates of group C, as indicated by the values of σ_2. Moreover, their activity is much more intense in the case of the plates of group C, as it can be concluded from the higher K_2 values measured. It is very likely that this behavior of the dual-phase plates of group C is related to the existence of a more heavily disturbed region at the ferrite-martensite interfaces as well as to the elongation of the second phase islets. The latter is responsible for a larger specific surface of those interfaces, at constant volume fraction of the second phase. This result is also partially enhanced by a somewhat larger amount of second phase in the C3 plate, compared to group B plates. Total interface length is greater by 30 percent in the C3 plate as compared to the other intercritically rolled dual-phase steels.

Using the measured microdeformation parameters it is also possible to show the greater ability to accommodate microdeformation presented by the dual-phase B plates as compared to the experimental low-carbon vanadium steel. It must also be added that the mild steel is exhibiting an even greater ability to develop microdeformation than the dual-phase steel B. Comparing now the microdeformation parameters of steels B and C, it must be emphasized that an increase of the yield and tensile strength of dual-phase steels does not imply a reduced ability for microdeformation. Moreover, when the same comparison is made for steels exhibiting the same UTS level (structural steel and the group C1-C2) a considerable gain relevant to the dual-phase steels is observed. Such a comment can also be made when comparing the same structural steel and the dual-phase C3, both caracterized by the same yield strength.

Finally, the elongated shape of the martensite islets observed in the plates of the group C are likely to induce a larger resistance to initial elastic deformation ; but this tendency is too weak to be quantified in the cases studied.

It is also of interest to discuss the Jaoul-Crussard analysis of the results of both microplasticity and standard mechanical tests (Fig. 8). Although experimental data are presently missing in the intermediate region, we assume that the three groups of plates behave there in a quite different way.

For group B, the behavior can be assumed to be quite continuous, as the portions of the corresponding straight lines can be matched when extrapolated. Group A, which exhibits a heterogeneous deformation of the Lüders type in this intermediate region, cannot be depicted rigorously and precisely when assuming an entirely homogeneous process as in Figure 8. Nevertheless, it can be predicted that, in the region of the Lüders band formation, a sudden drop in ($d\sigma/d\varepsilon_p$) will appear, so that both regions of the curve will join together. For group C, some degree of heterogeneous deformation is also to be expected, as evidenced for other steels by Matlock et al (10).

Conclusions

On the basis of the experimental results obtained by microplasticity tests, the following conclusions can be drawn :

1. Dual-phase steels obtained in the as-rolled condition are characterized by the existence of two parabolic laws of microdeformation, contrary to the general case where only one parabolic law exists.
   These laws are related to different types of dislocation sources, located in the ferrite for the first law, and at the ferrite-martensite interfaces for the second one.

2. The dislocation sources operating in the ferrite grains come into action at a rather low stress level in the case of the dual-phase microstructure.

3. The activity of these sources is enhanced the more the ferrite phase is substructured. This behavior is also associated with an increase of the standard yield and tensile strength.

4. The desired substructured ferrite is promoted by intercritical rolling, heavy reduction in the last rolling pass and final accelerated cooling.

5. The ability of ferrite to accommodate strains in the macroscopic deformation range is revealed by the appearance of deformation bands and stair-like configuration of grain boundaries.

References.

1. S. Hayami and T. Furukawa, "A Family of High-Strength, Cold-Rolled Steels"; pp. 311-321, in Microalloying 75, Union Carbide Corporation, ed.; New-York, N.Y., 1977.

2. A.P. Coldren, G. Tither, A. Cornford and J.R. Hiam, "Development and Mill Trial of As-Rolled Dual-Phase Steel", pp. 207-230 in Formable HSLA and Dual-Phase Steels, A.T. Davenport, ed. ; AIME, New-York, N.Y., 1979.

3. T. Gréday, H. Mathy and P. Messien, "About Different Ways to Obtain Multi-phase Steels", pp. 260-280 in Structure and Properties of Dual-Phase Steels, R.A. Kot and J.W. Morris, ed. ; AIME New York, N.Y., 1979.

4. J. Gouzou, "Microplasticity Phenomena. Their Significance in the Study of Deformation Mechanisms in Steel", CRM Metallurgical Reports, 26 (1971) pp. 39-50.

5. P. Ludwik, Elemente der Technologischen Mechanik, p. 32 ; Springer, Berlin, 1909.

6. J.H. Hollomon, "Tensile Deformation", Transactions of TMS-AIME, 162 (1945) pp. 268-290.

7. B. Jaoul, "Etude de la forme des courbes de déformation plastique", Journal of the Mechanics and Physics of Solids, 5 (1957) pp.95-114.

8. C. Crussard, "Rapport entre la forme exacte des courbes de traction des métaux et les modifications concomitantes de leur structure", Rev.Met., 10 (1953) pp. 697-710.

9. J. Gouzou, Le problème de la déformation plastique en traction monoaxée dans l'acier doux, CRM Publication, Liège, 1969.

10. D.K. Matlock, G. Krauss, L.F. Ramos and G.S. Huppi, "A Correlation of Processing Variables with Deformation Behavior of Dual-Phase Steels", pp. 62-90 in Structure and Properties of Dual-Phase Steels, R.A. Kot and J.W. Morris, ed. ; AIME, New York, N.Y., 1979.

CONTROL OF STRENGTH AND R-VALUE IN BOX-ANNEALED DUAL-PHASE STEEL SHEET

A. Okamoto and M. Takahashi

Central Research Laboratories, Sumitomo Metal Industries, Ltd.
Amagasaki, Japan

Abstract

A formable cold-rolled dual-phase steel sheet with an ultimate tensile strength of 400 MPa was developed by box-annealing a steel containing 0.02% carbon and 2.1% manganese. The effects of carbon content and the annealing conditions on the mechanical properties were examined and correlated with microstructure. The r-value was decreased by the presence of martensite. The steel contained pearlite, twinned martensite and retained austenite, which were enriched in carbon and manganese. The retained austenite can be transformed to martensite by a sub-zero temperature treatment, however it was stable even after tempering. Since the emergence of yield point elongation with the tempering treatment was greatly influenced by a sub-zero treatment before tempering, the role of retained austenite in the mechanical properties of the box-annealed dual-phase steel is discussed.

Introduction

As weight reduction is one of means to achieve fuel-efficient cars, many attempts have been practiced to apply high strength cold-rolled steel sheet to body panels. The main problem in these attempts was the lack of shape-fixability of panels, which was attributed to the high yield strength property of the steel sheet. Dual-phase steels have become of practical interest because they show a low yield strength, and a high ultimate tensile strength with a large work-hardening exponent.

The production of cold-rolled dual-phase steel sheet by box-annealing was primarily developed by Matsuoka and Yamamori (1). They found that inter-critical annealing of the cold-rolled steel containing manganese of more than 2% exhibits a ferrite-martensite mixed microstructure and the steel shows continuous yielding at low strength and high ultimate tensile strength. Recently, similar results were obtained by Mould and Skena (2). The problems in the mass production of dual-phase steel by box-annealing process have been the following: 1) high manganese content is required, 2) large strength variations are found with varying annealing conditions, and 3) the plastic strain ratio, i.e. r-value, is low.

The production of dual-phase steels by continuous annealing process was exploited by Hayami and Furukawa (3,4). Since a lower manganese content is sufficient to form a dual-phase microstructure in this process, dual-phase steels have extensively produced by the continuous annealing process and research efforts on continuously annealed dual-phase steels are seen to have overshadowed those on box-annealed dual-phase steels. However, in their applications to the panels of auto-body, they were not formable enough in comparison with conventional mild steels, because of their high strength and low r-value properties.

The present experiment was conducted to address some defects on conventional dual-phase steels, and to develop a highly formable dual-phase steel sheet with the ultimate tensile strength of 400 MPa by a conventional box-annealing process. The paper reports the metallurgical factors controlling the strength and r-value, and the role of retained austenite in the mechanical properties of box-annealed dual-phase steel.

Experimental Procedure

Materials

Two high manganese steels were melted by oxygen converter, vacuum-degassed by the DH process and subsequently cast into ingots with bottom-pouring. The slabs were soaked at 1280°C and hot-rolled to a final thickness of 2.8 mm or 2.3 mm in Wakayama Steel Works. The chemical compositions and the hot-rolling conditions are shown in Table I. They were hot-rolled in the austenite region and coiled at a low temperature in order to prevent the precipitation of aluminum nitrides. Steel A is the newly developed steel with low-carbon content and steel B is a companion steel containing normal carbon. Most of the experiments were processed with steel A.

Table I. Chemical composition and hot-rolling conditions of steels (wt %)

Steel	C	Si	Mn	P	S	sol. Al	N	F.T. (°C)	C.T. (°C)	Thickness (mm)
A	0.022	0.07	2.06	0.009	0.004	0.056	0.0047	850	570	2.8
B	0.047	0.11	2.44	0.017	0.010	0.037	0.0074	850	600	2.3

(F.T.: Finishing temperature, C.T.: Coiling temperature)

Cold-Rolling and Annealing

The hot-rolled steels were pickled and cold-rolled to the final thickness of 0.7 mm. The cold reductions were 75% and 70% for steel A and B, respectively.

Annealing was conducted in a laboratory furnace under an argon atmosphere. The heating rate, annealing temperature, soaking time and cooling rate were varied. In the base annealing condition, they were 40 °C/h, 710 °C, 4 hr and 80 °C/h, respectively. The base annealing condition simulated the heat cycle of an open-coil annealing furnace.

Tensile Testing

The mechanical properties of the annealed steels were evaluated by tensile testing of JIS No. 5 specimens with gauge sections 50 mm long by 25 mm wide. The r-value was measured at a tensile elongation of 15 or 20%, in three directions; longitudinal (0°), diagonal (45°) and transverse (90°). The mean r-value, $\bar{r}$, was calculated by $\bar{r} = (r_0 + 2r_{45} + r_{90})/4$ in usual manner. The yield strength was determined by the 0.2 % off-set strength method.

Analyses of Metallography

The microstructures of the annealed steels were examined by optical microscopy and by electron microscopy. The volume fractions of second phases were measured in electron micrographs of two-stage replicas at a magnification of 2,000 times. The enrichments of carbon and manganese in the second phases of the dual-phase steel were examined by an electron probe micro analyser (E.P.M.A.).

The textures of the annealed steels were examined by X-ray diffraction intensity of {222} planes parallel to the rolling plane and by a {200} pole figure. The amount of retained austenite was determined by Mössbauer spectroscopy. The method is preferred to that of X-ray analysis, because a small amount of austenite can be measured without the influence of crystalline texture. For the measurements of X-ray diffraction and Mössbauer spectroscopy, the steels were chemically polished to the thickness less than 0.05 mm.

Sub-zero Treatment and Tempering

In order to investigate the role of retained austenite on the mechanical properties, a dual-phase steel of composition A annealed in the base condition was sub-zero treated at ice water or at liquid nitrogen temperatures for 30 min and then reheated to a temperature between 150 °C and 710 °C within 3 min, soaked at the temperature for 1 hr and subsequently cooled at the rate of 80 °C/h. The changes in mechanical properties and in the amount of austenite resulting from these treatments were examined. The dual-phase steel was also strained in tension to an elongation of 5 % after the annealing. The amount of austenite was measured for the strained specimen and tensile testing was conducted for the specimen after tempering at 350 °C.

Results

Mechanical Properties and Textures

Figure 1 shows the effect of annealing temperature on the mechanical properties of steel A and B. Annealing in the intercritical range, that is above 690 °C or 700 °C, eliminates the yield point elongation and decreases the yield strength, indicating the formation of a dual-phase structure. Provided the steel is annealed in the intercritical range, the mechanical properties of steel A are less affected by annealing temperature than those of steel B, containing a normal amount of carbon. The stability in the properties of steel A is attributed to its lower carbon content. The decrease in r̄-value with intercritical annealing shown in Fig. 1 will be discussed later.

The effects of soaking time at two annealing temperatures in steel A are shown in Fig. 2. Though the soaking for more than 10 hr is required at 690 °C, 25 min is enough to develop dual-phase properties at 710 °C. The required soaking time is presumed to be the time for manganese atoms to diffuse into the austenite phase. Figure 2 also shows that the r̄-value decreases as the yield strength decreases.

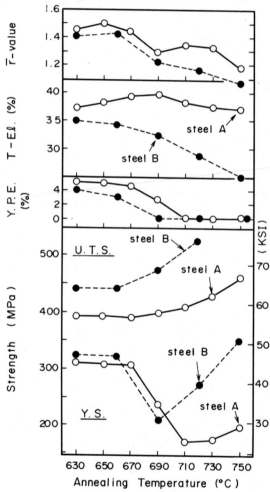

Figure 1 - Effect of annealing temperature on mechanical properties of steel A and B (Heating: 40 °C/h, Soaking: 4 hr, Cooling: 80 °C/h).

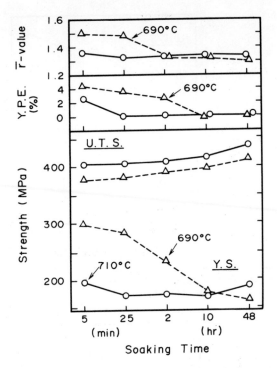

Figure 2 - Effect of soaking time at two annealing temperatures
on mechanical properties of steel A. (Heating: 40 °C/h,
Cooling: 80 °C/h).

The effect of cooling rate in steel A is shown in Fig. 3. The strength
does not change within the cooling rates of box-annealing (slower than 320
°C/h). However, both the yield and tensile strength increase at faster
cooling rates. This is due to the increase in volume fraction of martensite.

Figure 4 shows the effect of heating rate on r̄-value and texture of
steel A. A slowly heated aluminum-killed cold-rolled steel sheet is well
known to exhibit a high r̄-value (5) and a preferred recrystallization texture with <111> axis parallel to normal direction of sheet plane, owing to
the formation of aluminum nitride clusters at the recovery and recrystallization stages (6). The same effects of heating rate on r̄-value and on the
<111>// N.D. components are observed even in a high-manganese dual-phase
steel.

A {200} pole figure of steel A annealed in the base condition is shown
in Fig. 5. The figure indicates the strong {111}<011> + {554}<225> textures
(6) which are also the predominant components in a commercial aluminum-killed
steel sheet of deep drawing quality.

The results of the laboratory experiments indicates that the box-annealing of steel A at 710 °C, just above the A₁ temperature, with a slow heating rate offers a highly formable dual-phase steel. The mill productions
were conducted with steel A in Kashima and Wakayama Steel Works. After open
coil annealing, the steel was temper rolled to an elongation of 0.5 %. An
example of the results is shown in Table II. The steel has superior elonga-

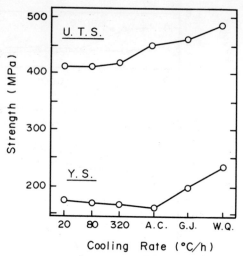

Figure 3 - Effect of cooling rate after soaking at 710 °C for
4 hr on strength of steel A (Heating: 40 °C/h, G.J.:
Gas jet cooling).

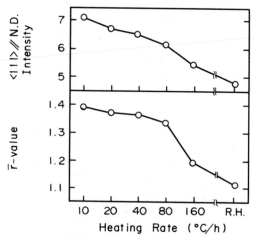

Figure 4 - Effect of heating rate to 710 °C on r̄-value and
X-ray diffraction intensity of steel A (soaking:
4 hr, Cooling: 80 °C/h).

tion to commercial rephosphorized steel of the same ultimate tensile
strength and has a higher r̄-value than conventional dual-phase steels whose
r̄-value is less than 1.0. Almost no variation in the mechanical properties
along the width and the length of the coils was observed.

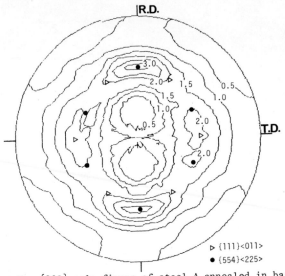

Figure 5 - {200} pole figure of steel A annealed in base
condition (710 °C, 4 hr).

Table II.   Mechanical properties of mill-produced dual-phase steel sheet

Y.S. (MPa)	U.T.S. (MPa)	T-El. (%)	Y.P.E. (%)	$\bar{r}$ value	C.C.V. (mm)	Er. (mm)	2%-WH (MPa)
172	409	39.7	0	1.38	27.5	10.4	92

(C.C.V.: Conical cup value, Er.: Erichsen value, 2%-WH: Work-hardening
by 2%-strain, JIS No. 5 specimen, G.L. = 50 mm, Thickness = 0.7 mm)

Microstructures

Figure 6 shows light micrographs of the dual-phase steels annealed in
laboratory furnace. Ferrite grains are elongated, which indicates the precipitation of aluminum nitride during their recovery and recrystallization
stages (5). Martensite and pearlite are located at their grain boundaries.
When steel B is annealed at 690 °C, cementite particles are observed in the
ferrite grains. These particles are absorbed in the austenite regions when
the steel is annealed at 720 °C or 750 °C and, as a result, the volume fraction of martensite increases. The microstructure of steel A is less affected by annealing temperature than that of steel B, which results in the
small variation in the mechanical properties of steel A shown in Fig. 1.

Figure 7 shows an example of electron micrograph of two-stage replicas
of steel A annealed at 710 °C. The volume fraction of pearlite and martensite were determined as 2.0 % and 1.8 %, respectively, by averaging the results obtained in 8 micrographs. The diameter of the martensite ranges
from 1 to 3 µm.

Transmission electron microscopic study revealed that most of the
martensite was twinned martensite, as shown in Fig. 8. This implies that
the martensite is highly enriched in carbon and was formed at below 300 °C.
Retained austenite could not be determined by transmission electron microscopy. It was believed that the retained austenite transformed to martensite
during electrolytic polishing.

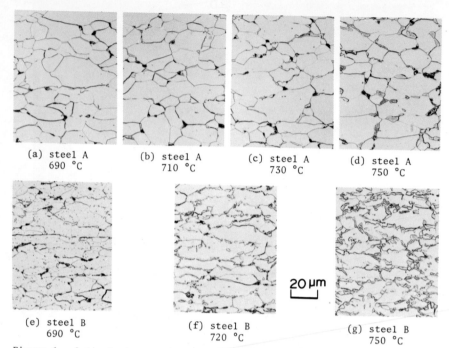

(a) steel A
690 °C

(b) steel A
710 °C

(c) steel A
730 °C

(d) steel A
750 °C

20 μm

(e) steel B
690 °C

(f) steel B
720 °C

(g) steel B
750 °C

Figure 6 - Optical micrographs of dual-phase steels (Heating: 40 °C/h, Soaking: 4 hr at indicated temperature, Cooling: 80 °C/h).

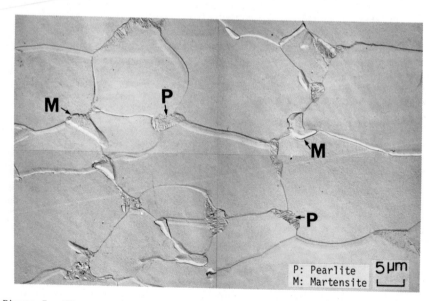

P: Pearlite
M: Martensite
5μm

Figure 7 - Electron micrograph of two-stage replicas of steel A annealed in base condition (710 °C, 4 hr).

434

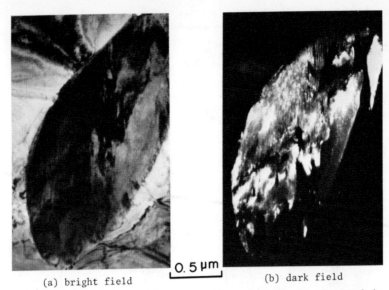

(a) bright field   0.5 μm   (b) dark field

Figure 8 - Transmission electron micrographs of martensite in steel A
annealed in base condition, showing twinned structure.

Figure 9 shows the relation between the mechanical properties and the
volume fraction of martensite of steel A and B annealed at various temperatures. As the volume fraction of martensite was measured in two-stage
replicas, it includes that of retained austenite. A martensite volume fraction of 2 % is enough to display the dual-phase properties with no yield
elongation. The minimum volume fraction of martensite to show no yield behavior is lower than the reported value for continuously annealed dual-
phase steels (7,8,9). The ultimate tensile strength of both steels increases linearly with the martensite volume fraction, as shown by many researchers (1,2,7,9,10). The different correlations between two steels can
be attributed to the difference in hardness of ferrite. The r̄-value, on the
other hand, initially decreases with increasing martensite volume fraction
but increasing the volume fraction above 4 % has little additional effect.
The figure indicates that both lowering carbon content from 0.05 % to 0.02 %
and controlling the volume fraction of martensite to less than 4 % are important to manufacture a dual-phase steel with an ultimate tensile strength of
400 MPa and a high r-value.

Enrichment of Carbon and Manganese

Segregation of carbon and manganese in steel B annealed for 4 hr at 720
°C was examined by E.P.M.A.. The result is shown in Fig. 10. Manganese and
carbon are enriched in the same place, which should be martensite, pearlite
or retained austenite in Fig. 6 (f). The figure also shows that the amount
of segregation for manganese is about twice as large as the bulk concentration.

Retained Austenite

The amount of retained austenite in steel A annealed in base condition
(710 °C, 4 hr) was measured by Mössbauer spectroscopy. Figure 11 shows an
example of the spectrum, indicating the existence of 1 % austenite by the
peak at the center. Thus the volume fractions of the second phases may be

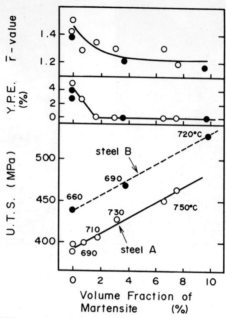

Figure 9 - Effect of volume fraction of martensite on mechanical properties of steel A and B. Annealing temperatures are indicated.

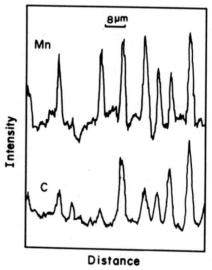

Figure 10 - Enrichment of C and Mn in second phases of steel B, detected by E.P.M.A. (Heating: 40 °C/h, Soaking: 720 °C, 4 hr, Cooling: 80 °C/h).

summarized as shown in Table III, by using the results of two-stage replicas and Mössbauer analysis.

The changes in the amount of austenite by sub-zero treatments or deformation are shown in Table IV. The retained austenite in box-annealed dual-phase steel is stable at 0 °C and it transforms to martensite at -196 °C. Moreover, the austenite shows a deformation-induced transformation within a strain of 5 % at room temperature.

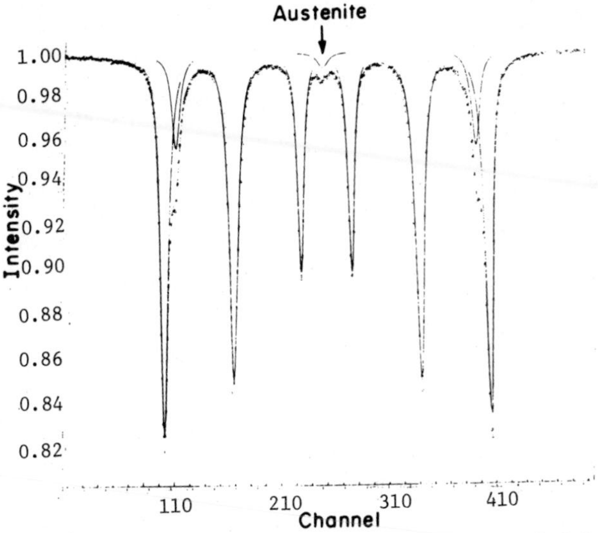

Figure 11 - Mössbauer spectroscopy of steel A annealed in base condition.

Table III. Volume fraction of second phases in steel A annealed in base condition.

Pearlite	Martensite	Retained Austenite
2.0 %	0.8 %	1.0 %

Table IV. Volume fraction of austenite measured by Mössbauer spectroscopy

Treatment	Volume Fraction of Austenite (%)
As-annealed (710 °C, 4 hr)	1.0
→ Ice water (0 °C, 30 min)	0.9
→ Liquid N_2 (-196 °C, 30 min)	<0.5
→ Strained (5 % in tension)	<0.5

Tempering

The annealed or the sub-zero treated dual-phase steels were tempered by reheating to various temperatures, soaked for 1 hr and subsequently cooled to room temperature at 80 °C/h. The effects of tempering on the mechanical properties and on the amount of austenite are shown in Fig. 12; the change in yield behavior with tempering is shown in Fig. 13.

When steel is not sub-zero treated before tempering, it contains austenite above the detectable limit and exhibits low yield strength even after any tempering treatment. It must be noted, however, that a small yield point elongation of 0.3 % is observed in Fig. 13 for steel tempered at 350 °C, 550 °C and 650 °C. The steel dipped in ice water before tempering showed almost the same behavior as the steel without sub-zero treatment. While, when steel is sub-zero treated at liquid nitrogen temperature, it shows a large yield point elongation and increased yield strength after tempering above 250 °C. A trough in yield point elongation and yield strength is observed at the tempering temperature of 500 °C. The same trough is also recognized in steel without sub-zero treatment. The ultimate tensile strength of the sub-zero treated steel decreases slightly by tempering above 500 °C. If the steel is reheated at 710 °C, which is the annealing temperature before tempering and also in the intercritical temperature range, the mechanical properties of the steel revert to the former properties, irrespective of the sub-zero treatment.

The pre-strained steel, shown in Table Ⅳ, was also tempered at 350 °C.

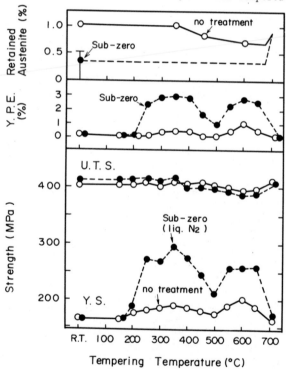

Figure 12 - Effect of sub-zero and tempering treatments on mechanical properties and volume fraction of retained austenite of steel A annealed in base condition.

438

Tensile testing of the steel resulted in a yield point elongation of 3 %. These results suggest that lack of retained austenite or martensite, not tempered, can not offer a continuous yielding characteristic to a dual-phase steel. The changes in r̄-value and total elongation by the sub-zero treatment and by the tempering were examined also, but their variations were within the experimental error.

Figure 14 shows the variation of n-value with strain, calculated from

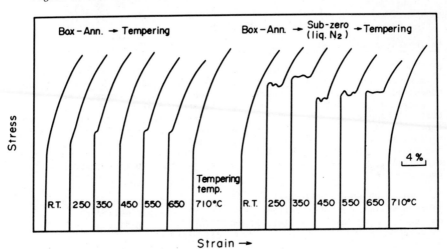

Figure 13 - Initial portion of stress-strain curves for specimens of steel A after tempering for 1 hr at the indicated temperatures.

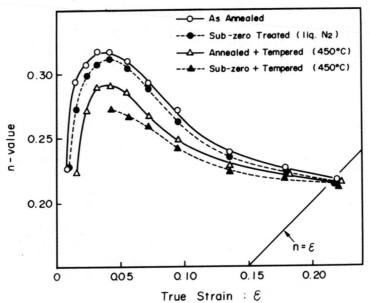

Figure 14 - Variations of n-value with strain for specimens of steel A box-annealed in base condition, sub-zero treated at liquid nitrogen temperature and tempered for 1 hr at 450 °C.

439

stress-strain curves for several conditions. The variation of n-value with strain is scarcely influenced by the sub-zero treatment, indicating that the retained austenite of 1 % itself does not affect the mechanical properties. The n-value at small strain is decreased by the tempering treatment, however, the n-value at which n is equal to true strain is not changed by tempering. The results imply that the uniform elongation is not influenced by the state of the second phase whether it is the retained austenite, martensite or tempered martensite.

Discussion

Enrichment of Manganese in Austenite

One of the features of box-annealed dual-phase steel is the enrichment of manganese in the second phases (11). The same analysis by E.P.M.A. was conducted for continuously annealed dual-phase steel, however no segregation of manganese was recognized because of the short soaking time at the inter-critical temperature (12). Recently, Koo et al. (13) reported that no pre-ferential partitioning of manganese between the phases is detected by X-ray micro-analysis for an AISI 1010 steel intercritically annealed for 10 min. The required soaking time for the enrichment of manganese is presumed from experimental data (see e.g. Fig. 2) to be more than 20 min for soaking temperature close to Ac_1.

The diffusion coefficient, D, of manganese in ferrite is not reported. If that of another substitutional element, for example molybdenum ($D_0 = 0.3$ cm^2/sec, Q = 49 kcal/mol) (14), is used, the diffusion distance of manganese in ferrite, x, may be estimated by the calculation of $x = A\sqrt{Dt}$, where A is a constant which is more than 1 and t is the soaking time. If the steel is soaked at 710 °C for 20 min and A = 3, x is calculated as 2.1 µm. This diffusion distance is about a quarter of the inter-austenite spacing at soaking, as inferred from Figs. 6, 7 and 10. As the austenite is located on the ferrite grain boundaries, the grain boundary diffusion must be considered also. Thus, the diffusion of manganese in ferrite to the austenite phase is believed to occur during the soaking period of intercritical box-annealing. This segregation phenomenon stabilizes the austenite phase remarkably and gives a unique behavior of retained austenite in the box-annealed dual-phase steel.

The r̄-value of Dual-Phase Steel

It is well known that the r̄-value is controlled by the texture of the sheet. When a steel contains much crystals whose <111> axis is normal to the sheet plane, the r̄-value increases, while, when it contains many <100> crystals, it decreases. A relation between the r̄-value and the ratio in diffraction intensity of <111> to <100> has been reported (15,16).

The present examination showed that the r̄-value is not only controlled by the crystalline texture but also by the existence of martensite. This was indicated in Fig. 9, and is also shown in Fig. 15. The textures in Fig. 15 were varied by varying the heating rate to the soaking temperature. It is clear that the r̄-value of a dual-phase steel varies with texture as in mild steel, however the r̄-values of a dual-phase steel are lower than those of mild steels with the same texture components. Similar effects of the hard phase on the r̄-value is also reported for water-quenched dual-phase steel by Hosoya et al. (17).

As the r̄-value decreases with the initial increase in the volume fraction of martensite in Fig. 9, the decrease in r̄-value is attributed to the activation of different slip systems in the ferrite adjacent to the marten-

440

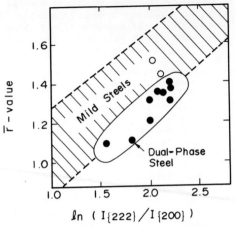

Figure 15 - Relation between the ratio of X-ray diffraction
intensities and r̄-value of steel A.
(○ : Subcritically annealed steel)
(● : Intercritically annealed steel)

site phase than those which would have operated in the absence of martensite, because slip system of the ferrite is constrained by the hard phase.
The yielding behavior itself does not affect the r̄-value, because the r̄-
value of sub-zero treated steel was not varied by tempering treatment at
above 250 °C.

Role of Retained Austenite

The existence of retained austenite in dual-phase steel has been reported by Marder (18), Rigsbee et al. (7) and Furukawa et al. (19). All of
them concluded that the austenite does not transform to martensite with the
sub-zero treatment but that it easily transforms with the plastic deformation. However, the dual-phase steels used in their studies have been produced by continuous annealing (19) or by a short-time soaking and cooling
process (7, 18).

The result of the present experiment indicates that a small amount of
austenite exists in the box-annealed dual-phase steels and that it transforms to martensite during a sub-zero treatment at liquid nitrogen temperature but that it is stable on reheating from room to intercritical temperature. The astonishing stability of the retained austenite on reheating is
possibly due to the enrichment of manganese in the austenite during soaking
and slow cooling.

The cause of the low ratios of yield/tensile strength and continuous
yielding characteristics of dual-phase steel has been discussed by many researchers (1,7,20). The internal residual stresses and mobile dislocations
produced by the volume expansion accompanying the transformation of a portion of the austenite to martensite during cooling from the annealing temperature have been thought to be the main causes of the above characteristics (7).

In the box-annealing process the retained austenite or the martensite
formed below 200 °C is presumed to take an important role in determining the

low yield/tensile ratio and continuous yielding characteristics. Because, when the martensite is formed at above 200 °C, it must be tempered during the slow cooling stage, as suggested in Fig. 12 and 13. Though the small amount of retained austenite itself does not affect the mechanical properties, its existence is important in suppressing the tempering of the second phase.

Davies reports (20) that the yield point elongation (yield plateau) emerges when the brine-quenched dual-phase steel is reheated at above 200 °C, and he concludes that it is due to the pinning of the dislocations in the ferrite by interstitial carbon atoms. However, if dislocation pinning is the cause of the yield plateau, it is surprising that the box-annealed and sub-zero treated dual-phase steel exhibits the yield plateau by the similar tempering treatment of 250 °C as the brine-quenched steel, because far smaller amount of carbon is in solution for box-annealed steel.

The difference in the yield behavior after tempering above 250 °C by the sub-zero treatment is attributed to the difference in the thermal stability above room temperature between the martensite and the retained austenite. Though both phases contain the same amount of carbon and manganese, the martensite phase is easier to decompose with shrinkage to ferrite plus cementite with tempering above 250 °C and, as a result, the internal stress in ferrite matrix is reduced. The volume shrinkage accompanied by the decomposition of martensite is presumed to be approximately 1 %. While the austenite phase hardly decomposes by tempering treatment, as previously mentioned, and can offer a new martensite phase with a large internal stress in the early stage of deformation.

It is well known that retained austenite in a martensitic steel decomposes by tempering treatment at a temperature between 230 and 280 °C (21). As an additional experiment, steel A was continuously annealed at 750 °C for 2 min. Though the steel contained retained austenite of 2.9 % to show a continuous yielding behavior after the annealing, it yielded discontinuously after a tempering treatment at 350 °C. This result is consistent with that of Davies (20), but differs from the present result on box-annealed dual-phase steel. The retained austenite in a continuously annealed steel is thought to decompose by tempering as in a martensitic steel because the austenite is not enriched in manganese. Thus the enrichment of manganese in retained austenite is believed to be indispensable to hold its unexpected stability to tempering.

The reason of the trough in yield strength at 500 °C tempering for the sub-zero treated steel, shown in Fig. 12, is uncertain, however it can not be interpreted with the presence of mobile dislocations. The internal stress is thought to be generated by the tempering at 500 °C. The authors believe that the internal stress formed with the transformation of austenite at or before the tensile testing is the predominant cause of the low yield/tensile ratio and continuous yielding characteristics of dual-phase steels.

Conclusions

1) The low yield dual-phase steel with the ultimate tensile strength of 400 MPa and the r̄-value of 1.4 was manufactured by box-annealing. The stability of the mechanical properties was achieved by lowering carbon content to 0.02 %.

2) The r̄-value of the dual-phase is controlled not only by crystalline textures but also by the volume fraction of martensite.

3) The dual-phase steel contains small amount of pearlite, twinned martensite and retained austenite. These second phases are enriched in carbon and manganese.

4) The martensite plus austenite volume fraction of 1 or 2 % is enough for box-annealed steel to exhibit low yield/tensile ratio and continuous yielding.

5) The retained austenite transforms to martensite by sub-zero treatment or by deformation, but it is stable to reheating.

6) The tempering of the box-annealed dual-phase steel does not influence the mechanical properties largely, because the steel contains retained austenite even after the tempering treatment. While, when the sub-zero treatment is operated for the steel before the tempering, the large yield point elongation is observed after the tempering at above 250 °C.

7) The existence of the retained austenite is important for the box-annealed dual-phase steel in the sense that it hardly decomposes at slow cooling because of the enrichment in manganese.

8) The internal stress formed at or before tensile testing was believed to be the predominant cause of the continuous yielding characteristic of dual-phase steel.

Acknowledgements

Grateful acknowledgement is given to Dr. Y. Ishida and Dr. S. Umeyama of the Institute of Industrial Science, University of Tokyo, for the measurement and analysis of Mössbauer spectroscopy.

References

1. T. Matsuoka and K. Yamamori, "Metallurgical Aspects in Cold Rolled High Strength Steel Sheets," Metallurgical Transactions, 6A (1975) pp. 1613-1622.

2. P.R. Mould and C.C. Skena, "Structure and Properties of Cold-Rolled Ferrite-Martensite (Dual-Phase) Steel Sheets," pp. 181-204 in Formable HSLA and Dual-Phase Steels, A.T. Davenport, ed.; AIME, New York, N.Y., 1979.

3. S. Hayami and T. Furukawa, "A Family of High-Strength, Cold-Rolled Steels," pp. 78-87 in Proceedings of Microalloying 75, 2A, Washington D.C., 1975.

4. S. Hayami, T. Furukawa, H. Gondoh, and H. Takechi, "Recent Developments in Formable Hot- and Cold-Rolled HSLA Including Dual-Phase Sheet Steels," pp. 167-179 in Formable HSLA and Dual-Phase Steels, A.T. Davenport, ed.; AIME, New York, N.Y., 1979.

5. R.L. Rickett, S.H. Kalyn and J.T. Mackenzie, "Recrystallization and Microstructure of Aluminum Killed Deep Drawing Steel," Transactions of the Metallurgical Society of AIME, 185 (1949) pp. 242-251.

6. J.T. Michalak and R.D. Schoone, "Recrystallization and Texture Development in a Low-Carbon Aluminum-Killed Steel," Transactions of the Metallurgical Society of AIME, 242 (1968) pp. 1149-1160.

7. J.M. Rigsbee and P.J. Vander Arend, "Laboratory Studies of Microstructures and Structure-Property Relationships in "Dual-Phase" HSLA Steels," pp. 56-86 in Formable HSLA and Dual-Phase Steels, A.T. Davenport, ed.; AIME, New York, N.Y., 1979.

8. K. Nakaoka, K. Araki and K. Kurihara, "Strength, Ductility and Aging Properties of Continuously-Annealed Dual-Phase High-Strength Sheet Steels," pp. 126-141 in Formable HSLA and Dual-Phase Steels, A.T. Davenport, ed.; AIME, New York, N.Y., 1979.

9. N. Ohashi, I. Takahashi and K. Hashiguchi, "Processing Techniques and Formability of Age-Hardenable, Low Yield, High Tensile Strength Cold-Rolled Sheet Steels," Transactions of Iron and Steel Institute of Japan, 18 (1978) pp. 321-329.

10. R.G. Davies, "Influence of Martensite Composition and Content on the Properties of Dual-Phase Steels," Metallurgical Transactions A, 9A (1978) pp. 671-679.

11. T. Matsuoka, M. Takahashi and A. Okamoto, "Production of Cold-Rolled Dual-Phase Steel Sheet for Outer Panel," pp. 62-63 in VANITEC Seminar on Dual-Phase and Cold Pressing Vanadium Steels in the Automobile Industries, West Berlin, 1978.

12. M. Takahashi, K. Kunishige and A. Okamoto, "Dual-Phase Steel Sheet Characterized in Low Yield Ratio and Excellent Formability," Bulletin of Japan Institute of Metals, 19 (1980) pp. 10-16.

13. J.Y. Koo, M. Raghavan and G. Thomas, "Compositional Analysis of Dual-Phase Steels by Transmission Electron Microscopy," Metallurgical Transactions A, 11A (1980) pp. 351-355.

14. T. Kunitake, "Diffusion of Iron and Carbon in Steel," Bulletin of Japan Institute of Metals, 3 (1964) pp. 466-476.

15. F.A. Hultgren, "Reversion and Re-precipitation of Aluminum Nitride in Aluminum Killed Drawing Quality Steel," Blast Furnace and Steel Plant, (1968) pp. 149-156.

16. K. Matsudo and T. Shimomura, "Effect of Carbon Content on Deep Drawability of Rimmed Steel Sheet," Transactions of Iron and Steel Institute of Japan, 10 (1970) pp. 448-458.

17. Y. Hosoya, K. Kurihara and K. Nakaoka, "Effect of Strength Ratio of Secondary Phase/Matrix on R-Value of Dual-Phase Steel Sheet," paper presented at Fall Meeting of the Iron and Steel Institute of Japan, Toyama, Japan, Oct. 1978.

18. A.R. Marder, "Factors Affecting the Ductility of "Dual-Phase" Alloys," pp. 87-98 in Formable HSLA and Dual-Phase Steels, A.T. Davenport, ed.; AIME, New York, N.Y., 1979.

19. T. Furukawa, H. Morikawa, H. Takechi and K. Koyama, "Process Factors for Highly Ductile Dual-Phase Steel Sheets," pp. 281-303 in Structure and Properties of Highly Formable Dual-Phase Steels, R.A. Kot and J.W. Morris, ed.; AIME, New Orleans, La, 1979.

20. R.G. Davies, "Early Stage of Yielding and Strain Aging of a Vanadium-Containing Dual-Phase Steel," Metallurgical Transactions A, 10A (1979) pp. 1549-1555.

444

21. W. Hume-Rothery, The Structure of Alloys of Iron: An Elementary Intro-
     duction; Pergamon Press Ltd., England, 1966.

RELATION BETWEEN HOLE EXPANSION FORMABILITY AND

METALLURGICAL FACTORS IN DUAL-PHASE STEEL SHEET

A. Nishimoto, Y. Hosoya and K. Nakaoka

Technical Research Center, Nippon Kokan K. K.

1-1, Minamiwatarida-cho, Kawasaki-ku
Kawasaki, 210 Japan

Results of a systematic investigation on the hole expansion formability of sheared edge of dual-phase steel sheet are presented in terms of metallurgical factors, such as volume fraction of second phase, degree of tempering, morphology of martensite, retained austenite, solute carbon and sulfur content. The expanding limit of the milled hole is closely related to the total-elongation. However, the limit of the punched hole is less than that of the milled hole and does not show any correlation with the tensile properties. Main factors affecting the deterioration due to punching are volume fraction of second phase, degree of tempering, banded-structure and sulfur content. The experimental results show that the air-cooled high-manganese steel is more sensitive to the punching deterioration as compared to the water-quenched and tempered low-manganese steel ; this tendency can be accounted for by the difference of the hardness ratio of the martensite to the ferrite matrix.

Introduction

Dual-phase steels composed of ferrite and martensite have been studied by numerous researchers world-wide because of their superior strength-ductility characteristics. These products are being supplied by some steel-makers and their application is increasing gradually partly as a result of the demand for vehicle weight reduction in the automotive industry.

Most of metallurgical studies up to now have been performed on strength and ductility, however only a few papers have been published on the sheared edge ductility that is important in practice.

Butler and Bucher (1) reported that uniform-elongation and stretch formability of dual-phase VAN-QN were superior to those of the conventional VAN steels and also that the flange stretch performance of the VAN-QN was slightly inferior to that of the VAN steels. That means flange stretch performance of dual-phase steel is decidedly inferior to most of conventional steels with same elongation. Also, Shimomura et al. (2) reported that hole expansion limits of cold-rolled high-manganese dual-phase steel sheet with low yield tensile ratio were inferior to those of notched tensile elongation.

The present authors have made a systematic investigation on the hole expansion formability of a sheared edge in terms of metallurgical factors. Three types of steel, 1.5%Mn-V, 2.1%Mn, 0.5%Mn-P, were principally used in this study, but other steels were also used in some aspects of this investigation.

Experiment

The chemical compositions of the steels used in this investigation are given in Table I. These steels were prepared by vacuum induction melting of 150 kg laboratory heats that were fully aluminum-killed prior to teeming. After soaking at 1250 °C the ingots were hot-rolled to 30 mm slabs, and then these slabs were cut, reheated at 1250 °C and hot-rolled to 3.2 mm strips. After the final pass at 900 °C, the strips were also cooled to 500 °C at a rate of approximately 5 °C/sec, and subsequently furnace-cooled to simulate coiling.

Table I. Chemical composition of experimental steels

Steel		Chemical compositions (wt %)									Note
Type	No	C	Si	Mn	P	S	sol.Al	V	Cr	Ni	
1.5Mn-V	1	.08	.29	1.43	.004	.003	.053	.074	–	–	Fig.2,3,5,6,9
	2	.07	.30	1.47	.004	.009	.056	.074	–	–	Fig.2,3,4,5,6
	3	.06	.30	1.46	.004	.015	.061	.073	–	–	Fig.2,3,5,6,16
2.1Mn	4	.07	.29	2.14	.004	.004	.059	–	–	–	Fig.2,3,5,6,9
	5	.07	.29	2.13	.004	.010	.060	–	–	–	Fig.2,3,4,5,6
	6	.06	.29	2.10	.004	.015	.057	–	–	–	Fig.2,3,5,6,13-16
0.5Mn-P	7	.07	.48	.47	.081	.001	.049	–	–	–	Fig.2,3,5,6,12
	8	.07	.49	.48	.079	.007	.057	–	–	–	Fig.2,3,4,5,6
	9	.07	.50	.48	.077	.014	.057	–	–	–	Fig.2,3,5,6,13-16
1.5Mn-Cr	10	.06	.30	1.50	.005	.003	.060	–	1.01	1.01	Fig.9
1.0Mn	11	.05	tr	1.04	.008	.004	.040	–	–	–	Fig.8,12
1.5Mn	12	.06	tr	1.54	.008	.004	.039	–	–	–	Fig.7,11

The pickled strips were cold-rolled to 0.8 mm, cut into 200 mm x 200 mm specimens and heat-treated using a salt bath. The main heat-treatment cycles are shown in Fig. 1. The 1.5Mn-V and 2.1Mn steels were subjected to the air-cooling type annealing cycles, and the 0.5Mn-P steels to the water-quenching type annealing cycles. The specimens heat-treated by the water-quenching type cycles were temper-rolled to 1.0 % for flattening before making test pieces. In addition to the above standard samples, additional samples were prepared to vary metallurgical factors such as volume fraction or distribution of martensite, degree of tempering, and solute carbon ; a specific description will be made in the following sections dealing with each metallurgical factor.

The hole expansion test was conducted by expanding a 30.0 mm diameter hole in a square specimen using a 100.0 mm diameter hemi-spherical punch. The holes were made by both punching and milling. The punching clearance was set at 40 % from the standpoint of punch abrasion in practical use. The test was stopped when a crack approximately 2 mm in length was observed at the hole edge. The expanding limit was evaluated by means of the limiting expansion ratio, λ, defined as :

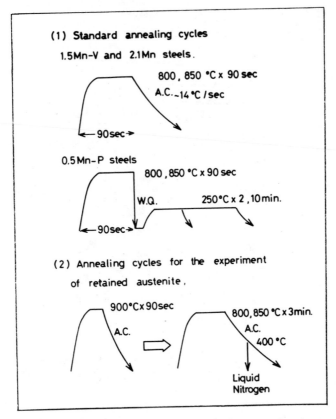

Fig. 1 - Annealing cycles used for experiment.

$$\lambda = \frac{D_f - D_O}{D_O} \times 100 \quad (\%) \text{ ----------- (1)}$$

where D_O and D_f are the initial and final hole diameters. The degree of deterioration due to punching, D , was defined as :

$$D\lambda = \frac{\lambda_M - \lambda_P}{\lambda_M} \times 100 \quad (\%) \text{ ----------- (2)}$$

where λ_M and λ_P are the limiting expansion ratio on milled and punched holes , respectively.

Results

Mechanical properties and hole expansion

Neither λ_P nor λ_M showed a dependence on tensile strength. As shown by Fig. 2, however, it became evident that λ_M is closely related to total-elongation, although the correlation between λ_P and elongation is very weak. From Fig. 3, it is seen that the degree of deterioration due to punching is not influenced by the elongation.

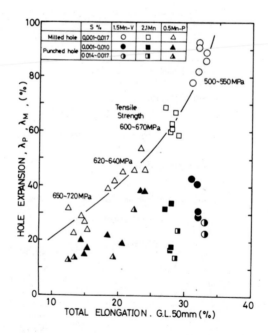

Fig. 2 - Relationship between total-elongation and hole expanding limit for three types of dual-phase steels.

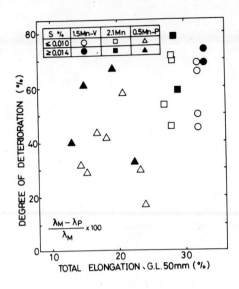

Fig. 3 - Relationship between total-elongation and
degree of deterioration for three types
of dual-phase steels.

Metallurgical factors and hole expansion

According to Figures 2 and 3, a general tendency is recognized that λ_p is smaller and $D\lambda$ is larger for higher sulfur content steels. However, there is a large scatter in the degree of deterioration due to punching and higher sulfur steels do not necessarily show a higher degree of deterioration for the same elongation. Other factors, for instance volume fraction and morphology of the second phase are assumed to have considerable influence since the microstructures show distinctive differences as shown by Figure 4.

The metallurgical factors selected for further study in this investigation are volume fraction and morphology of second phase, sulfur content, degree of tempering, retained austenite and solute carbon.

Effects of volume fraction of second phase and sulfur content

Figure 5 shows the results of the punched hole expanding limit, λ_p, and the degree of deterioration due to punching, $D\lambda$, as a function of volume fraction of the second phase. What is evident from the figure is that λ_p decreases and $D\lambda$ increases in proportion to the second phase, except for the case of low-manganese steels. For the same volume fraction of second phase, the degree of deterioration of high-manganese steel and/or high-sulfur content steels is higher, than that observed in low-manganese or low-sulfur content steels.

Figure 6 is a replot of the relationship between the deterioration and sulfur content. The deterioration increases in proportion to surfur content for a steel of a fixed martensite volume fraction. In the case of the air-cooled steels, the difference of the degree of deterioration between the lower and higher volume fractions of the second phase is larger when the

451

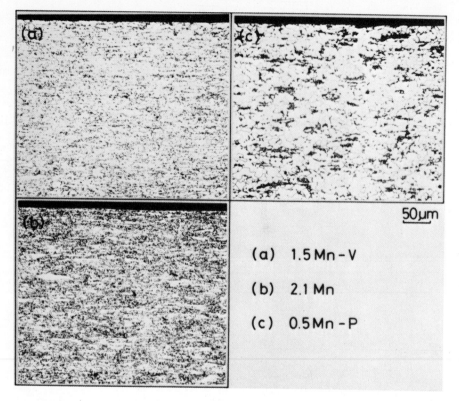

(a) 1.5 Mn - V

(b) 2.1 Mn

(c) 0.5 Mn - P

Fig. 4 - Optical microstructures for three types of dual-phase steels.

sulfur content is less. However, in the case of the water-quenched and tempered steels, the difference is roughly equal for all the sulfur contents covered in the experiment. In addition, general tendency is found for the degree of deterioration of the water-quenched and tempered steels due to punching to be smaller than that of the air-cooled steels.

Figure 7 illustrates the effect of martensite volume fraction on the 1.5Mn steel. After the cold-rolling to 0.8 mm, the steel was pre-annealed at 680 °C for 2 hours with box-annealing furnace in order to avoid morphology variations between samples and then reheated using a salt bath to a temperature between 730 °C and 850 °C for 2 minutes, followed by water-quenching. In the case of the as-quenched condition, the deterioration initially increases with an increase in the martensite volume fraction, but then reaches a maximun, and decreases gradually thereafter.
The deterioration of the samples after tempering remained at about 50 % up to a volume fraction of 25 % and then began to decrease.

Effect of tempering

The cold-rolled 1.0Mn steels were pre-annealed at 680 °C for 2 hours, reheated at 760 °C for 2 minutes, water-quenched, and then tempered at temperatures between 150 °C and 350 °C for 2 minutes.

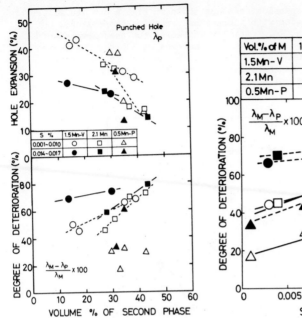

Fig. 5 - Punched hole expanding limit, λ_P, and the degree of deterioration due to punching, $D\lambda$, as a function of martensite volume fraction.

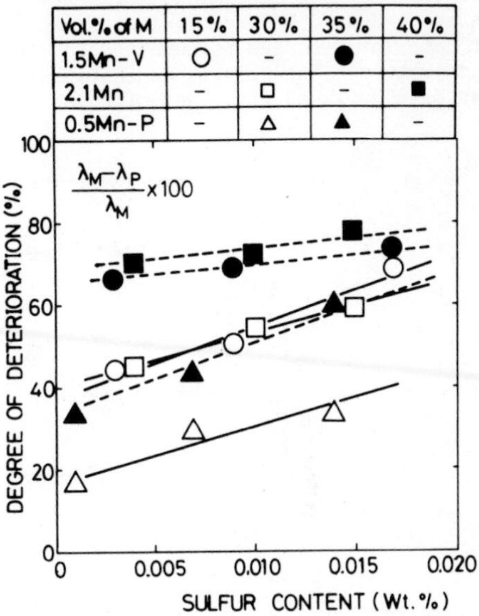

Fig. 6 - Relationship between the deterioration and sulfur content.

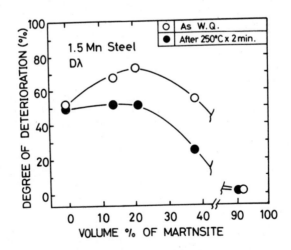

Fig. 7 - Effect of martensite volume fraction on the degree of deterioration in 1.5Mn steel.

The results of the hole expansion test are shown in Figure 8. The milled hole expansion, λ_M, increases with increasing tempering temperature, but the punched hole expansion, λ_P, initially increased but then leveled-off at about 200 °C before showing a dramatic increase at 300 °C. The degree of deterioration due to punching decreases with the rise in tempering temperature. The results suggest that the reduction of deterioration can be associated with the softening of martensite.

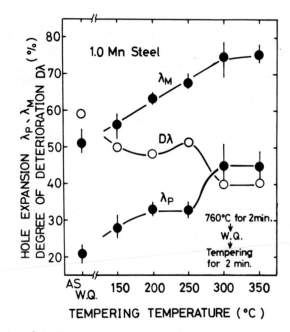

Fig. 8 - Effect of tempering temperature on the hole expanding limit and degree of deterioration.

Effect of retained austenite

An examination on the effect of retained austenite was conducted on three steels. As shown in Figure 1, these steels were normalized and re-heated to 800 °C or 850 °C, and subsequently air-cooled or quenched from 400 °C into liquid nitrogen. Retained austenite volume fractions ranging from 1 to 7 % were found in the resultant micro-structure. The results are presented in Figure 9. It is clear that the degree of deterioration due to punching is independent of the amount of retained austenite.

Effect of solute carbon

Two low-manganese steels were used in this experiment. The chemical compositions and the heating cycles to attain different level of solute carbon are shown in Figure 10, in addition to the results of hole expansion test. The two-stage tempering at 200 °C and 150 °C was undertaken to stabilize martensite by the first tempering and to vary the solute carbon by the second tempering. From Figure 10, it can be concluded that the

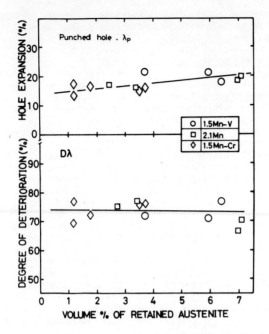

Fig. 9 - Effect of retained austenite on the hole
expanding limit and degree of deterioration.

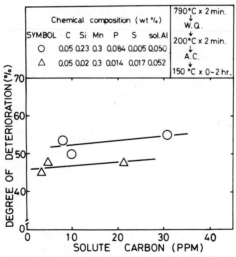

Fig. 10 - Effect of solute carbon on the
degree of deterioration.

deterioration increases slightly with increase in solute carbon, although there might also be an influence of martensite softening at 150 °C in the experimental data.

Effect of second phase morphology

In this study, the morphology of martensite was varied by means of several pre-heatings after cold-rolling, and subsequently, in order to achieve equal volume fraction and hardness of martensite, all the samples were re-heated to the same temperature and quenched.

The pre-heatings for the 1.5Mn steel (No.12) were a normalizing at 900 °C, a box annealing at 680 °C, and a quenching from 1050 °C to 700 °C followed by air cooling. The subsequent common re-heating at 790 °C for 2 minutes and quenching resulted in the microstructures of Figure 11. Volume fraction of the martensite and the result of the hole expansion test are shown on the right hand side of each microstructures. There are remarkable differences in the martensite spacing in these microstructures, however the hole expansion limits are almost the same.

Figure 12 shows the effect of the martensite banded-structure. As in the above case, different pre-heatings and a common final treatment was used. The pre-heatings for the 0.5Mn-P steel are a normalizing at 900 °C (upper) and a box annealing at 680 °C (lower), and those of the 1.0Mn steel are box annealings at 680 °C (upper left) and 750 °C (lower left). In both cases the latter pre-heatings gave pronounced banded-structures. Also in this case the volume fraction and the edge ductility are indicated on the side of the microstructures. It is clearly seen from the photographs that the banded-structures have undesirable influences on the sheared edge ductility.

Discussion

The deformation mode of the periphery of the hole in the hole expansion test is essentially a circumferential uniaxial stretching along the edge of the hole with radial stress gradient. Therefore, it is quite reasonable that the milled hole expanding limit is closely related to the total-elongation, as was seen in Figure 1. However the expanding limit of the punched hole can not be simply related to the elongation because of damage along the hole-edge. The edge region of punched hole contains both a fractured surface and a work-hardened zone. On the fractured surface, there are micro-cracks which, in the course of hole expansion, grow into cracks. These cracks connect with each other, and the resultant larger cracks eventually pass through the thickness, forming a notch at the edge of the hole, and leading to the circumferential stress concentration at the root of the notch, which results in the growth of the visible cracks. The presence of the work-hardened zone may promote the growth of the cracks since the stress required for the plastic deformation is higher as compared to the non work-hardened zone. It thus appears that metallurgical factors influence the number of micro-cracks on the fractured surface and their growth in the work-hardened zone.

In Figure 5 it was recognized that an increase in martensite volume fraction led to a decrease in the expanding limit of the punched hole and raised the degree of deterioration for the air-cooled high-manganese steel. This tendency was not clear in the low-manganese quenched and tempered steels. Generally speaking an increase in the hard second phase will lead to an increase in the number of micro-cracks on the fractured surface and it will also serve to promote the propagation of cracks in the hole expansion. The experimental data for the higher manganese steel agree with the above expectation. The results for the low-manganese steel seem to indicate that the influence of the martensite was weakened because of its

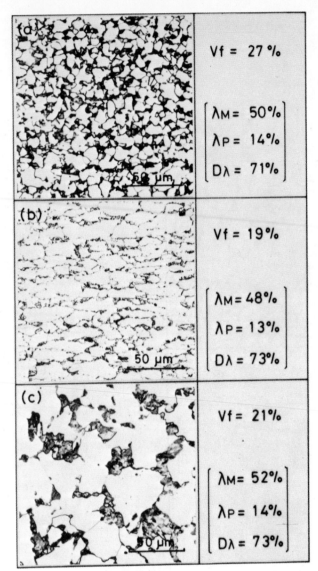

Fig. 11 – Optical micrographs and the results of hole expansion
test showing the effect of martensite-spacing on the
sheared edge ductility.

Pre-annealings : (a) Normalizing at 900°C,
                 (b) Box-annealing at 680°C,
                 (c) Normalizing at 1050°C.

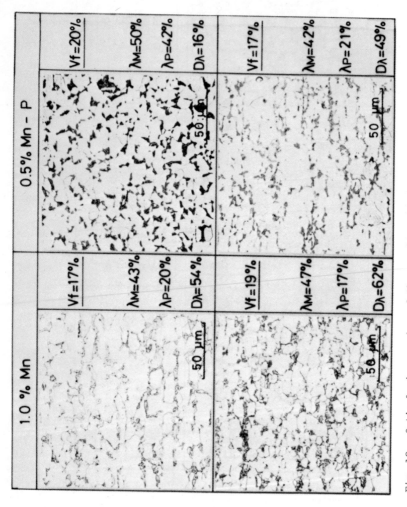

Fig. 12 – Optical micrographs and the results of hole expansion test showing the effect of the martensite banded-structures on the sheared edge ductility.

softening by tempering, as will be discussed later.

In Figure 7, a gradual decrease in the degree of deterioration was observed when the volume fraction of martensite exceeded 20 %. It may be attributed to the softer martensite in the larger volume fraction range, since the volume fraction was varied by means of quenching temperature, the higher quenching temperature giving martensite with less carbon. The favorable influence of the softening of martensite is demonstrated also in Figure 7.

In Figures 5 and 6, it was recognized that when the sulfur content increased, the influence of the martensite became less evident. Like the martensite, sulfides will act as a second phase and aggravate the edge ductility, although there are differences in the hardness and nature of phase interface.

When the difference in the hardness between matrix and second phase is large, formation of the micro-cracks during punching will be promoted and the harder second phase in the vicinity of the micro-cracks will cause a pronounced stress concentration, leading to an acceleration of the crack propagation, in the hole expansion. The influence of the tempering as observed in Figure 8 can be attributed to the fact that the softening of the martensite helps reduce the above tendencies.

Optical micrograpfs of the punched edge were taken over the section parallel to the sheet surface, and shown in Figure 13. In Figure 13-(a) which is the micrograph of the 2.1Mn steel air-cooled from 800 °C, micrographs were clearly observed at the interface of the ferrite-martensite in the work-hardened zone. However, in Figure 13-(b) which is the micrograph of the 0.5Mn-P steel quenched from 800 °C and subsequently tempered at 250 °C, no micro-cracks were observed. Figure 14 shows the similar micrographs of the two steels after hole expansion. In the case of the 2.1Mn steel, edge of the expanded hole took a zigzag form as shown in Figure 14-(a), while in the case of 0.5Mn-P steel the irregularity was less pronounced as can be seen in Figure 14-(b).

The regions at the tip of the crack were investigated by means of a scanning electron microscope, and the results are shown in Figure 15. The crack in the 0.5Mn-P steel propagated in a straight manner (Figure 15-(a)), and the enlarged picture revealed that the growth took place crossing through the martensite (Figure 15-(c)). In contrast, in the case of the 2.1Mn steel it was recognized that the crack propagated in a zigzag form passing the interface of the farrite and martensite (Figure 15-(b),-(d)). It was confirmed that crack propagation in the air-cooled 1.5Mn-V steel was very similar to the case of the 2.1Mn steel.

Hardness of the ferrite matrix and the martensite were measured by a micro-micro-Vickers hardness tester, which was specially designed to our specification several years ago, with a 0.5 g loading. The results are shown in Figure 16, where Hv_M and Hv_F are the hardness of martensite and that of ferrite, respectively. The micrographs attached to each set of data are examples of the scanning electron micrographs of the indented surface, by means of which the hardnesses were determined. It is evident that the difference of hardness between ferrite and martensite is larger in the air-cooled higher manganese steels than in the quenched and tempered low-manganese steel. This difference seems to be the principal reason of the difference in the formation of micro-cracks in the manner of crack propagation in the hole expansion.

In the experiment of Figure 11, influence of the spacing was not recognized, but we can not say that similar result would be obtained in a different experimental condition. Further investigation should be made to elucidate the influence of the morphology and spacing.

Qualitatively influence of retained austenite may be similar to that of martensite. Result of Figure 9 seems to be reasonable because of the

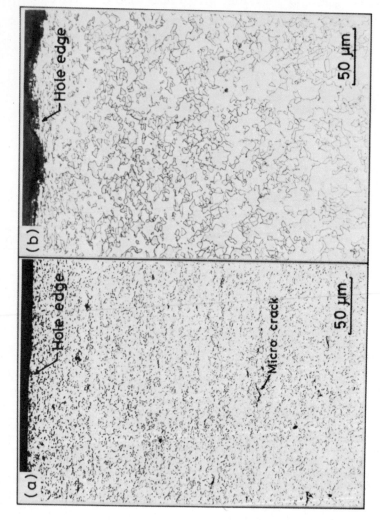

Fig. 13 – Optical micrographs of punched edge (plane-section view).

(a) : 2.1 Mn steel air-cooled from 800°C.

(b) : 0.5 Mn-P steel quenchen from 800°C and tempered at 250°C.

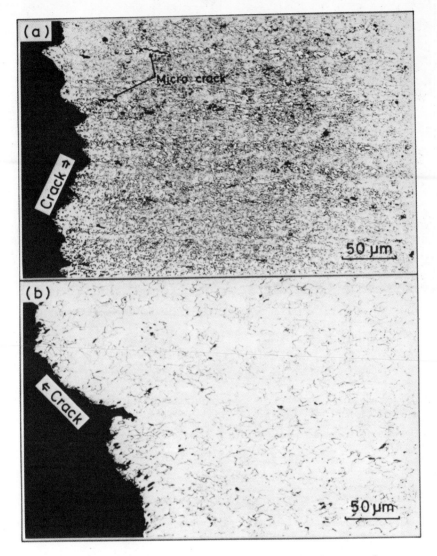

Fig. 14 – Optical micrographs of punched edge after hole expansion.

    (a) : 2.1 Mn steel air-cooled from 800°C.

    (b) : 0.5 Mn-P steel quenched from 800°C and tempered at 250°C.

                                   (plane-section view)

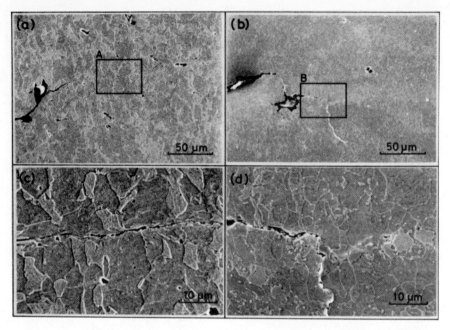

Fig. 15 - Scanning electron micrographs of the tip of cracks after
hole expanding.

(a): 0.5Mn-P steel   (b): 2.1Mn steel
(c): Enlarged picture from the region marked A in (a)
(d): Enlarged picture from the region marked B in (b)

smaller quantity as compared to the coexisting martensite and sulfide.
   Solute carbon will have an influence since it raises strength and
lowers ductility, but also in this case its effect seems to have been
very small in these experiments with the large influence of martensite
and sulfides.

Conclusions

1. The expanding limit of the milled hole is closely related to the
   total elongation. However, the limit of the punched hole does not
   show any correlation with the ordinary mechanical properties as
   determined by tensile testing.

2. Presence of martensite raises the degree of deterioration due to
   punching. The deterioration increases with the increases of the
   hardness and volume fraction of martensite.

3. The difference in the hardness between ferrite and martensite is
   larger in the air-cooled higher manganese steels as compared to
   the water-quenched and tempered lower manganese steel. The higher
   degree of deterioration due to punching in the higher manganese
   steel sample seems to be attributable to the above aspect.

1.5Mn - V	2.1 Mn	0.5 Mn - P
Hv_M : $\bar{x} = 684$ $\delta = 83$	Hv_M : $\bar{x} = 405$ $\delta = 65$	Hv_M : $\bar{x} = 307$ $\delta = 22$
Hv_F : $\bar{x} = 125$ $\delta = 1$	Hv_F : $\bar{x} = 127$ $\delta = 4$	Hv_F : $\bar{x} = 144$ $\delta = 4$
$Hv_M / Hv_F = 5.5$	$Hv_M / Hv_F = 3.2$	$Hv_M / Hv_F = 2.1$

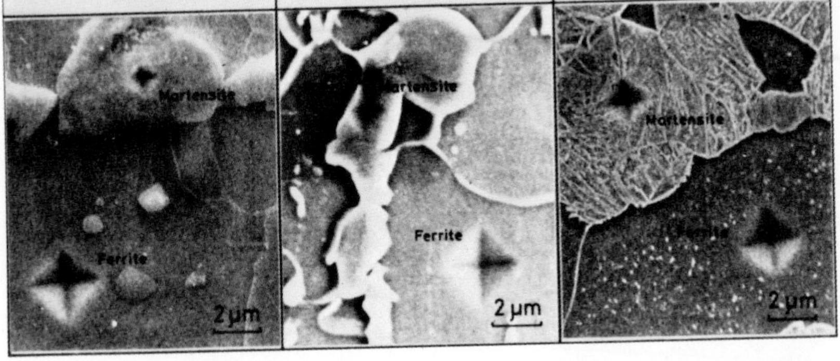

Fig. 16 - Hardness of ferrite matrix and martensite in dual-phase steels.

$$\left[\begin{array}{l} \text{1.5Mn-V and 2.1Mn steels : } 850°C \text{ x 90sec} \longrightarrow \text{A.C.} \\ \text{0.5Mn-P steel : } 850°c \text{ x 90sec} \longrightarrow \text{W.Q.} \longrightarrow 250°C \text{ x 2min} \longrightarrow \text{A.C.} \end{array} \right]$$

4. In this experiment, influence of spacing of the martensite was not observed.

5. Banded structure of martensite is one of the factors to raise the degree of deterioration due to punching.

6. When sulfur content increases, the influence of the martensite become less pronounced.

7. In this experiment, influence of retained austenite and solute carbon were not evident.

References

1. J.F.Butler and J.H.Bucher, "Development of Vanadium HSLA Steels with Improved Formability," Iron and Steel International, April (1979) pp. 85-94.

2. T.Shimomura, M.Yoshida, M.Sakoh and K.Matsudo, Proceeding of 12th IDDRG (1979).

STRETCH FORMABILITY STUDIES ON A DUAL-PHASE STEEL

H. Conrad[†], K. P. Datta, and K. Wongwiwat[††]
Metallurgical Engineering and Materials Science Department
University of Kentucky
Lexington, Kentucky 40506

Summary

Forming limit diagrams (FLD's) were determined on a vanadium-bearing dual-phase steel over the temperature range of 78-700K, which range provided a significant variation in the material parameters n and m determined in concurrent uniaxial tension tests. The failure mode at 78 and 200K was by quasi-cleavage, whereas that at 300K and above was by local necking followed by ductile fracture. The effect of temperature on the maximum principal failure strain e_1^* varied along the FLD. e_1 tended towards a maximum at 300K and a minimum at 400-600K, whereas for the right-hand side e_1^* tended to increase continuously with temperature. At all temperatures the FLD fit reasonably well to the expressions derived previously, namely

$$\bar{\varepsilon}*/Z \simeq (\bar{\varepsilon}*/Z)_{\varepsilon_2^*=0} + \beta(\varepsilon_2^*/\varepsilon_1^*) \tag{a}$$

$$\simeq \{[n_u + m_u(\alpha* + \varepsilon_u)]^o - \beta(\varepsilon_2^*/\varepsilon_1^*)^o\} + \beta(\varepsilon_2^*/\varepsilon_1^*) \tag{b}$$

where $\bar{\varepsilon}*$ is the effective failure strain, Z is the appropriate subtangent, $n = d\ln\sigma/d\ln\varepsilon$, $m = \partial\ln\sigma/\partial\ln\dot{\varepsilon}$, $\alpha = d\ln\varepsilon/d\ln\dot{\varepsilon}$ and β is a constant of the order of 0.1. The subscript u refers to the maximum load and the superscript o to values determined in uniaxial tension. FLD's calculated using Eqs. (a) and (b) and the results from a uniaxial tension and a dry, unlubricated cup test were in accord with those determined experimentally in the conventional manner. The superior formability of the dual-phase steel over other steels at a constant tensile strength can be explained by the appreciably higher values of the strain hardening coefficient $d\sigma/d\varepsilon$ and of the strain rate hardening coefficient $\partial\sigma/\partial\ln\dot{\varepsilon}$.

[†]Presently with Materials Engineering Department, North Carolina State University, Raleigh, North Carolina 27650.

[††]Presently with Fairchild Republic Company, Farmington, Long Island, New York.

Introduction

Dual-phase steels offer promise for weight reduction in automobiles because of their good formability at comparatively high tensile strengths (1, 2). Of both practical and theoretical interest is the effect of material parameters on the formability of such steels. Since forming limit diagrams (FLD's) have been found useful in predicting the stretch formability of sheet metals in practical forming operations (3-8), an expression for the FLD in terms of materials parameters determined in uniaxial tension is of special interest.

It is now generally accepted that the material parameters having the greatest influence on stretch formability are the strain hardening exponent $n = d\ln\sigma/d\ln\varepsilon$ and the strain rate hardening exponent $m = \partial\ln\sigma/\partial\ln\dot{\varepsilon}$, where σ is the true stress, ε the true strain and $\dot{\varepsilon}$ the true strain rate in uniaxial tension. The roles of n and m in plastic instability in uniaxial tension have been identified by computer calculations using specific constitutive equations (9-11). Along a different line, Conrad (12) derived an analytic expression for the effect of material parameters on the critical failure strain in stretch forming. Assuming that failure in stretch forming occurs when

$$\frac{d\bar{\sigma}}{d\bar{\varepsilon}} = \frac{\bar{\sigma}}{Z} \tag{1}$$

where $\bar{\sigma}$ and $\bar{\varepsilon}$ are the effective stress and strain respectively and Z the critical subtangent defined by Keeler and Backofen (13), he obtained the following expression for the critical effective failure strain $\bar{\varepsilon}*$ in stretch forming

$$\frac{\bar{\varepsilon}*}{Z} = \bar{n}* + \bar{m}*\bar{\alpha}* \tag{2}$$

$$\simeq n_u + m_u(\alpha* + \varepsilon_u) \quad \text{(uniaxial tension)} \tag{2a}$$

where $\bar{n} = \dfrac{d\ln\bar{\sigma}}{d\ln\bar{\varepsilon}}$, $\bar{m} = \dfrac{\partial\ln\bar{\sigma}}{\partial\ln\bar{\dot{\varepsilon}}}$ and $\bar{\alpha} = \dfrac{d\ln\dot{\bar{\varepsilon}}}{d\ln\bar{\varepsilon}}$ is the strain rate localization in the element which eventually failed.[†] The superscript * refers to the failure strain and u to the maximum load. Uniaxial tension test results were found to be in reasonable accord with Eq. 2a (12,14,15). Agreement also occurred for dry, unlubricated cup tests and for the FLD when certain corrections were made to Eq. 2a (16-18).

The objective of the present investigation was to determine the FLD's of a dual-phase steel by the method of Hecker (5-8) and to evaluate the effect of material parameters on the formability in the manner proposed by Conrad and coworkers (12,14-18) by performing tests over a sufficiently wide temperature range (78-700K) to provide a significant variation in the material parameters n and m.

[†] Use of the parameters n and m as here defined does not imply that the stress-strain curve obeys a power law with respect to strain and strain rate; rather n and m are simply the derivatives indicated.

466

Experimental

Material

The dual-phase steel (VAN-QN80) specimens used in this investigation were furnished by the Jones and Laughlin Steel Corporation and had the following chemical composition in wt.%:

C	S	Mn	P	Si	Mo	Al	V	Fe
0.15	0.007	1.39	–	0.70	–	0.04	0.05	balance

The microstructure of the material is shown in Fig. 1. The grain size measured on three specimens ranged between 4-6 μm determined by the linear intercept method.

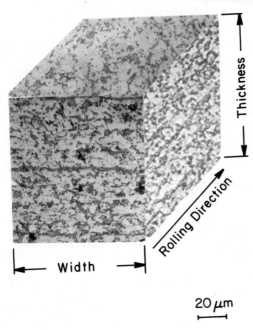

Fig. 1. Microstructure of the dual-phase steel.

Uniaxial Tension Tests

Standard sheet tensile specimens 12.5mm wide x 1.04mm thick with a gauge length of 63.5mm were machined parallel to the rolling direction. The tensile tests were carried out at 78-700K in an Instron machine at a nominal strain rate of $1.3 \times 10^{-4} s^{-1}$ on specimens onto which a square grid (1.27mm on a side) had been electroetched. The parameters n, m and α were determined in the manner described previously (14,15). n was determined as a function of strain from the slope of the true stress-true strain curve through the relation $n = (\varepsilon/\sigma)(d\sigma/d\varepsilon)$, and m was obtained as a function of strain by strain rate cycling tests using the relation $m = \ln(P_2/P_1)/\ln(v_2/$

467

v_1), where P_1 and P_2 are the load prior to and following a crosshead velocity change from v_1 to v_2. The true uniform strain ε_u was taken from the intersection of plots of σ versus ε and $d\sigma/d\varepsilon$ versus ε on the same graph. This value of ε_u generally agreed within 0.01 with that taken from the load-elongation chart at maximum load. The values of m_u and n_u were taken from plots of these parameters versus strain at $\varepsilon = \varepsilon_u$. The value of $\alpha = d\ln\dot{\varepsilon}/d\ln\varepsilon$ was derived from measurements of the increase in strain as a function of the time of the grid element which eventually failed. Details regarding the measurement of these parameters in the dual-phase steel and other observations on plastic instability in uniaxial tension will be presented in a separate paper.

Forming Limit Diagrams

The FLD's were determined by the method of Hecker (5-8) except that a 5cm dia. punch was employed here rather than 10cm. The present punch speed was 4×10^{-5}m/s. The grids employed were 2.54mm dia. circles electrochemically etched onto one side of each blank.

At 100°K the die-specimen-punch assembly was immersed in liquid nitrogen and the test initiated when iron-constantan thermocouples attached to both the specimen and die indicated that the desired temperature had been attained. Tests at 200°K were performed in the same manner except that a mixture of dry ice and methanol was used as the refrigerant. At 300°K the tests were conducted in air at ambient conditions. At elevated temperatures the die-specimen-punch assembly was heated in an electric resistance furnace to the desired temperature measured by chromel-alumel thermocouples. Generally, one hour was allowed to reach temperature. The surface of the specimen containing the electrochemically etched grid was protected from oxidation at 600 and 700°K by spraying with Ti-50 Formkote and in addition by passing argon gas over that surface. In all cases the temperatures were maintained within ±5K of those listed.

At elevated and low temperatures, Ti-50 Formkote and dry lubricant Formula FL-40 were used on 12.7x12.7cm specimen blanks to determine the positive minor strain portion of the FLD. For the negative minor strain side, the width of the blank was varied from 6cm to 0.95cm. They were tested in the dry, unlubricated condition at all temperatures for this side of the FLD.

To define the FLD, three types of grid ellipses were identified in directions parallel and transverse to the rolling direction in the immediate vicinity of a failed (fractured or local necked) region: (a) neck free, (b) neck-affected and (c) fractured. Measurements of the engineering radial strain (major, e_1) and circumferential strain (minor, e_2) using a standard measuring tape were made on grid ellipses immediately above and below the fracture and immediately above and below any local neck which may have existed in the other directions. The grid strain measurements are estimated to be accurate within ±2% strain. The FLD was taken as an average between the lower bound of the "neck-affected" data points and the upper bound of the "neck-free" points. The two bounding curves defined in this way were generally separated by a major strain difference of 7-15%. Figs. 2 and 3 show that FLD's determined by the present test method are in accord with those obtained by others in more conventional tests using a larger diameter punch.

Results and Discussion

Uniaxial Tension

The conventional tensile properties and the pertinent material parameters obtained in the uniaxial tension tests at the different test temperatures are given in Tables 1 and 2, respectively. When multiple data are listed, they reflect results by different investigators on separate samples. It is not clear whether the scatter is due to material variations or to experimental error. Due to a shortage of material the α^* values could not be determined at 500, 600 and 700K. The tensile properties listed in Table 1 at 300K are typical for this steel (20-22). The α^* values in Table 2 are

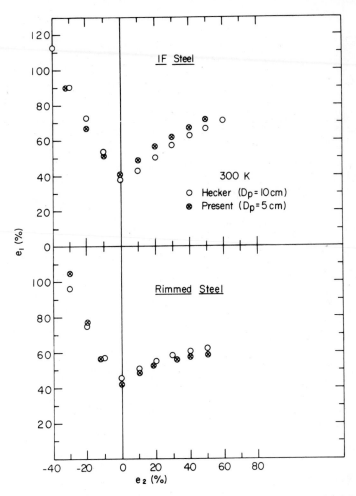

Fig. 2. Comparison of the FLD's for interstitial free and rimmed steels obtained in the present investigation using a 5 cm dia. punch with those by Hecker (19) using a 10 cm dia. punch

469

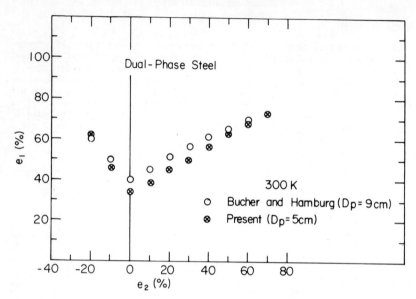

Fig. 3. Comparison of the FLD for a dual-phase steel obtained in the present investigation using a 5 cm dia. punch with that by Bucher and Hamburg (20) using a 9 cm dia. punch.

Table 1. Conventional tensile properties parallel to the rolling direction of the dual-phase steel at 78-700K.

T (K)	0.2% Y.S.* MPa	U.T.S.[†] MPa	Uniform El.[††] %	Total El. in 63.5cm %	r_o[†††]
78	530	1106	17.8	17.8	0.85
200	361	690	17.0	18.3	0.86
300	335	737	17.9	19.5	1.09
	357	754	23.3	28.4	0.95
	360	747	19.7	26.2	1.01
400	384	691	15.8	21.0	0.90
500	384	757	13.5	14.1	0.60
600	289	555	5.8	9.2	0.51
700	236	348	5.2	12.9	0.46

Notes:

[†]Engineering stress at maximum load.
[††]Engineering strain at maximum load.
[†††]Plastic anisotropy ratio: $\varepsilon_w/\varepsilon_{th}$.

similar to those obtained previously for conventional low carbon steels (14,18).

Worthy of note regarding Table 2 are: (a) the strain hardening exponent at maximum load (n_u) does not vary appreciably from 78 to about 500K (with perhaps a slight maximum at 300K), after which it decreases significantly,

Table 2. Formability parameters in uniaxial tension for the dual-phase steel at 78-700K.

T(K)	n_u	ε_u	m_u	$\alpha*$	$\varepsilon*$	$Z[n_u+m_u(\alpha*+\varepsilon_u)]$	Fracture Mode
78	0.17	0.17	0.0103	2.5	0.18	0.20[†]	Quasi-cleavage
200	0.14	0.14	0.0106	2.1	0.17	0.17[††]	Quasi-cleavage
300	0.16	0.16	0.0070	2.0	0.35	0.36[††]	Dimple
	0.19	0.21	0.0065	3.8	0.45	0.43[††]	
	0.18	0.18	0.0075	3.6	0.43	0.42[††]	
400	0.15	0.15	0.0034	3.0	0.35	0.32	
					0.32		
					0.40		
500	0.14	0.13	0.0120		0.39		
					0.38		
					0.42		
600	0.06	0.06	0.0190		0.41		
					0.33		
700	0.05	0.06	0.0430		0.47		

Notes: [†]$Z = 1$; [††]$Z = 2$.

(b) the strain rate hardening exponent at maximum load (m_u) is relatively constant between 78 and 200K, decreases at 300 to 400K, following which it increases again and (c) the relationship between the critical strain and the material parameters n_u and m_u proposed earlier (12,14,15) (i.e., $\varepsilon*/Z \approx n_u + m_u(\alpha*+\varepsilon_u)$) holds at all temperatures providing the appropriate value of the subtangent Z is employed. In this regard it must be pointed out that at 78 and 200K, the relationship holds for Z=1 (failure occurred near the maximum load with no evidence of local necking), whereas at 300K and above it holds for Z=2 (failure occurred by local necking). The basis for taking Z=1 at maximum load and Z=2 for local necking is reviewed by Backofen (23). Associated with the change in the Z value is a change in the fracture mode; see Fig. 4. To be noted is that the fracture mode at 78 and 200K is quasi-cleavage (25), whereas that at 300K is microvoid coalescence, i.e., dimpled.

Also to be noted in Table 2 is that the n_u values are in good accord with the values of the true strain at maximum load ε_u (uniform elongation), in keeping with the observation of Davies (24). This agreement between n_u and ε_u does not require that the true stress-strain curve obey the Hollomon power law, as is often presumed, but follows directly from the expression for the maximum load, i.e.,

$$(1/\sigma)(d\sigma/d\varepsilon) = 1 \qquad (3)$$

Multiplying both sides of Eq. 3 by ε_u gives

$$(\varepsilon/\sigma)(d\sigma/d\varepsilon)\big|_{\varepsilon=\varepsilon_u} = \varepsilon_u \qquad (4)$$

or

$$n_u = d\ln\sigma/d\ln\varepsilon\big|_{\varepsilon=\varepsilon_u} = \varepsilon_u \qquad (4a)$$

Worthy of mention is that dynamic strain-aging, evidenced by slight serrations in the load-elongation curve, occurred in the tests at 400 to 700K. The low value of m_u at 400K is in keeping with the occurrence of dynamic strain aging (26). The increase in m_u with temperature at 500K and above might then be due to a recovery type of plastic flow mechanism coming

471

into play, which ultimately leads to a decrease in strain hardening and in n_u at 600 and 700K.

Punch Tests

Comparison of the Dual-Phase Steel with Other Steels.

A comparison of the FLD of the dual-phase (DP) steel at 300K with those of several other

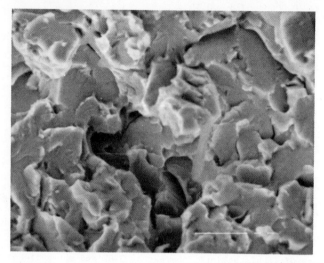

Quasi-
Cleavage

(a) 78K

|——— 10µm ———|

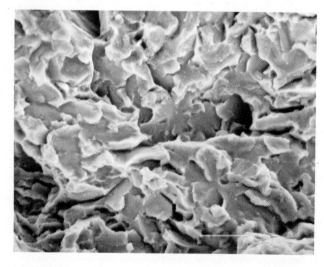

Quasi-
Cleavage

(b) 200K

|——— 10µm ———|

472

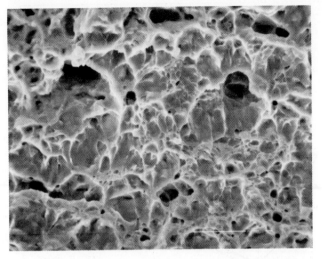

Micro-Void
Coalescence

(c) 300K   ├──── 10μm ────┤

Fig. 4. SEM fractographs of dual-phase steel specimens
tested in uniaxial tension at: (a) 78K, (b)
200K and (c) 300K.

steels, namely rimmed (R), interstitial free (IF), enameling iron (E-iron)
and a high strength low alloy steel (HSLA, Republic 50XK) is presented in
Fig. 5. The thickness of the R, IF, E-iron and DP steels is approximately
1mm; that of the HSLA is 0.7mm. It is seen that for $e_2 \leqslant 0$, the three unalloyed steels show similar behavior, being superior to that of the dual-phase
steel, which in turn is slightly better than that of the HSLA steel. For
$e_2 > 0$, the same trend exists for the IF, dual-phase and HSLA steels. However, e_1 for the rimmed and E-iron steels does not increase as rapidly with
e_2 as for the other three steels, and as a result the FLD's for the rimmed
and E-iron steels cross that for the dual-phase steel at $e_2 \simeq 40\%$. Since
the FLD is generally lowered with decreasing thickness, the difference between the HSLA and the DP steel at constant thickness would be less than
that shown in Fig. 5.

Effect of Temperature on the FLD of the Dual-Phase Steel. Fig. 6 shows
the FLD's of the dual-phase steel obtained at various temperatures. Similar
to uniaxial tension, the failure mode for a dry, unlubricated cup at 78K was
by quasi-cleavage. Time did not permit examination of the failure mode at
other temperatures, but based on the uniaxial tension test results it is
presumed that failure at 200K was by cleavage and that at higher temperatures by ductile fracture.

The effects of temperature on the maximum principal strain e_1 at various minor strains e_2 along the FLD and on the ratio of dome height h to
punch diameter D for the dry, unlubricated cup test, $(h/D)_d$, are presented
in Fig. 7. No clear, consistent effect of temperature on the FLD is evident.
For $e_2 < 0$ there is a tendency for a maximum in e_1 to occur at 300K followed
by a decrease at 400K and then a slight increase with temperature to 700K.

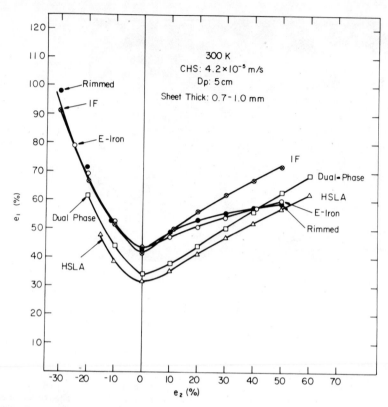

Fig. 5. Comparison of the FLD of the dual-phase steel with those for rimmed, E-iron, IF, HSLA (Republic 50XK) steels. The thickness of the dual-phase, rimmed, E-iron and IF steels is ∿1.0 mm; that of the HSLA is 0.7mm.

For e_2=0, the maximum in e_1 at 300K is followed by a minimum at 500K before increasing again at higher temperatures. For e_2>0 the fluctuations in e_1 with temperature are reduced, and e_1 appears to increase slightly with temperature over the entire range from 100 to 700K. These variations in the temperature dependence of e_1 with the value of e_2 indicate that the deformation behavior is influenced by the stress state. Similar trends in the effect of temperature on the FLD were noted for rimmed steel (18). This suggests that the formability of the dual-phase steel may be largely governed by the ferrite phase. The exception to this idea is the results for large negative minor strains at 500K, where a minimum in ε_1 occurs for rimmed steel, whereas for the dual-phase steel ε_1 exhibits a maximum at this temperature, suggesting a difference in the effect of temperature on dynamic strain aging and work hardening in the two steels at 500K.

Worthy of mention is that the parameter $(h/D)_d$ shows the same variation with temperature as does e_1 at e_2=+10%, these values of e_1 and e_2 being those associated with the dry, unlubricated cup test. Thus, for a dry,

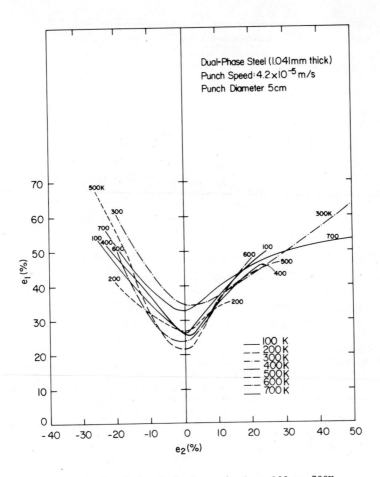

Fig. 6.   FLD's of the dual-phase steel at 100 to 700K.

unlubricated cup test h/D exhibits the same general behavior as does e_1.
This was also found to be the case for punch tests on rimmed steel (18).

The effect of temperature on the load at which fracture occurred in a
dry, unlubricated cup test divided by the sheet thickness is presented in
Fig. 8. Also included for comparison are results for rimmed steel. It is
seen that the fracture load for the dual-phase steel decreases with increase
in temperature to about 500-600K and then increases again, with a plateau
occurring at 200-300K. In comparison, the fracture load of the rimmed steel
decreases continuously and more steeply with temperature to about 500K and
then increases again at higher temperatures. The fracture loads for the
rimmed steel are about 1/3 to 1/2 those of the dual-phase steel, in keeping
with the difference in tensile strength between the two materials.

Limiting dome height (LDH) measurements were made on dry, unlubricated
blanks of varying width and plotted versus e_2 to yield a limiting dome

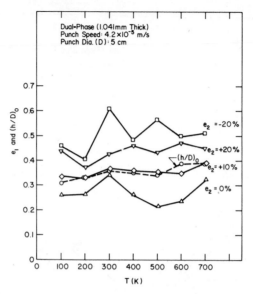

Fig. 7. e_1 at several values of e_2 and $(h/D)_d$ for punch tests on the dual-phase steel versus temperature.

height diagram (LDHD). The effect of temperature on $(h/D)_d$ at $e_2=0$ is shown in Fig. 9; included for comparison are the values for rimmed steel (18). To be noted is that $(h/D)_d$ is slightly larger for the dual-phase steel, even though the FLD for this material tends to be below that of the rimmed steel at each temperature. This indicates a more uniform distribution of strain for the dual-phase steel compared to the rimmed steel. Also to be noted in Fig. 9 is that the temperature dependence of $(h/D)_d$ at $e_2=0$ for the dual-phase steel is similar to that for the rimmed steel, going through a maximum at 200-300K and a minimum at 500-600K. This again suggests that the formability of the dual-phase steel may be largely controlled by the ferrite phase.

Similar to the behavior noted for rimmed steel (18), $(h/D)_d$ for the dual-phase steel was found to correlate with e_1 from the FLD, both taken at $e_2=0$; see Fig. 10. To be noted is that $(h/D)_d$ is roughly proportional to e_1 and that for a given value of e_1 the limiting dome height is larger for the dual-phase steel, in keeping with a more uniform distribution of strain in the dual-phase steel.

Analytical Expression for the FLD

In an analysis of the FLD's of a number of metals (17,18) it was found that $\bar{\varepsilon}*/Z$ increases slightly with the strain ratio $Y = \varepsilon_2^*/\varepsilon_1^*$ and that for the strain states normally represented by the FLD a good approximation to the FLD is given by

$$\bar{\varepsilon}*/Z = (\varepsilon*/Z)_{\varepsilon_2^*=0} + \beta(\varepsilon_2^*/\varepsilon_1^*) \qquad (5)$$

476

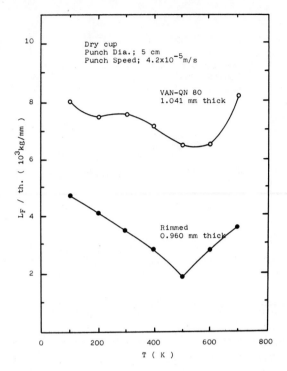

Fig. 8.   The effect of temperature on the fracture load divided
          by the sheet thickness for dry, unlubricated cup tests
          on the dual-phase steel and on rimmed steel.

where $\bar{\varepsilon}^*$ is the effective critical strain, Z the appropriate subtangent (13)
and β a constant of the order of 0.1 reflecting the effect of stress state.
Fig. 11 shows that Eq. 5 is a reasonable approximation to the FLD's for the
dual-phase steel considered here, with however some deviation tending to occur at large positive values of $\varepsilon_2^*/\varepsilon_1^*$.

The values of the intercepts and slopes of the straight lines in Fig.
11 are listed in Table 3. The slopes (β) exhibit no clear temperature dependence, although a minimum is suggested at 500-600K. The intercepts
$((\bar{\varepsilon}^*/Z)_{\varepsilon_2^*=0})$ have a temperature dependence similar to that of the maximum
principal strain at $e_2=0$ in Fig. 7. The β values are similar in magnitude
(~ 0.1) as those for a number of metals (17,18).

The present values of the intercepts and slopes for the dual-phase
steel are compared with those for rimmed steel (18) in Fig. 12. It is here
seen that the effects of temperature on both β and $(\bar{\varepsilon}^*/Z)_{\varepsilon_2^*=0}$ are similar
for the two steels; the values of β are approximately the same, whereas the
values of $(\bar{\varepsilon}^*/Z)_{\varepsilon_2^*=0}$ are lower for the dual-phase steel. Again the similarity in the effect of temperature suggests that the behavior mainly reflects
the plastic deformation of the ferrite phase.

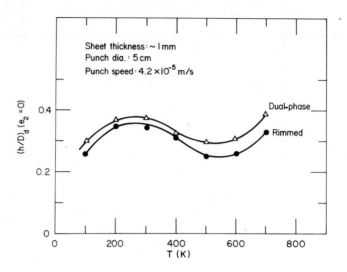

Fig. 9. Effect of temperature on the limiting dome height at $e_2=0$ for dry, unlubricated blanks of rimmed and dual-phase steels

Table 3. Values of $(\bar{\varepsilon}*/Z)_{\varepsilon_2^*=0}$ and β as a function of temperature (taken from Fig. 11) and formability parameters obtained in uniaxial tension

T (K)	$(\bar{\varepsilon}*/Z)_{\varepsilon_2^*=0}$	Slope (β)	$[n_u+m_u(\alpha*+\varepsilon_u)]^{o\dagger\dagger}$	$(\varepsilon_2^*/\varepsilon_1^*)^o$	$[n_u+m_u(\alpha*+\varepsilon_u)]^o-\beta(\varepsilon_2^*/\varepsilon_1^*)^o$
100	0.22	0.11	0.18	−0.45	0.23
200	0.22	0.13	0.17	−0.45	0.23
300	0.28	0.11	0.21	−0.33	0.25
400	0.23	0.14	0.16	−0.33	0.21
500	0.22	0.07		−0.40	
600	0.20	0.08		−0.38	
700	0.26	0.15		−0.33	

Notes: $\dagger\dagger$Superscript ^{o} refers to values in uniaxial tension.
$\dagger$From Table 2.

Considering Eqs. 2 and 5 one obtains

$$(\bar{\varepsilon}*/Z)^o = (\bar{\varepsilon}*/Z)_{\varepsilon_2^*=0} + \beta(\varepsilon_2^*/\varepsilon_1^*)^o \qquad (6)$$

$$= [n_u + m_u(\alpha*+\varepsilon_u)]^o \qquad (6a)$$

where the superscript ^{o} refers to uniaxial tension. Rearranging Eq. 6 gives

$$(\bar{\varepsilon}*/Z)_{\varepsilon_2^*=0} = [n_u+m_u(\alpha*+\varepsilon_u)]^o - \beta(\varepsilon_2^*/\varepsilon_1^*)^o \qquad (7)$$

in which all of the parameters on the right-hand side (RHS) of Eq. 7 except

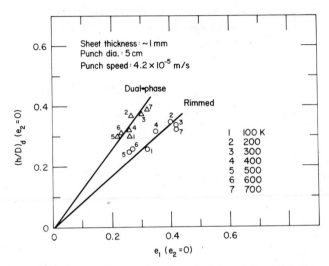

Fig. 10. Correlation of limiting dome height with critical maximum principal strain e_1 (FLD) at $e_2=0$. Data for rimmed steel from Ref. 18.

β are obtained in uniaxial tension. Table 3 provides a comparison of the values of $(\bar{\epsilon}*/Z)_{\epsilon_2^*=0}$ taken from Fig. 11 (second column) with those calculated from the RHS of Eq. 7 (last column). It is seen that reasonably good agreement occurs between the measured and calculated values.

Eq. 5 indicates that the entire FLD might be calculated from the values of the critical principal strains ϵ_1^* and ϵ_2^* for only two tests, e.g., a uniaxial tension test and a dry, unlubricated cup test. Such a calculation was carried out for rimmed steel (18), and reasonable agreement occurred between the calculated FLD and that determined experimentally for tests at temperatures of 100-700K. Furthermore, combining Eqs. 5 and 7 one obtains

$$\bar{\epsilon}*/Z = \{[n_u+m_u(\alpha*+\epsilon_u)]^o - \beta(\epsilon_2^*/\epsilon_1^*)^o\} + \beta(\epsilon_2^*/\epsilon_1^*) \qquad (8)$$

which expresses the FLD in terms of: (a) formability parameters determined in uniaxial tension and (b) the constant β, which can be obtained from the results in uniaxial tension and a dry, unlubricated cup test. A comparison of the FLD at 300K for the dual-phase steel calculated based on Eq. 8 (using the results from a uniaxial tension test and a dry, unlubricated cup test) with that determined experimentally in the conventional manner is presented in Fig. 13. It is seen that fair agreement exists over most of the strain range, the agreement tending however to break down at large positive values of the minor strain, in keeping with the behavior noted in Fig. 11 for this temperature. Similar agreement between the calculated and experimentally determined FLD was obtained for the other temperatures for which $\alpha*$ values were available.

479

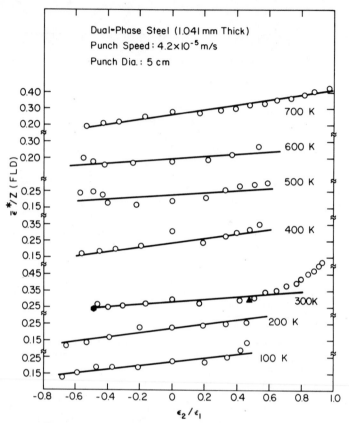

Fig. 11. $\bar{\epsilon}^*/Z$ versus $\epsilon_2^*/\epsilon_1^*$ for the dual-phase steel as a function of temperature (derived from LFD's in Fig. 6). ● Uniaxial tension test; ▲ cup test.

Effect of Material Parameters on Formability

Considering the data of Tables 1 to 3 it is clear that the conventional tensile ductility parameters such as uniform elongation e_u, total elongation e_t or post-uniform elongation (e_t-e_u) do not by themselves correlate directly with the critical failure strain ϵ^* in uniaxial tension or with stretch formability given by the FLD. On the other hand, reasonable correlation is obtained through the combined parameter $Z[n_u+m_u(\alpha^*+\epsilon_u)]^o$, which includes the material parameters n_u and m_u and the strain rate localization parameter α^*, all determined in uniaxial tension. Similar behavior has been noted for sheet metals in general (14–18). Hence, on the basis of Eqs. 2a and 8 one can understand why at the same tensile strength dual-phase steel has better formability than unalloyed low carbon steels and conventional HSLA steels. Since these steels all have similar α^* values (14,18), the formability is determined to a large extent by the material parameters n_u and m_u, which are given by

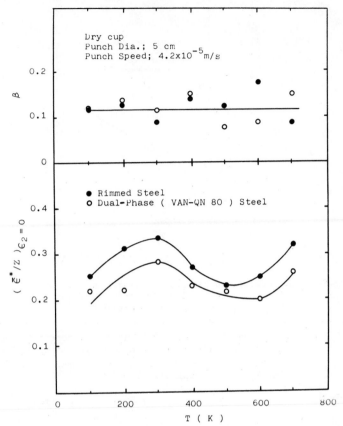

Fig. 12.   Comparison of the values of $(\varepsilon*/Z)_{\varepsilon_2^*=0}$ and β as a function
of temperature for the dual-phase steel and rimmed
steel.   Data for rimmed steel from Ref. 18.

$$n_u = \left.\frac{d\ln\sigma}{d\ln\varepsilon}\right|_{\varepsilon=\varepsilon_u} = \frac{\varepsilon_u}{\sigma_u}\left(\frac{d\sigma}{d\varepsilon}\right)_{\varepsilon=\varepsilon_u} \qquad (9)$$

and

$$n_u = \left.\frac{\partial\ln\sigma}{\partial\ln\dot\varepsilon}\right|_{\varepsilon=\varepsilon_u} = \frac{1}{\sigma_u}\left(\frac{\partial\sigma}{\partial\ln\dot\varepsilon}\right)_{\varepsilon=\varepsilon_u} \qquad (10)$$

Thus, at a constant value of the true stress at maximum load σ_u, the material
parameters which govern formability are $(d\sigma/d\varepsilon)_{\varepsilon=\varepsilon_u} = \sigma_u$ and $(\partial\sigma/\partial\ln\dot\varepsilon)_{\varepsilon=\varepsilon_u}$.
A comparison of the values of $(d\sigma/d\varepsilon)$ and $(\partial\sigma/\partial\ln\dot\varepsilon)$ at maximum load for the
dual-phase steel with those for a HSLA steel and rimmed steel is presented in
Table 4.   It is here seen that both the strain hardening coefficient $(d\sigma/d\varepsilon)$
and the strain rate hardening coefficient $(\partial\sigma/\partial\ln\dot\varepsilon)$ for the dual-phase steel
are about twice those for the HSLA and the rimmed steels, which accounts for
the better formability of the dual-phase steel.

481

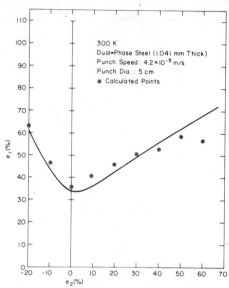

Fig. 13. Comparison between the experimentally determined
FLD and that calculated from Eq. 8 in the text
for the dual-phase steel at 300K.

Table 4. Comparison of the uniaxial tension parameters ($d\sigma/d\varepsilon$) and
($\partial\sigma/\partial\ln\dot{\varepsilon}$) at maximum load and α^* at the critical failure
strain for the dual-phase steel with those for rimmed
steel and a HSLA steel at 300K.

Steel	$(\frac{d\sigma}{d\varepsilon})_{\varepsilon=\varepsilon_u}$	$(\frac{\partial\sigma}{\partial\ln\varepsilon})_{\varepsilon=\varepsilon_u}$	$\alpha^*=(\frac{d\ln\dot{\varepsilon}}{d\ln\varepsilon})_{\varepsilon=\varepsilon^*}$
Dual-Phase	869	6.08	2.0
(VAN–QN80)	930	6.05	3.8
	894	6.71	3.6
HSLA	508	3.42	4.0
(Republic 50XK)			
Rimmed	404	3.64	3.3
	413	3.51	3.0

Conclusions

The formability of a vanadium-bearing dual-phase steel (VAN-QN80) was
investigated in uniaxial tension and by punch tests over the temperature
range of 78-700K, which range yields significant variations in the material
parameters n and m. The following is a summary of the results and conclusions reached:

1. The failure mode at 78 and 200K was found to be quasi-cleavage fracture,
    whereas at 300K and above failure occurred by local necking followed by
    ductile fracture.

482

2. The effect of temperature on formability varies with the stress state, greater variations occurring for the negative side of the FLD than for the positive side.

3. The effect of temperature on the formability of the dual-phase steel exhibits trends similar to those for rimmed steel, suggesting that the behavior of the dual-phase steel may be governed to a large extent by the ferrite phase.

4. The following phenomena appear to play an important role in the formability of dual-phase steel at 400 to 700K, where serrations occurred in the stress-strain curves: (a) 400 K, dynamic strain aging, (b) 500 and 600K, dynamic strain aging plus dynamic recovery and (c) 700K, dynamic recovery.

5. The critical effective failure strain $\bar{\varepsilon}*$ along the FLD is described reasonably well by the equation derived earlier, namely

$$\varepsilon*/Z = \{[n_u + m_u(\alpha* + \varepsilon_u)]^o - \beta(\varepsilon_2^*/\varepsilon_1^*)^o\} + \beta(\varepsilon_2^*/\varepsilon_1^*)$$

where the superscript ^{o} refers to uniaxial tension and u to the maximum load. FLD's calculated on the basis of this equation were in good accord with those determined experimentally.

6. The superior formability of the dual-phase steel over other steels at a constant tensile strength can be explained by the appreciably higher values of the strain hardening coefficient $d\sigma/d\varepsilon$ and of the strain rate hardening coefficient $d\sigma/d\ln\dot{\varepsilon}$ of the dual-phase steel.

Acknowledgements

This research was sponsored by the American Iron and Steel Institute under Project No. 57-367. The authors also wish to thank Dr. J. H. Bucher and the Jones and Laughlin Steel Corp. for providing the dual-phase steel specimens used in this study.

References

1. M. S. Rashid, SAE Preprint 760206 (February, 1976).
2. K. Arabi, K. Nabaska, M. Abe, and N. Ohashi, Proc. 9th Biennial Conf. International Deep Drawing Research Group (IDDRG), p. 39, 1976.
3. S. P. Keeler, Machinery (March-July 1968).
4. R. Pearce, Sheet Metal Industries (1971) p. 943.
5. S. S. Hecker, Proc. 7th Biennial Conf. International Deep Drawing Research Group (IDDRG), Amsterdam, Netherlands, 1972.
6. S. S. Hecker, Metals Engineering Quarterly, Vol. 13 (1973), p. 92.
7. S. S. Hecker, Sheet Metal Industries (November 1975) p. 671.
8. A. K. Ghosh and S. S. Hecker, Met. Trans., Vol. 5 (1974) p. 2161.
9. A. K. Ghosh, Met. Trans., Vol. 8A (1977) p. 1221.
10. A. K. Ghosh, Acta Met., Vol. 25 (1977) p. 1413.
11. J. W. Hutchinson and K. W. Neale, Acta Met., Vol. 25 (1977) p. 839.
12. H. Conrad, J. Mechwork. Tech., Vol. 2 (1978) p. 67.
13. S. P. Keeler and W. A. Backofen, Trans. ASM, Vol. 56 (1963) p. 25.
14. H. Conrad, M. Y. Demeri and D. Bhatt, Formability: Analyses, Modeling and Experimentation, TMS-AIME, p. 208, 1978.
15. K. Okazaki, M. Kagawa, and H. Conrad, Acta Met., Vol. 27 (1979) p. 301.
16. H. Conrad, K. Okazaki, and C. Yin, Proc. 6th North American Mfg. Research Conf. (NAMRC VI), p. 264, 1978.

17. H. Conrad and C. Yin, <u>Mechanical Behavior</u> of <u>Materials</u>, Vol. 2, ICM 3, p. 595; Pergamon Press, New York, 1979.
18. H. Conrad and K. Wongwiwat, <u>Proc. 8th North American Mfg. Research Conf.</u> (NAMRC VIII), Rolla, Mo., May 18-22, 1980.
19. S. S. Hecker, <u>Formability: Analyses, Modeling</u> and <u>Experimentation</u>, TMS-AIME, p. 150, 1978.
20. J. H. Bucher and E. G. Hamburg, <u>SAE Int. Automotive Eng. and Expo.</u>, paper 770164, Detroit, February 28-March 4, 1977.
21. J. H. Bucher and E. G. Hamburg, <u>Formable HSLA and Dual-Phase Steels</u>, TMS-AIME, p. 142, 1979.
22. J. H. Bucher, E. G. Hamburg, and J. F. Butler, <u>Structure and Properties of Dual-Phase Steels</u>, TMS-AIME, p. 346, 1979.
23. Walter A. Backofen, <u>Deformation Processing</u>, pp. 199-213; Addison-Wesley, Reading, Massachusetts, 1972.
24. R. G. Davies, <u>Met. Trans.</u>, Vol. 7A (1978) p. 451.
25. <u>ASM Metal Handbook</u>, Vol. 9, p. 65, 1974.
26. R. E. Reed-Hill, <u>Mat. Reviews</u>, Vol. 2 (1974), p. 218.

EFFECT OF MARTENSITE BANDS AND ELONGATED MANGANESE SULFIDE

INCLUSIONS ON THE FORMABILITY OF DUAL-PHASE STEELS

Y.J. Park,[1] A.P. Coldren[1] and J.W. Morrow[2]
Climax Molybdenum Company
A Subsidiary of AMAX, Inc.

[1]Ann Arbor, Michigan
[2]Pittsburgh, Pennsylvania

The formability of dual-phase steels was characterized by zero-T bend tests, hole-expansion tests, and Olsen cup tests. The formability test results were correlated with two microstructural features; that is, martensite bands and elongated manganese sulfide inclusions.

The results clearly indicate that martensite bands and elongated inclusions significantly influence the formability of dual-phase steels. Only materials that were free from both of these microstructural features passed the zero-T bend test. These materials also exhibited high hole-expansion values.

Introduction

The current effort in the automotive industry to improve fuel economy through vehicle weight reduction is prompting the substitution of reduced-gage high-strength steel for conventional low carbon steel. A promising group of recently developed high strength materials are the heat treated and as-rolled dual-phase steels, which show continuous yielding at a relatively low level of stress with a rapid work hardening rate and good tensile ductility. To completely explore the possibility of materials substitution, a necessary step is to investigate the formability of any candidate replacement material. While the formability of dual-phase steels has been rather extensively studied in recent years (1-4), the effects of banded microstructure and inclusion shape, which were found to greatly influence the formability of conventional HSLA steels (5), have not been fully evaluated in the case of dual-phase steels. The purpose of the present study was therefore to investigate these microstructural effects on the formability of commercially produced dual-phase steels by simple laboratory tests; that is, bend, hole-expansion and Olsen cup tests. Dual-phase steels satisfying the specification of a minimum tensile strength of 620 MPa (90 ksi) and a total elongation value of 27% in 51 mm (2 in.) were selected for the present study. Since the tensile characteristics of these steels were similar, no attempt was made to correlate the tensile parameters (e.g., elongation and strain hardening exponent)with the formability data.

Materials

Nine heat treated dual-phase (HTDP) steels and eight as-rolled dual-phase (ARDP) steels were evaluated. Thicknesses of the as-received strips and compositions of these steels are presented in Table I. HTDP and ARDP

Table I. Type, Thickness and Composition of the Steels Investigated

Steel No.	Type of Steel	Thickness mm	Composition, Weight Percent									
			C	Mn	Si	Cr	Mo	V	Al	N	P	S
1		2.6	0.09	1.36	0.56	–	–	0.050	0.04	0.008	0.015	0.004
2		4.3	0.12	1.62	0.68	–	–	0.051	0.06	0.008	0.006	0.011
3		4.4	0.11	1.66	0.70	–	–	0.060	0.05	0.009	0.010	0.008
4		4.4	0.10	1.65	0.69	–	–	0.060	0.05	0.008	0.006	0.007
5	HTDP	4.3	0.11	1.71	0.61	–	–	0.066	0.06	0.008	0.006	0.010
6		4.3	0.11	1.63	0.74	–	–	0.069	0.08	0.008	0.014	0.003
7		4.3	0.11	1.64	0.84	–	–	0.060	0.07	0.008	0.006	0.011
8		2.5	0.11	1.79	0.63	–	–	0.033	0.06	0.008	0.001	0.024
9		3.1	0.10	1.50	0.48	–	0.10	–	0.04	0.015	0.020	0.015
10		2.7	0.04	1.00	1.28	0.59	0.40	–	0.08	0.006	0.02	0.015
11		4.8	0.09	0.89	1.11	0.39	0.38	–	0.08	0.010	0.012	0.007
12		3.4	0.05	1.37	1.37	0.52	0.39	–	0.26	0.017	0.020	0.009
13	ARDP	3.1	0.04	0.84	1.05	0.49	0.35	–	0.04	0.009	0.012	0.001
14		4.3	0.09	1.00	1.32	0.45	0.41	–	0.14	0.012	0.013	0.008
15		3.4	0.06	0.91	1.52	0.54	0.40	–	0.003	0.002	0.004	0.009
16		2.8	0.04	0.84	1.05	0.49	0.35	–	0.04	0.009	0.012	0.004
17		1.6	0.06	1.56	1.46	0.56	0.24	–	0.03	0.004	0.018	0.008

486

steels exhibit similar microstructures that consist of relatively hard islands of martensite plus retained austenite (MA constituent) dispersed in a matrix of soft and highly ductile ferrite. The primary difference in microstructure between HTDP and ARDP is that the latter generally has a lower MA content and a finer ferrite grain size than the former. In addition, many HTDP steels exhibit martensite bands, while this condition has never been observed in ARDP steels. The microstructural differences between HTDP and ARDP are a consequence of their respective compositions and manufacturing processes, which are briefly described below.

Heat Treated Dual-Phase Steel

HTDP steel is produced by intercritically annealing hot-rolled or cold-rolled strips. The steel with an initial ferrite plus carbide structure is heated into the intercritical temperature region where the carbides dissolve, leaving a mixture of ferrite plus carbon-rich austenite. On subsequent cooling at relatively moderate rates, depending on the steel alloy content, the carbon-rich austenite has sufficient hardenability to transform to martensite.

The manganese content of HTDP steels produced commercially in the United States varies from 1.3 to 1.8%. Manganese segregation occurs during solidification and the high-manganese regions can become greatly elongated in the hot-rolled strip. During subsequent intercritical annealing, martensite bands may form in the regions of higher manganese content (6).

As-Rolled Dual-Phase Steel

The dual-phase microstructure can also be obtained directly by conventional rolling on a hot-strip mill without heat treatment (7). Designated as-rolled dual-phase steel, this high-strength steel has been produced on a trial basis in many hot-strip mills. ARDP steel produced on a hot-strip mill undergoes two stages of cooling after the last rolling pass. First, the strip is spray cooled for about 10 seconds while traveling on a runout table. Second, after the strip is coiled the cooling rate is reduced to approximately 22 C/hr (40 F/hr).

Obtaining a dual-phase microstructure under these production conditions requires a steel composition with transformation characteristics relatively insensitive to variation in thermomechanical treatment. The recommended composition range for obtaining reproducible properties in commercial production of ARDP steel is 0.04/0.07% C, 1.2/1.5% Si, 0.8/1.0% Mn, 0.4/0.5% Cr, and 0.33/0.38% Mo (7,8). This optimized composition is highly effective in making the ARDP steel almost totally insensitive to the normal variations in hot-mill finish-rolling temperature, runout table cooling rate, and coiling temperature.

Because ARDP steel contains less manganese than HTDP steel, manganese segregation is less severe in the ARDP steel than in the HTDP steel. In addition, ARDP steel has a lower carbon content and lower volume fraction of MA islands because a large amount of polygonal ferrite (more than 85%) forms during rapid cooling on the runout table. For example, to achieve a 620 MPa (90 ksi) minimum tensile strength, only 8 to 12% of the MA constituent is required in the ARDP steels compared to a level of 18 to 23% MA required for the HTDP steels. Due to the minimal manganese segregation and the lower volume fraction of MA constituent, the ARDP steels do not exhibit martensite bands.

Experimental Procedures

Longitudinal tensile properties were obtained using specimens with a 13 mm (0.5 in.) wide by 51 mm (2 in.) long gage section. The tensile tests were performed at room temperature with a cross head speed of 5.1 mm (0.2 in.) per minute.

To determine the existence of martensite bands and the adequacy of inclusion shape control, a metallographic specimen was prepared for each steel. The surface examined was parallel to the direction of rolling and perpendicular to the plane of rolling (longitudinal face). Each specimen was scanned at 100X and 250X, and individual features were examined in greater detail at 1000X.

The formability tests performed were the bend, hole-expansion, and Olsen cup tests. Experimental procedures for each test are described below.

Bend Test

Duplicate bend tests were performed using sheared transverse specimens (crease of bend parallel to rolling direction) which measured 25 to 38 mm (1 to 1.5 in.) wide by 125 to 150 mm (5 to 6 in.) long. Each specimen was bent 180 degrees flat upon itself (zero-T radius) with the shear burr in tension. When the bending was completed, the convex surface of the specimen was inspected for cracks. If an edge crack propagated beyond the raised edge of the specimen, the specimen was considered to have failed the bend test.

Hole-Expansion Test

The hole-expansion test consisted of enlarging a 12.7 mm (0.5 in.) diameter punched hole with a conical ram having a 20 degree apex angle. Replicate tests were performed on each material using a sheet metal testing machine which has a 36-tonne capacity for the primary ram and an 11-tonne capacity for its clamping ram. The test was interrupted when a crack was observed at the free edge of the hole. The percentage hole expansion was calculated as:

$$\text{percentage hole-expansion} = \frac{D_f - D_o}{D_o} \times 100\%$$

where D_o and D_f are the initial and final hole diameters, respectively.

During the hole-expansion testing program, a tendency was noticed that percentage hole-expansion values decreased with increasing sheet thickness. This is believed to be a consequence of the thicker sheets having a larger sheared surface area. A shear burr and a cold-worked layer at a sheared edge would promote splitting in subsequent forming operations and thus result in decreased percentage hole-expansion values. To account for the effect of thickness, the hole-expansion data were corrected to a uniform thickness of 3.3 mm (0.130 in.) by a factor of -9.5%/mm (-241%/in.) of sheet thickness. This correction factor was developed by a study of dual-phase steels which were surface ground to various thicknesses (9).

Olsen Cup Test

The Olsen cup tests were performed on the same sheet metal testing machine as the hole-expansion tests. The test consisted of forcing a 22 mm (7/8 in.) steel ball into a sheet sample which was held firmly by smooth circular die plates. The load was applied at a rate of 25 mm/min (1 in./min). The ball-sheet interface was lubricated by a 0.1 mm (0.004 in.) thick polyethylene sheet wetted with SAE 50-weight oil. A plot of load vs. cup depth was automatically drawn and used to determine the load and cup depth at maximum load and at fracture for each test.

Results

Tensile properties of the steels are presented in Table II. Table III summarizes the results of the metallographic examination. The existence of martensite bands was judged at 250X. As mentioned above, many HTDP steels exhibited martensite bands, while no ARDP steel exhibited the banded structure. Optical micrographs of steels 1, 4, and 10 are shown in Figures 1 and 2. Most steels were inclusion (sulfide) shape controlled either by a rare earth metal (REM) addition or by a Zr addition as shown in Table III. In some steels, the inclusion shape control was not adequate. Here adequate control means complete absence of elongated manganese sulfide (MnS) inclusions as judged at 1000X. Micrographs illustrating "adequate" and "inadequate" inclusion shape control are shown in Figure 3.

From this metallographic examination, the steels were divided into three microstructural groups. Group A exhibited neither martensite bands nor elongated MnS inclusions (adequate inclusion shape control). Group B exhibited either martensite bands or elongated MnS inclusions. Group C exhibited both martensite bands and elongated inclusions. These microstructural groups could be correlated with the formability test results as mentioned below.

Table II. Room Temperature Tensile Properties of the Steels

Steel No.	Type of Steel	Offset Yield Strength, MPa (ksi) 0.2%	3%	Ultimate Tensile Strength, MPa (ksi)	Total Elongation in 51 mm (2 in.), %
1-7[a]		<379 (<55)	>483 (>70)	>620 (>90)	>27
8	HTDP	429 (62.2)	593 (86.0)	693 (100.5)	28
9		373 (54.1)	ND[b]	656 (95.1)	27
10		341 (49.4)	485 (70.2)	622 (90.2)	31
11		410 (59.4)	532 (77.2)	631 (91.5)	31
12		365 (53.0)	521 (75.5)	636 (92.2)	30
13	ARDP	318 (46.1)	512 (74.2)	641 (93.0)	28
14		424 (61.5)	577 (83.7)	675 (97.9)	30
15		346 (50.1)	538 (78.0)	669 (97.1)	29
16		325 (47.1)	ND[b]	647 (93.8)	27
17		392 (56.9)	ND[b]	690 (100.1)	27

[a]Steels are from seven different lots of a commercially produced HTDP steel. Tensile properties of each lot were not determined but are believed to meet the GM 980X requirements which are listed in the table.

[b]ND = not determined.

Table III. Martensite Bands and Inclusion Shape
Control of the Steels

Steel No.	Type of Steel	Continuous Martensite Bands?	Inclusion Shape Control	
			Method	Adequate*
1		No	REM	Yes
2		No	Zr	Yes
3		No	Zr	No
4		Yes	Zr	Yes
5	HTDP	Yes	REM	Yes
6		Yes	Zr	No
7		Yes	Zr	No
8		Yes	REM	No
9		Yes	None	No
10		No	REM	Yes
11		No	REM	Yes
12		No	REM	Yes
13		No	Zr	Yes
14	ARDP	No	REM	Yes
15		No	REM	No
16		No	None	No
17		No	None	No

*As judged by absence of elongated MnS
inclusions at 1000X.

The bend test results shown in Table IV clearly indicate that only
steels in Group A passed the zero-T bend test. Steels in Groups B and C
failed to pass the test. Correlation between percentage hole-expansion
values and the microstructural groups is less distinct. Steels in Group A
generally exhibited higher percentage hole-expansion values than those in
either Groups B or C. The difference between Group B and Group C was rather
small, although the average hole-expansion value of steels in Group B was
slightly higher than that of steels in Group C. These results are illustrated in Figure 4. Neither the bend test results nor the hole-expansion
values could be correlated with the strength variations of the steels. However it should be recalled that all the dual phase steels in this study
exhibited tensile strengths in excess of 620 MPa (90 ksi).

Olsen cup test results are presented in Figure 5. It was observed that
the Olsen cup height at maximum load is somewhat correlated with sheet thickness but not with the microstructure. When cup height at fracture is
considered, differences related to inclusion shape control can be seen. The
steels showing the highest values of the quantity cup height at fracture
minus cup height at maximum load, are also those steels (1, 10, 12 and 13)
which have the best metallographic rating (Group A), pass the zero-T bend
test, and show high hole-expansion values.

Discussion

The above formability test results indicate that martensite banding and
elongated MnS inclusions significantly reduce the formability of dual-phase

490

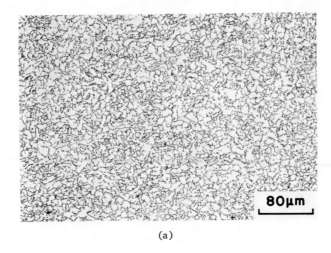

(a)

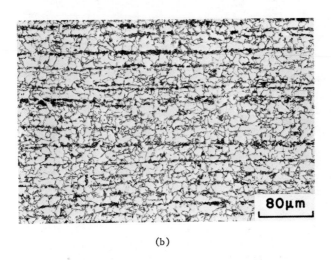

(b)

Fig. 1 - Microstructure of Heat Treated Dual-Phase Steel
(a) 1 and (b) 4. Note the banded structure of
Steel 4.

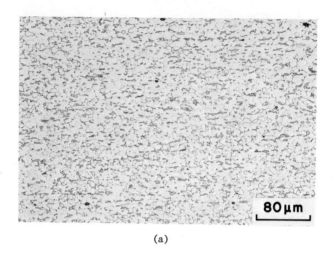

(a)

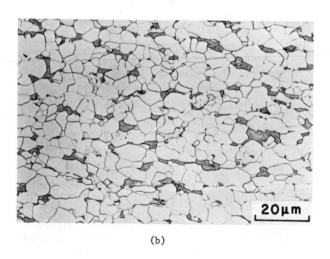

(b)

Fig. 2 - Microstructure of As-Rolled Dual-Phase Steel 10 at
(a) Low and (b) High Magnification.

492

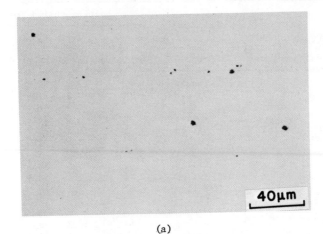

(a)

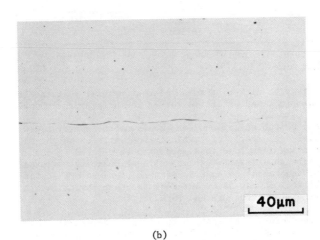

(b)

Fig. 3 - Optical Micrographs Illustrating "Adequate" and
"Inadequate" Inclusion Shape Control: (a) Round
Inclusions Obtained by Adequate Shape Control
(Steel 10); (b) An Elongated Inclusion Resulted
from Inadequate Shape Control (Steel 15).

Table IV. Bend Test and Hole-Expansion Test Results

Steel No.	Type of Steel	Micro-structural Group	Zero-T Bend Test	Number of Tests	Hole-Expansion Test $(D_f - D_o)/D_o$ x 100, %* Range	Mean	Corrected
1		A	Passed	2	12.4	52.4	45.4
2			Passed	2	3.2	17.6	27.0
3			Failed	2	3.0	16.5	27.1
4		B	Failed	2	1.4	15.7	25.8
5	HTDP		Failed	2	5.0	14.9	24.3
6			Failed	2	5.8	18.1	28.0
7		C	Failed	2	2.4	16.4	26.5
8			Failed	17	7.1	20.7	13.5
9			Failed	2	2.0	9.0	7.1
10			Passed	17	26.3	58.9	53.1
11			Passed	2	0.4	27.8	42.0
12		A	Passed	2	6.0	33.8	34.8
13	ARDP		Passed	3	5.6	31.1	29.4
14			Passed	2	0.4	20.0	28.9
15			Failed	2	1.0	29.2	30.3
16		B	Failed	2	10.2	25.3	20.3
17			Failed	2	4.2	26.3	10.5

*Corrected to a uniform thickness of 3.3 mm (0.130 in.) by a factor of -9.5%/mm (-241%/in.)(9).

steels. Steels with either of these microstructural features failed to pass the zero-T bend test and exhibited rather low hole-expansion values.

Olsen cup height at maximum load could not be correlated with these microstructural features. Interestingly however, cup height at fracture minus height at maximum load does appear to be related to the microstructural features. While this is not the conventional quantity measured in the Olsen cut test (10) and cannot be considered as a standard method for quantifying steel formability, it may be of interest to other investigators as a means for correlating microstructure and formability.

During the zero-T bend test, a specimen is deformed rather severely, i.e., beyond its necking point. This test therefore can be used to predict the formability of dual-phase steels for components experiencing rather severe deformation (e.g., automotive wheel rims). However, the zero-T bend test is too severe for predicting performance in many commercial forming operations. In these cases, a series of less severe bend tests, i.e., 1/2T,1T, 2T, and 3T (11) can probably be used to evaluate the relative formability of the steels with either martensite bands or elongated MnS inclusions.

Since martensite bands and elongated inclusions strongly influence the formability of dual-phase steels in certain forming modes, the manufacturing process of HTDP steels should be designed to avoid the formation of martensite

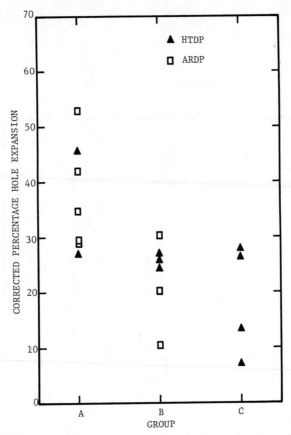

Fig. 4 - Correlation Between Hole-Expansion Test Results
and Microstructural Group.

bands. Adequate sulfide shape control is also necessary for both types of
dual-phase steels to maximize formability. The lack of a correlation between sulfur content (Table I) and percent hole-expansion (Table IV) for the
steels studied suggests that adequate sulfide shape control is more important
than the overall sulfur content.

It is difficult to predict how accurately the bend test and hole-expansion test results can be correlated with the commercial part-forming
characteristics of dual-phase steels. It has been reported that martensite
bands and elongated MnS inclusions in HSLA steels strongly influence the
forming characteristics of automotive wheels (5). Therefore it seems
reasonable that the simple laboratory tests (bend and hole-expansion) can be
used to estimate certain aspects of the formability of dual-phase steels in
commercial parts-production.

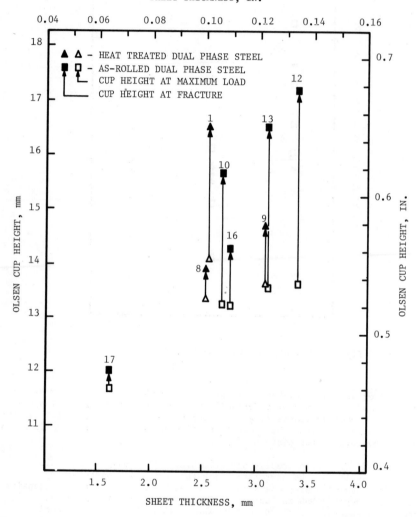

Fig. 5 - Olsen Cup Test Results. Numbers correspond to the steel numbers listed in tables.

Summary and Conclusions

1. Many HTDP steels exhibit martensite bands, while no ARDP steels exhibit martensite bands.

2. Only steels free from both martensite bands and elongated MnS inclusions passed the zero-T bend test. These steels also exhibited high hole-expansion values.

From the above observations, it is concluded that martensite bands and elongated MnS inclusions strongly influence the formability of dual-phase steels for certain forming modes. The zero-T bend test may be used to identify whether or not a steel is free from both of these microstructural features.

References

(1) M.S. Rashid, "GM 980X - A Unique High Strength Sheet Steel with Superior Formability," SAE Paper No. 760206, February 1976.

(2) W.G. Brazier and R. Stevenson, "Evaluation of a New Dual Phase, Cold Rolled Steel - Formability," SAE Paper No. 780137, February 1978.

(3) A.E. Cornford, J.R. Hiam, and R.M. Hobbs, "Properties of As-Rolled Dual Phase Steels," SAE Paper No. 790007, February 1979.

(4) S.S. Hansen, "The Formability of Dual-Phase Steels," presented at 1980 AIME Fall Meeting in Pittsburgh, Pennsylvanis, October 1980.

(5) J.A. Lumm, G.M. Hughes, and B.J. Bastien, "Application of High Strength Steels to Wheel Manufacturing," SAE Paper No. 740179, February 1974.

(6) G.R. Speich and R.L. Miller, "Mechanical Properties of Ferrite-Martensite Steels," pp. 145-182 in Structure and Properties of Dual-Phase Steels, R.A. Kot and J.W. Morris, Ed.; AIME, New York, NY, 1979.

(7) A.P. Coldren and G.T. Eldis, "Using CCT Diagrams to Optimize the Composition of an As-Rolled Dual-Phase Steel," Journal of Metals, March 1980, pp. 41-48.

(8) G.T. Eldis, A.P. Coldren and F.B. Fletcher, "Alloying and Transformation Control in Mn-Si-Cr-Mo As-Rolled Dual-Phase Steels," Paper No. 4 in Alloys for the 80s, Climax Molybdenum Company, Ann Arbor, MI, 1981.

(9) Y.J. Park, J.W. Morrow, and A.P. Coldren, "Formability of Dual Phase Steels," Climax Report L-176-206, January 1980.

(10) ASTM E643-78, "Standard Method for Conducting a Ball Punch Deformation Test for Metallic Sheet Materials," 1980 Annual Book of ASTM Standards, Part 10, Philadelphia, PA, 1980.

(11) J.D. Grozier, "Sheared Edge Bend Test (Hutchinson Bend Test)," Journal of Materials, JMLSA, Vol. 7, No. 1 March 1971, pp. 307.

Subject Index

498